Jichuan Wang · Haiyi Xie · James H. Fisher
Multilevel Models

Jichuan Wang
Haiyi Xie
James H. Fisher

Multilevel Models

Applications using SAS®

De Gruyter 高等教育出版社 HIGHER EDUCATION PRESS

Mathematics Subject Classification 2010: 97K80, 93E24, 62J12, 91G70, 91B62, 62H10, 62N01, 62G09.

ISBN 978-3-11-026759-4
e-ISBN 978-3-11-026770-9

Library of Congress Cataloging-in-Publication Data

Wang, Jichuan.
[880-01 Duo ceng tong ji fen xi mo xing]
Multilevel models : applications using SAS / by Jichuan Wang, Haiyi Xie, James H. Fischer.
p. cm.
Includes bibliographical references and index.
ISBN 978-3-11-026759-4 (alk. paper)
1. Social sciences – Research – Mathematical models. 2. Multilevel models (Statistics) 3. SAS (Computer file) I. Xie, Haiyi. II. Fischer, James H. III. Title.
H61.W36 2011
005.5'5–dc23
2011037278

Bibliographic information published by the Deutsche Nationalbibliothek

The Deutsche Nationalbibliothek lists this publication in the Deutsche Nationalbibliografie; detailed bibliographic data are available in the Internet at http://dnb.d-nb.de.

Printing and binding: Hubert & Co. GmbH & Co. KG, Göttingen
♾ Printed on acid-free paper

Printed in Germany

www.degruyter.com

Preface

Interest in multilevel statistical models for social science and public health studies has been aroused dramatically since the mid-1980s. New multilevel modeling techniques are giving researchers tools for analyzing data that have a hierarchical structure. Multilevel models are now applied to a wide range of studies in sociology, population studies, education studies, psychology, economics, epidemiology, and public health.

Individuals and social contexts (e.g., communities, schools, organizations, or geographic locations) to which individuals belong are conceptualized as a hierarchical system, in which individuals are micro units and contexts are macro units. Research interest often centers on whether and how individual outcome varies across contexts, and how the variation is explained by contextual factors; what and how the relationships between the outcome measures and individual characteristics vary across contexts, and how the relationships are influenced or moderated by contextual factors. To address these questions, studies often employ data collected from more than one level of observation units, i.e., observations collected at both individual level (e.g., students) and contextual levels (e.g., schools, cities). As a result, the data are characterized by a hierarchical structure in which individuals are nested within units at the higher levels. This kind of data is called hierarchically structured data or multilevel data. The conventional single-level statistical methods, such as ordinary least square(OLS) regression are inappropriate for analysis of multilevel data because observations are nonindependent and the contextual effects cannot be addressed appropriately in such models. Multilevel modeling not only takes into account observation dependence in the multilevel data, but also provides a more meaningful conceptual framework by allowing assessment of both individual and contextual effects, as well as cross-level interaction effects.

This book covers a broad range of topics about multilevel modeling. Our goal is to help students and researchers who are interested in analysis of multilevel data to understand the basic concepts, theoretical frameworks and application methods of multilevel modeling. This book is written in non-mathematical terms, focusing on the methods and applications of various multilevel models, using the internationally widely used statistical software, the *Statistical Analysis System* (SAS®). Examples are drawn from analysis of real-world research data. We focus on two-level models in this book because it is most frequently encountered situation in real research. These models can be readily expanded to models with three or more levels when applicable. A wide range of linear and non-linear multilevel models are introduced and demonstrated.

There are six chapters in this book.

Chapter 1 presents a brief introduction and overview of multilevel modeling. In this chapter, we discuss the problems inherent in applying traditional analytical methods to hierarchically structured data or multilevel data; we explain why multilevel models are needed for analyzing such data; and discuss the conceptual framework, its advantages, and limitations of multilevel modeling. Chapter 1 concludes with a brief overview of computer software for multilevel modeling.

Chapter 2 summarizes basic concepts of multilevel models, including intra-class correlation (ICC), model formulation, statistical assumptions, model estimation, model fit and model comparison, explained micro and macro level variances, and strategies of model building. Expansion of the 2-level model to 3-level models is also discussed.

Chapter 3 demonstrates linear multilevel models, also known as hierarchical linear models (HLM) using cross-sectional data. This chapter presents detailed model building strategies and illustrates model development and statistical testing procedures step by step.

Chapter 4 extends multilevel models to longitudinal data. The chapter covers both linear and curvilinear growth models. Some complex modeling strategies such as orthogonal polynomial modeling and piecewise modeling techniques are also presented.

Chapter 5 discusses advanced multilevel models for discrete outcome measures, such as binary, ordinal, nominal and count outcomes. The chapter starts with introduction of the generalized linear models. Then we present the model formulation for each type of discrete outcomes: multilevel logistic regression for binary outcome, multilevel cumulative logistic regression for ordinal outcomes, multilevel multinomial logit models for nominal outcome, and multilevel Poisson model, as well as multilevel negative binomial model, for count data. Alternative SAS procedures are used to analyze different types of discrete outcomes, and detailed count of model specifications and interpretations of model results are presented.

Chapter 6 discusses some special issues that are often encountered in multilevel modeling, including approaches for modeling count data with extra zeros, semi-continuous outcome measures, and multilevel data with a small number of groups (i.e., level-2 units). We demonstrate multilevel or random effects zero-inflated Poisson (RE-ZIP) models, random effect zero-inflated negative binomial models (RE-ZINB), mixed-effect mixed-distribution models, and bootstrapping multilevel models using SAS procedures. In addition, group-based models are introduced to assess growth trajectories of various outcome measures using longitudinal data. A special SAS procedure, SAS *PROC TRAJ* is used to demonstrate group-based logit models, group-based ZIP models, group-based Poisson models, group-based censored normal models, and group-based normal models. Finally, missing values and sample size/statistical power estimation for multilevel modeling are discussed.

While many computer programs are available for multilevel modeling, we have chosen the internationally distributed statistics package *Statistical Analysis System* (SAS) to demonstrate multilevel models in this book. SAS is a suitable package for many analysts because of its powerful data manipulation and modeling capabilities. The models demonstrated in this book are intended to show readers, step by step, how to build multilevel models using SAS for both cross-sectional and longitudinal data. SAS syntax for all of the models covered in the book are provided in each corresponding chapter of the book. The data used, as well as SAS syntax for all examples, can be downloaded from the website of Higher Education Press (academic.hep.com.cn). Although data used for these examples are drawn from public health studies, the methods and analytical techniques are applicable to other fields of social sciences.

Washington, DC — Jichuan Wang
Lebanon, NH — Haiyi Xie
EUsworth, ME — James H. Fisher

August, 2011

Contents

Chapter 1
Introduction

Over the past two decades *multilevel models* (Mason et al., 1983; Bryk & Raudenbush, 1992; Raudenbush & Bryk, 2002; Goldstein, 1987, 1995) have gained popularity in various research fields including education, psychology, sociology, economics, and public health. Multilevel models extend ordinary least square (OLS) regression to analyze multiple level data or hierarchically structured data that involve both micro and macro observation information. Multilevel models also appear under different names in the literature, including *hierarchical linear models* (Bryk & Raudenbush, 1992; Raudenbush & Bryk, 2002), *random-effect models* (Laird & Ware, 1982), *random coefficient models* (DeLeeuw & Kreft, 1986), *variance component models* (Dempster et al., 1981), *mixed models* (Longford, 1987), and *empirical Bayes models* (Strenio et al., 1983).

Prior to the development of formal statistical methodology for multilevel models, sociologists engaged in contextual or multilevel analysis of hierarchically structured data. In the late 1950s and early 1960s Lazarsfeld (1961) and Merton (1957) at Columbia University began to assess contextual effects on individual behavior. The 1970s witnessed a significant jump in analysis of multilevel data in education (Barr & Dreeben, 1977; Block & Burns, 1976; Bronfenbrenner, 1976; Burstein, 1980; Cronbach, 1976; Herriot & Muse, 1973; Pedhazur, 1975; Snow, 1976; Spady, 1973; Walberg, 1976). In a systematic study of contextual analysis, Boyd and Iversen (1979) discussed how to model multilevel data with micro-macro models, i.e., to formulate within group regression model at individual level, then relate the within group regression coefficients to contextual variables that describe the groups. Although multilevel observations are discussed in their models, their estimation was conducted using ordinary least square (OLS) techniques that were inappropriate for multilevel analysis.

Statistical theories of multilevel models and corresponding computer programs were developed in early 1980s by sociologists and demographers. Models were applied to analyze the large scale multilevel data of the United Nation's World Fertility Survey (WFS) (Hermalin & Mason, 1980; Mason et al., 1983). Further methodological and substantive work in educational studies and the user-friendly windows-based computer programs by Bryk & Raudenbush (1992) and Goldstein (1987, 1995) have popularized the multilevel models. Multilevel models are now applied in a wide range of studies in the social sciences.

1.1 Conceptual framework of multilevel modeling

A key concept in social sciences is that a society can be described in hierarchical structures. By hierarchy, we mean that units at a lower level are nested within or grouped into units at a higher level. People cannot be treated as isolated individuals but as social beings. Individuals are members of many different types of groups and are embedded in different social contexts. For example, individuals belong to families, neighborhoods, organizations and communities. Awareness has been mounting that individual behaviors and outcomes are affected not only by individual characteristics, but also shaped by the social contexts in which they are imbedded (Lazarsfeld, 1961; Merton, 1957; Bronfenbrenner, 1976; Blalock, 1984; Iversen, 1991). Davis' so-called "frog-pond" theory proposes that individual students evaluate personal ability relative to in-groups and pay little attention to out-groups (Davis, 1966). A moderately intelligent student (a medium size "frog") in a highly intelligent school (a large "pond") may become discouraged and thus become an under-achiever, while the same student in a considerably less intelligent school (a small "pond") may gain confidence and become an over-achiever. On the contrary, a moderately intelligent student might be motivated to study harder in a highly intelligent school and become more successful. In other words, the effect of an individual student's intelligence on his/her achievement may be influenced by specific features in the school he/she attends. In addition to composite measures (e.g., average intelligence level), student academic achievement may also be influenced by a variety of school level variables such as student/teacher ratio, teachers' work experience, school facilities, budget, etc.

The relationships between academic achievement and individual level variables vary across schools. For example, differences in academic achievement among ethnic groups may be larger in some schools and smaller in others. In such cases the extent of the effect of ethnicity on academic achievement may relate to identifiable school-level characteristics. On the other hand, the school's effect on student academic achievement may also vary among individuals. For example, while students usually benefit from smaller student/teacher ratios, these ratios and other school features are unlikely to influence all types of students equally. Cross-level interactions in multilevel modeling enable us to assess the degree to which relationships between individual explanatory and outcome variables are moderated by group level variables.

Good examples of this class of multilevel studies can be drawn from population studies. It is well-known that fertility levels vary among countries. In general, fertility is low in developed countries and high in developing countries. Fertility has multi-level determinants. Individual fertility behavior is determined not only by individual characteristics, such as a couple's preference for children, ethnicity, education, and income at the micro level. Features of the social contexts or social environments where the individuals live, such as culture or subculture, GDP, average education level, and in particular, the intensity/efficacy of the family planning programs (FPP) at the macro-level can also produce measurable effects. Assessing cross-level interactions is very important in fertility studies. FPP analysts and officers are interested in knowing: What individual characteristics influence individual fertility behaviors? Do family planning programs work? How do differences in program implementation among various locations or macro-level units affect individual fertility behaviors? And, for what classes of people are family planning programs most effective? Multilevel modeling helps us to gauge how family planning programs interact with individual characteristics to affect fertility behavior.

Public health studies indicate that individual health behaviors and outcomes are jointly determined by individual and environmental factors (Von Korff et al., 1992; Duncan et al., 1996; Diez-Roux, 1998; Wang et al., 1998). For example, initiation of smoking among adolescents may be associated with gender, ethnicity, school achievement and family background, as well as the social setting in which the individual is imbedded, such as geographic location, prevalence of smoking, and restrictions on smoking in public areas.

From these examples we can see that research interest in social science studies often centers on questions like: 1) what and how explanatory variables measured at the individual level affect the individual-level dependent variable, 2) what and how variables measured at the context or group level affect the individual-level dependent variables, 3) how the relationships between the individual-level explanatory and dependent variables vary across contexts or groups, and 4) what and how group-level variables moderate the effects of individual level variables on the individual-level dependent variable.

To answer these questions, both micro and macro data are needed. A common challenge in multilevel data is within-group observation dependence. That is, individuals in the same group tend to be alike and share similar attitudes and behaviors relative to individuals from other groups. For example, people living in the same neighborhood may share similarities with each other because they are influenced by the same neighborhood socio-economic characteristics. This may be true even for groupings that are only recently established. For example, students who are in the same school may not be associated with each other before they get into the same school. Once students enter a school, they become members of the same group. Once groupings are established, individuals in the group will tend to share traits that differentiate them from members of other groups. In statistical terms we say there exist within-group homogeneity and between-group heterogeneity in the hierarchically structured data.

Traditional analytical methods such as Ordinary Least Squares (OLS) Regression assume that observations are independently, identically distributed (IID). The same assumption is required for generalized linear models. Violation of this assumption will result in incorrect inference in statistical analysis. Chapter 2 demonstrates how observation dependence can be measured using an *Intraclass Correlation* Coefficient (ICC). Studies show that even a small ICC can lead to substantial Type-I errors in statistical testing, thus falsely rejecting a true null hypothesis. Dealing with ICC has been a challenge in statistical analysis of multilevel data for many years.

Multilevel models provide an appropriate analytical framework to deal with observation dependence in multilevel data. More importantly, multilevel models permit us to explore the nature and extent of the relationships at both micro and macro levels, as well as across levels.

1.2 Hierarchically structured data

Hierarchical social structures naturally give rise to hierarchical or multilevel data in which the lower level units are nested or grouped in the next higher level units. Such hierarchically structured data exist in many real life situations. The simplest and the most often used multilevel data are collected at two levels (i.e., one micro level and one macro level). For example, a study on student academic achievement may collect information at the student level and at the school level for multilevel modeling.

Multilevel designs can be readily extended to more than two levels. For example, students are nested in classes, and classes nested in schools; thus observation units lie at three levels of a hierarchy: the level-1 units are students; the level-2 units are classes; and the level-3 units are schools. The lowest level units (e.g., students) are the micro-level units or individual units, while the higher level units are the macro level units or context/group units.

Hierarchically structured data may arise in a variety of forms and from a variety of situations, either observed or by design. Survey data obtained from a complex sampling design are hierarchically structured. Multi-stage or cluster sampling is conducted to take full advantage of information from a hierarchy of study units. The first stage or "*Primary Sampling Unit*" (PSU) is often a well-defined geographic unit (e.g., county in a state). Once the PSUs are randomly selected, further stages of random selection are carried out within the PSUs (e.g., districts in a county) until the final units (e.g., households or individuals) are selected (Kalton, 1983). As a result, the survey data collected from cluster sampling design have a hierarchical structure in which individuals are nested within higher level sampling units.

Hierarchical data also frequently arise from experimental designs. For example, clinical trails may be carried-out in randomly selected clinics or medical centers, thus creating data that have a hierarchical structure. However, in practice clinics and medical centers are often not randomly selected. This is also true in many multi-site research projects. For example, a national multi-site research project on public health is often conducted with many project sites located in different regions, states, or cities. Very often, rather than being randomly selected, project sites are selected based on the quality of the grant proposals, the level of seriousness of the health problems under study, or the feasibility of conducting a successful study in a specific site. Although the distribution of the project sites may be carefully taken into consideration, they are not randomly selected, thus they are not representative of the corresponding higher level units in the targeted population. As a result, inferences based on the multilevel analysis for non-randomly selected study sites should be interpreted with caution.

Hierarchical data structures are not confined to cross-sectional settings with multiple units. Individuals may also be higher level units. For example, in longitudinal or panel studies individuals are followed up over time. Data are collected repeatedly from the same individuals. Such longitudinal data can be considered hierarchically structured. The repeated measures for each individual at different times are level-1 observation units, and individuals become the level-2 units. A third level can be introduced into the data structure, if the higher level units (e.g., clinics) in which individuals are nested are available.

Depending on the situation, some individuals may be considered as level-1 units while other individuals are higher level units. For example, patients and doctors can form a multilevel data structure. As a doctor treats multiple patients, the doctor may be considered as a level-2 unit, while patients are level-1 units. Similar situations include teachers and students, coaches and athletes, as well as interviewers and interviewees.

Finally, a special type of hierarchical data arise from meta-analysis in which results or findings from a series of related studies are summarized quantitatively to assess consistency or inconsistency of study results (Glass, 1976). In meta-analysis data, individuals are nested within specific studies. However, it is usually difficult or impossible for researchers to obtain the raw data from all the studies of interest. As such, a special approach is required for multilevel modeling of meta-analysis data. Detailed examples of formulating multilevel models for meta-analysis are available in many studies (Goldstein et al., 2000; Raudenbush & Bryk, 2002).

1.3 Variables in multilevel data

In addition to the format of multilevel data, choosing the variables that describe the features of the distinct levels of the hierarchical structure is another important consideration. For multilevel analysis, the dependent variable is measured at the individual level and explanatory variables are measured at both individual and group level or at both micro and macro levels. As in regular statistical analysis, individual explanatory variables usually include socio-demographic characteristics (e.g., gender, ethnicity, education, age) and other measures such as psychological status and behaviors, depending upon the analyst's conceptual model.

Contextual variables are macro level measures. They can be aggregate measures, such as mean values of some individual measures (e.g., average family income) or proportion of individuals for a particular characteristic within a particular context (e.g., percentage of minority population). These contextual variables represent the collective social characteristics of contexts/groups. They can be derived from either the sample or obtained separately from other sources such as census or government statistics.

Many contextual variables are not aggregations of individual information. Some characteristics are unique to contexts/groups and cannot be captured at the individual level. For example, in studies of student school performance, contextual variables could include aggregate measures such as student gender ratio or average enrollment test scores, and school feature measures such as school ranking, student-teacher ratio or teacher's level of experience. The former is an aggregation of student data and can be generated from the sample; the latter represents contextual aspects of the schools that must be collected from other sources at the school level.

Contextual variables can also be categorical measures. For example, in a multilevel study on childhood obesity in which children are level-1 units and neighborhoods are the level-2 units. The researcher may include a dummy variable (1-yes; 0-no) to indicate whether there are fast-food restaurants in the neighborhood because easy access to fast food may have a significant impact on children's diet, thus on their obesity level.

Conceptually, one might use J-1 dummy variables to represent all the contextual features of the J groups. This approach, however, is not feasible even with a moderately large number of groups because too many dummy variables would be needed to represent the groups.

The following tables illustrate a fictional two-level data structure. Table 1.3.1 shows the individual level outcome variable y_{ij} and independent variable x_{ij} for the ith individual in the jth group. There are a total of n_j individuals in the jth group, and individuals in all the groups sum up to the total sample size $\sum n_j = N$. z_j is a contextual variable describing the group. The values of the variable z_j for specific groups ($j = 1, 2, \ldots, J$ groups) are shown in Table 1.3.2.

Table 1.3.1 Individual level data

Unit		Variable	
Group	Individual	y_{ij}	x_{ij}
1	1	5	11
1	2	3	8
⋮	⋮	⋮	⋮
1	n_1	2	7
2	1	6	12
2	2	9	10
⋮	⋮	⋮	⋮
2	n_2	10	15
3	1	11	15
3	2	15	18
⋮	⋮	⋮	⋮
3	n_3	16	20
⋮	⋮	⋮	⋮
J	1	4	7
J	2	5	9
⋮	⋮	⋮	⋮
J	n_J	6	8

Note:
n_1, n_2, and n_j — Number of individuals in the first, second, and the jth groups, respectively. $\sum n_j=N$.
y_{ij} and x_{ij}—Individual level outcome and independent variable, respectively.

Individual level data and group level data shown in Tables 1.3.1 and 1.3.2 are integrated into a mixed data set and shown in Table 1.3.3. When merging the data sets, both individual ID and group ID must be matched for every individual. As such, the same value of the contextual variable z_j of group j is assigned to each individual in this group. Consequently, the value of z_j does not vary across individuals within the same group (see Table 1.3.3).

Table 1.3.2 Group level data

Group	z_j
1	8.7
2	12.3
3	17.6
⋮	⋮
J	8.0

Note:
z_j — Contextual variable at the group level.

Table 1.3.3 Individual and group level mixed data

Unit		Variable		
Group	Individual	y_{ij}	x_{ij}	z_j^*
1	1	5	11	8.7
1	2	3	8	8.7
⋮	⋮	⋮	⋮	⋮
1	n_1	2	7	8.7
2	1	6	12	12.3
2	2	9	10	12.3
⋮	⋮	⋮	⋮	⋮
2	n_2	10	15	12.3
3	1	11	15	17.6
3	2	15	18	17.6
⋮	⋮	⋮	⋮	⋮
3	n_3	16	20	17.6
⋮	⋮	⋮	⋮	⋮
J	1	4	7	8.0
J	2	5	9	8.0
⋮	⋮	⋮	⋮	⋮
J	n_J	6	8	8.0

Note:
*—The same value of contextual variable z_j of the jth group is assigned to each individual in the group.

The data format for multilevel modeling varies slightly by computer programs. Some programs require separate individual and groups data sets while others work with mixed data formats like the one shown in Table 1.3.3.

1.4 Analytical problems with multilevel data

Prior to the availability of multilevel analytical techniques and computer programs, multilevel data were analyzed separately at a single level, either the individual level or the group level[1]:

Individual level model:

$$y_{ij} = \beta_0 + \beta_1 x_{ij} + \varepsilon_{ij} \quad (1.4.1)$$

[1] For simplicity, only one independent variable is included in each model.

Group level model:

$$\bar{y}_j = \gamma_0 + \gamma_1 \bar{x}_j + \varepsilon_j \tag{1.4.2}$$

Equation 1.4.1 is a model at the individual level in which both dependent and explanatory variables are measured at the individual level. Equation 1.4.2 is a model at the aggregate or group level in which both dependent and explanatory variables are measured as the mean values of the corresponding individual level variables. The underlying problem encountered in such an approach is that it ignores the multilevel structure of the data. Model 1.4.1 ignores group membership and focuses exclusively on individual-level characteristics and inter-individual variation and thus ignores the potential importance of group-level features in influencing individual-level outcomes. Another serious problem with this model is that it assumes the independence of observations. As discussed in Section 1.1, generally individuals within each group are more alike compared with those in other groups. Thus the within-group observations are unlikely to be independent.

Model 1.4.1 cannot control the intraclass correlation coefficient (ICC), it ignores the within-group observation dependence, and thus violates the basic assumption underlying traditional regression models. As a result, standard errors of parameter estimates would be biased downwards, resulting in a large Type I error — falsely rejecting a true null hypothesis in statistical significance testing (De Leeuw & Kreft, 1986; Snijders & Bosker, 1999; Hox, 1998, 2002). Even a small ICC can lead to a Type I errors. (Hox, 1998; Barcikowski, 1981). Consequently, analyzing multilevel data with traditional regression models can produce misleading conclusions.

Model 1.4.2 focuses exclusively on the inter-group variation and on the data aggregated to the group level. The group-level model eliminates the observation dependence problem, but ignores the role of individual-level variables in shaping the outcome on one hand; and on the other hand, it substantially reduces statistical power by using a group level sample with a much smaller sample size.

Traditionally, researchers tended to use model results at one level to draw statistical inferences at another level. This has proven incorrect. The results from the two single level models frequently differ either in magnitude or in sign. The relationships found at the group level are not reliable predictors for relationships at the individual level, and vice versa. This phenomenon is known as the *ecological fallacy, aggregation bias,* or the *Robinson effect* (Robinson, 1950).

What causes *Robinsin effect*? Model 1.4.2 analyzes the variation in variable $\bar{y}_{ij}$ at the group level. *Aggregating* individual measures changes their meaning. If x_{ij} is a continuous measure (e.g., age), then $\bar{x}_j$ would be the average or mean value of the x_{ij} (e.g., mean age) in the jth group of individuals. If x_{ij} is a dichotomous variable, denoting gender (e.g., 1-male; 0-female), then $\bar{x}_j$ would be the proportion of males in the jth group. Clearly, x_{ij} and $\bar{x}_j$ are different measures, and we should not expect them to have the same effect in separate models based on either individual or group data.

A critical analytical problem with multilevel data is the *heterogeneity of relationships* of independent variables with the dependent variable. The relationship between individual level dependent and independent variables may vary across groups. For example, suppose we were studying academic performance of minority students in high schools. The average academic performance score for the minority students may vary across schools. The effect of minority status on the dependent variable may vary across schools for a variety of reasons. The proportion of minority students in a school, a "sample composition contextual variable" might partially account for the variation in performance in addition to other contextual variables at the school level.

In the past, heterogeneity of micro level relationships was often studied using the following fixed-effect regression model:

$$y_{ij} = \beta_0 + \beta_1 x_{ij} + \beta_2 \overline{x}_j + \beta_3 x_{ij} \cdot \overline{x}_j + \varepsilon_{ij} \tag{1.4.3}$$

where y_{ij} denotes the performance score for the ith student in the jth school; x_{ij} is a dummy variable (1-yes; 0-no) indicating the minority status at the student level; and $\overline{x}_j$ denotes the proportion of minority students in the jth school. In this model the macro-level (e.g., school level) variables were disaggregated to the micro-level (e.g., student level). In this example, students are assigned various school-level variables and all students in the same school are assigned the same value on the school-level variable (e.g., $\overline{x}_j$). The model is then run at the student level. Slope coefficients of β_1 and β_2 are the main effects of the individual level variable x_{ij} and school level variable $\overline{x}_j$ respectively. The slope coefficient β_3 is the interaction of these two variables. If the cross-level interaction is statistically significant, we conclude that the relation between student's minority status and the performance score is influenced or moderated by the proportion of minority students at the school level. This kind of model takes into account the effects of contextual variables on the relationships of individual explanatory variables and the dependent variable at individual level.

One serious problem with this model is that it treats observations as independent though they are not, thus leading to biased standard error estimates. In addition, in this fixed-effect model, the variation in the intercept and slope coefficients are assumed to be perfectly explained by group level variables without error, which is highly unlikely.

Van de Eeden (1988) and others have examined the *heterogeneity of relationships* problem using a two-step approach. In Step 1, they estimated the individual level regression models for each group separately. The assumption of invariance in the intercept and slope coefficients is tested by running multi-group regression models with and without equality restrictions on the coefficients across groups, using structural equation modeling software such as LISREL. If the coefficients show significant variance across groups, then the second step is to regress each of the regression coefficients on the contextual variables at the group level.

Although this approach enables analysts to test the significance of variations in the regression coefficients estimated in Step 1, it has several limitations. OLS models are used at both Steps 1 and 2, even though it is technically incorrect to use OLS to estimate the standard errors in the second step (De Leeuw & Kreft, 1986, p. 61). It is also impractical to run separate regression for each group when the number of groups is large, and particularly when the number of observations per group is small. This approach treats the groups as unrelated and ignores the likelihood that the groups are drawn from a larger population of groups that share common attributes.

Given the shortcomings of traditional methods, a new statistical method, called multilevel modeling is needed. Multilevel models are explicitly designed to analyze hierarchically structured data, modeling variables at both micro and macro levels simultaneously without aggregation or disaggregation. In the following section we will discuss the advantages as well as the limitations of multilevel models for multilevel data analysis.

1.5 Advantages and limitations of multilevel modeling

The problems encountered in traditional multilevel data analysis can be readily solved by multilevel modeling. First, the *Robinson effect* is avoided because multilevel models simultaneously analyze data obtained from both the individual level as well as the context/group level. The effects of individual and contextual variables are simultaneously examined in a multilevel model. In addition, the assumption of observation independence is not required in multilevel modeling because multilevel models are designed to measure and thus account for ICC in hierarchically structured data. The inflated type-I error in statistical testing that results from observation dependence is corrected in multilevel modeling. We will discuss this issue in detail in the section on model estimation in Chapter 2.

Multilevel modeling advances the theoretical and methodological aspects of social science studies. Multilevel models provide a convenient analytical framework with concordance between theoretical approaches and statistical analysis for studying data that have a hierarchical structure. This framework enables researchers to test multilevel theories statistically by systematically analyzing the effects of micro and macro factors on outcome measures, as well as testing cross-level interactions among macro and micro level variables. With multilevel modeling, researchers are able to decompose variation in the outcome measure into within and between group variations, and to understand where effects on the outcome measure are occurring, and how the effects of individual level variables on outcome measures are moderated by the group level variables.

These features of multilevel models facilitate a wide range of applications in social science studies. For example, women's child preference is one of the most important micro level variables in population studies, while the intensity/efficacy of family planning programs is one of the most important macro level variables that influence individual fertility or the number of children ever born (CEB) to a woman. While both the micro and the macro variables affect women's fertility, the strength of the effect of women's child preference on fertility is usually strongly moderated by the intensity/efficacy of family planning programs in the social setting where the women live. At the same time, the strength of family planning programs on individual fertility may vary among individuals. Program may be more effective for women who have lower child preference.

Multilevel modeling is a particularly useful analytical approach when data are sparse. For example, in a study of minority student performance, regression models cannot provide valid statistical inferences if the number of minority students in the sample from a school is too small. However, if a number of schools from which the students are sampled are available, then multilevel models can utilize information from all sampled schools, thus compensating for sparseness of data in particular schools with a small number of minority students. In Chapter 2 we will discuss the "*empirical Bayes*" estimator used in multilevel model estimation in which data from each group are combined with the data summed over all groups. Model parameters are estimated using individual group data separately, as well as using the entire sample with all the groups. This approach is called "*shrinkage estimation*" that enables the model to "*borrowing strength*" from data drawn from all the groups to "support" statistical estimation for groups with too few cases. Sizes of groups vary considerably in most quasi-experimental studies. Therefore, sparseness is likely to be the case for some groups. Multilevel modeling is an effective method for analysis of such data.

Finally, multilevel models can be readily extended to study growth or change trajectories of outcome measures over time using longitudinal data. In this case the multilevel model becomes a *growth model* (GM) which is clearly superior to traditional "repeated measure" analysis methods (Raudenbush & Bryk, 2002). The GM examines not only intra-individual changes over time, but also inter-individual variations in these changes. Other benefits of GM include: 1) GM does not require the data to be balanced (i.e. equal numbers of cases at each time) nor equal intervals between time points. 2) Missing values caused by attrition are allowed under the assumption of "missing at random" (MAR). 3) In a GM, time-varying covariates can be readily included. 4) The association between the rate of outcome change and the initial level of the outcome measure can be assessed. And 5) GM can be readily expanded to a three or more level model by including more layers or higher level units to which individuals belong. We will provide a detailed discussion of the GM in Chapter 4.

Of course, no statistical method is a panacea for data analysis in social science studies. Limitations of multilevel models include, but are not limited to:

1) Multilevel models are more complex than ordinary regression models because outcome variation at both micro and macro levels are modeled simultaneously. As a result, the number of model parameters can become large so that the model may become less parsimonious.

2) Very often the higher level units or groups are selected on the basis of convenience rather than being randomly sampled from a well-defined population. In these cases it is potentially incorrect to infer model parameter estimates to other groups.

3) Very often, contextual variables are measured by aggregating individual data within groups in the sample rather than the groups in the target population. If the number of observations in a group is not large enough, then the aggregated composition measures of the group may be biased and result in misleading group information.

4) Individuals are mobile, and it is reasonable to expect that members of a group may not have entered the group at the same time. Length of exposure to group influences may have systematic effects upon individuals. One possible solution to this problem is to control for duration of group membership in data analysis. Unfortunately, information on membership duration is frequently unavailable. We are then left with no alternative but to assume that everyone in a group is affected by the group in the same way and to the same degree.

5) Like traditional regression analysis, all explanatory variables are typically assumed to have no measurement error. In fact, very often variable measurement includes errors and may vary from measurement to measurement. For example, measurements of a person's blood pressure are likely to vary, even under the same conditions. Some of this variation is simply random measurement error. Structural equation models (SEM) are designed to assess underlying conceptual relationships while controlling for measurement error (Jöreskog, 1971, 1977; Jöreskog & Sörbom, 1979; Bentler, 1983; Bollen, 1989). However, multilevel models we introduce in this book assume that all explanatory variables are measured without errors. Readers who are interested in multilevel structural equation modeling, which deals with multilevel data and measurement errors simultaneously, are referred to Muthén & Muthén (1998—2010).

6) In multilevel modeling, researchers often encounter data with a relatively small number of higher level units or groups. As a result of this and/or non-normality of the residuals, model parameter estimates, particularly the variance components and standard errors of parameter estimates at the group level, may be biased (Bussing, 1993; Van der Leeden & Bussing, 1994; Van der Leeden et al., 1997; Hox, 1998). We will discuss how to deal with this issue in Chapter 6.

7) The multilevel data considered in this book are assumed to be completely nested. That is, each individual belongs to only one group. If individuals are nested within more than one group, then mixed models with crossed random effects should be applied (Raudenbush, 1993). This topic is beyond the scope of this book.

1.6 Computer software for multilevel modeling

It was technically impossible to fit multilevel models until the early 1980s when Dr. William Mason and colleagues at the University of Michigan Population Studies Center developed the GENMOD computer program. This software was developed to run in DOS environment and was unfortunately never upgraded to MS Windows. However, a growing number of computer programs have been developed to fit multilevel models in the past two decades. In recent years the major statistical software companies, such Statistical Analysis System (SAS) and Statistical Package for Social Sciences (SPSS) have incorporated procedures or modules for multilevel modeling. Each of the available programs has strengths and weaknesses. The choice of computer software is a matter of personal preference. The major computer software currently available for multilevel modeling, includes, but is not limited to:

- HLM: This is the first commercial computer software designed for multilevel modeling. This user-friendly program was initially developed in the mid-1980s and has been in active development ever since. HLM was a leading package during the development of multilevel modeling in the 1990s, and has been widely used since. The program is developed by Drs S. W. Raudenbush and A. S. Bryk and distributed by the Scientific Software International (SSI) in U.S.A. (www.ssicentral.com).
- MLwin: This is a popular special-purpose computer program for multilevel modeling developed by Dr. Harvey Goldstein and his colleagues in the Centre for Multilevel Modelling and other institutes in UK. MLwin was first released in 1997 and Version 2 was released in beta form in 2003 (www.mlwin.com). The program provides a system for the specification and analysis of a wide range of multilevel models. A graphical user interface (GUI) is available for model specification, along with plotting, diagnostic and data manipulation tools.
- Mixed-Up Suite and SuperMix: This is a family of standalone programs including MIXREG, MIXOR, MIXNO, MIXPREG, and MIXGSUR for multilevel modeling for continuous, binary, ordinal, nominal, count, or survival outcome measures. The software was developed by Drs. Donald Hedeker and Robert D. Gibbons of the University of Illinois at Chicago, U.S.A., and Version 1.0 was released by SSI in 2008. (www.ssicentral. com/supermix).
- aML: Computer software developed by American economists Lee Lillard and Stan Panis became commercially available in 2000. The software offers a wide range of models for multilevel data analysis. It extends multilevel modeling to fitting econometric models such as simultaneous equation models with multilevel data. aML is a product of EconWare, a California corporation, U.S.A. Full details on ordering aML can be found at www.applied-ml.com.
- EGRET: Computer software designed for analyzing biomedical and epidemiological data. It has been widely used by epidemiologists and biostatisticians for fitting generalized linear models with and without random effects and survival models. The software was originally developed in 1999 at the School of Public Health of University of Washington U.S.A. The current Window version is developed by CYTEL Software Corporation of Cambridge, U.S.A. (www.cytel.com/products/egret).
- LISREL: is the earliest and still very popular statistical package for structural equation modeling (SEM)[2]. LISREL began implementing multilevel modeling in 1999, with version 8.30. MULTILEV fits multilevel linear and nonlinear models to multilevel data from simple random and complex survey designs. It allows for models with continuous and categorical response variables. LISREL is distributed by the Scientific Software International (SSI) in U.S.A. (www.ssicentral.com).
- Mplus: Initially released in 1998, Mplus provides a generalized modeling framework for structural equation modeling with continuous and categorical observed variables, as well as continuous and categorical latent variables. Mplus V.3. and subsequent versions include a

[2] Structural equation modeling is a more generalized statistical analysis approach. It provides a powerful analytic framework that allows for the simultaneous estimation of the relations between a set of observed variables and a smaller set of underlying latent constructs, as well as the relations among the latent constructs (Bentler, 1980, 1983; Jöreskog, 1971a, 1971b; Jöreskog & Sörbom, 1979; Bollen, 1989).

multilevel extension of the full modeling framework. Multilevel structural equation models can be readily run in Mplus. Detail information about Mplus is available at www.statmodel.com.

- STATA: A general purpose statistical program has been used increasingly in recent years (www.stata.com). STATA provides a broad statistical base for data analysis. Its treatment of generalized linear mixed models has capacity of implementing multilevel modeling for continuous, binary, and count outcomes, as well as crossed random effects model.
- SPSS: Statistical Package for Social Sciences (SPSS), another major statistical package supports multilevel modeling with the Linear Mixed Models procedure in the Advanced Models module since version 11.5. With SPSS, most commands are available either through the graphical user interface or through the use of command syntax. SPSS is distributed by the Scientific Software International (SSI) in U.S.A. (www.ssicentral.com).
- SAS: The internationally recognized Statistical Analysis System (SAS or SAS/STAT) has provided several procedures adaptable to multilevel modeling since Release 6.0. The *PROC MIXED* has been improved and now is widely used for multilevel modeling for continuous outcome measures. Two other procedures, *PROC GLIMMIX* and *PROC NLMIXED*, are designed to fit multilevel models for categorical outcome measures and count data (e.g., multilevel logit, probit, ordered logit, multinomial logit, Poisson, and ZIP models). Information about obtaining SAS is available at www.sas.com.
- A variety of other statistical packages, such as LIMDEP (www.limdep.com), GenStat (sue.welham@lshtm.ac.uk), SYSTAT (www.systat.com), S-Plus (www.insightful.com), WINBUGS (www.mrc-bsu.cam.ac.uk/bugs), and R (cran.r-project.org) also have functions for conducting multilevel modeling.

Throughout this book, we use SAS, specifically SAS *PROC MIXED*, *PROC NLMIXED* and *PROC GLIMMIX* for analysis of continuous and categorical data in multi-level models.

Chapter 2
Basics of linear multilevel models

This chapter introduces the basics of multilevel models. Readers will gain a basic understanding of the conceptual framework, methods, and strategies of multilevel modeling. Although multilevel modeling is designed to analyze data with two or more levels, we focus on the most commonly used two-level model. The expansion from two-level models to three-level models is introduced at the end of the chapter.

2.1 Intraclass correlation coefficient (ICC)

As noted in Chapter 1, multilevel modeling enables analysts to accommodate "within-group observation dependence" in analysis of the hierarchically structured data. Within-group observation dependence is usually measured by the intraclass correlation coefficient (ICC). In this section, we will define ICC, describe how to calculate ICC, and explain why ICC is a measure of both within-group homogeneity and between-group heterogeneity.

Shrout and Fleiss (1979) provide a classical definition of ICC as the ratio of the *between-group variance* to the *total variance* (i.e., the sum of the within-group and the between-group variances):

$$\text{ICC} = \frac{\sigma^2_b}{\sigma^2_b + \sigma^2_w} \tag{2.1.1}$$

where σ_w^2 and σ_b^2 are the within-group variance (or variance at the individual level) and the between-group variance (or variance at the group level), respectively; and ($\sigma_w^2 + \sigma_b^2$) is the total variance of the outcome measure. As such, ICC is defined as the ratio of the between-group variance to the total variance, representing the proportion of the total variance that is due to the between-group variance.

From eq. 2.1.1, we know that ICC can be readily estimated if σ_w^2 and σ_b^2 are known. Usually, σ_w^2 and σ_b^2 are estimated from the *empty model* (*intercept-only model* or *unconditional means model*), which is identical to a *one-way random effect analysis of variance* (ANOVA):

$$y_{ij} = \beta_{0j} + e_{ij} \tag{2.1.2}$$

$$\beta_{0j} = \gamma_{00} + u_{0j} \tag{2.1.3}$$

$$y_{ij} = \gamma_{00} + u_{0j} + e_{ij} \tag{2.1.4}$$

where explanatory variables are included in neither level-1 nor level-2 models. In eq. 2.1.2 β_{0j} and e_{ij} represent mean outcome of group j and the random individual variation around this mean. In eq. 2.1.3 γ_{00} denotes the overall intercept representing the grand mean of y_{ij}; and u_{0j} captures the variation between group means (i.e., the deviation of the mean of the jth group from the grand mean). Equation 2.1.4 is a *combined model* in which the outcome measure y_{ij} is a linear combination of two parts: 1) the fixed part (i.e., γ_{00}), which is the grand mean across all the individuals in the sample; and 2) the random part, consisting of the two random effects: u_{0j} for the group-specific variation from the grand mean and e_{ij} for the within-group individual variation around group-specific means.

Because no covariates are involved, the outcome variance in the empty model is not explained at all. It can be decomposed into two components: within-group variance and between-group variance. The total outcome variance is described as:

$$\begin{aligned}\text{Var}(y_{ij}) &= \text{Var}(\gamma_{00} + u_{0j} + e_{ij}) \\ &= \text{Var}(u_{0j}) + \text{Var}(e_{ij}) \\ &= \sigma_{u0}^2 + \sigma^2\end{aligned} \tag{2.1.5}$$

where $\sigma_{u0}^2 = \sigma_b^2$ representing the between-group variance, and σ^2 (or σ_w^2) is the within-group variance. Then σ_b^2 and σ_w^2 can be used to calculate ICC using eq. 2.1.1.

From another perspective, ICC can also be considered as a measure of the within-group observation dependence. Suppose e_{ij} and $e_{i'j}$ are independent level-1 residuals in group j, and the level-1 and level-2 residuals (i.e., e_{ij} and u_j) are also independent of each other. The covariance between individual observations y_{ij} and $y_{i'j}$ $(i \neq i')$ in the same group can be expressed as:

$$\text{Cov}(y_{ij}, y_{i'j}) = \text{Cov}[(u_{0j} + e_{ij}), (u_{0j} + e_{i'j})] = \sigma_{u0}^2 \tag{2.1.6}$$

As such, within-group observation covariance is equal to the between-group variance σ_{u0}^2. That is, the *within-group homogeneity* indicates the *between-group heterogeneity* and vice versa.

Further, the correlation between individual observations y_{ij} and $y_{i'j}$ in the same group j can be estimated as:

$$\begin{aligned}\rho(y_{ij}, y_{i'j}) &= \frac{\text{Cov}(y_{ij}, y_{i'j})}{\text{Var}(y_{ij})^{1/2}\,\text{Var}(y_{i'j})^{1/2}} = \frac{\text{Cov}(y_{ij}, y_{i'j})}{\text{Var}(y_{ij})} \\ &= \frac{\sigma_{u0}^2}{\sigma_{u0}^2 + \sigma^2} = \text{ICC}\end{aligned} \tag{2.1.7}$$

From eq. 2.1.7, we know that ICC can be considered a correlation among individuals within the same group. However, the values of ICC do not range between−1 and 1, but between 0 and 1 since both the numerator and denominator in eq. 2.1.7 are variances whose values should be equal to or greater than 0 by definition. In addition, ICC is interpreted as the proportion of total variance explained by the between-group variation. The proportion of the explained variance can never be a negative number. Although a negative ICC is considered to be possible in some studies (Tammemagi et al., 1995), and some computer programs (e.g., LISREL) provide an option for multilevel modeling with a negative ICC, Giraudeau et al. (1996) proved that negative values of ICC are not theoretically possible. Empirical estimates of ICC may sometimes be negative when a model is mis-specified or when the between-group variance is close to zero. When the between-group variation is very large, relative to the within-group variation, ICC will approach 1. On the contrary, when the within-group observations are independent of each other, ICC will approach 0, indicating no within-group homogeneity. In this case the multilevel model would be reduced to a fixed-effect model (i.e., eq. 1.4.3).

As within-group homogeneity is also an indication of the between-group heterogeneity, the significance testing for ICC is equivalent to testing for the null hypothesis that the between-group variance is zero. If the intercept variance in eq. 2.1.2 or variance of u_{0j} in eq. 2.1.3 is statistically significant, we conclude that ICC is statistically significant and the multilevel modeling should be considered for explaining the between-group heterogeneity in data analysis. If

ICC is not statistically significant, then the multilevel modeling is not necessary and a traditional multiple regression model is appropriate for data analysis.

2.2 Formulation of two-level multilevel models

The empty model specified in eqs. 2.1.2—2.1.4 is a fundamental two-level multilevel model, which provides important preliminary information about within-group and between-group variations in the outcome measure under study. Once ICC has been tested to be statistically significant, explanatory variables are included in the model. For simplicity, we start with a two-level multilevel model which involves only two level-1 explanatory variables and one level-2 explanatory variable in the following equations:

$$y_{ij} = \beta_{0j} + \alpha_1 x_{1ij} + \beta_{1j} z_{1ij} + e_{ij} \tag{2.2.1}$$

$$\beta_{0j} = \gamma_{00} + \gamma_{01} w_{1j} + u_{0j} \tag{2.2.2}$$

$$\beta_{1j} = \gamma_{10} + \gamma_{11} w_{1j} + u_{1j} \tag{2.2.3}$$

where y_{ij} represents level-1 outcome measure of the ith individual in the jth level-2 unit (e.g., the jth context or group); $i = 1, 2, \ldots, N$ (N is the total sample size), and $j = 1, 2, \ldots, J$ (J is the number of level-2 units). Equation 2.2.1 is the level-1 equation that models the variation of y_{ij}. The level-1 intercept β_{0j} has subscript "j" indicating that the level-1 intercept varies across level-2 units or groups. In other words, the expected value of y_{ij} corresponding to zero values of all level-1 explanatory variables, x_{1ij} and z_{1ij} in this model, varies across groups. Assuming the effect of level-1 variables x_{1ij} on y_{ij} does not vary across groups, while the effect of z_{1ij} on y_{ij} does, we have a fixed level-1 slope coefficient α_1 and a random level-1 slope coefficient β_{1j}, respectively.

Level-2 equations model the variation of the level-1 random regression coefficients (β_{0j} and β_{1j} in this model). Equations 2.2.2 and 2.2.3[1] illustrate the case of two level-2 equations corresponding to two level-1 random regression coefficients. The level-1 random regression coefficients become the dependent variables in the level-2 models.

Conceptually, the two-level model can be estimated in two stages: First, the same regression model is run using level-1 observations within each of the level-2 units. A total of J sets of regression coefficients would be estimated, which result in a set of J level-1 intercepts, J level-1 slope coefficients for independent variables, respectively. In the second step the estimated level-1 regression coefficients are treated as functions of level-2 variables. The level-2 model is constructed by regressing level-1 regression coefficients on appropriate level-2 variables. Before multilevel modeling computer packages became available, some multilevel data were analyzed as these separate stages. As we discussed in Chapter 1, the two separate-stage analysis approach is problematic. In multilevel modeling, two-stage modeling is conducted simultaneously.

Substituting the macro eqs. 2.2.2 and 2.2.3 into the micro eq. 2.2.1, we have:

$$y_{ij} = \gamma_{00} + \gamma_{01} w_{1j} + \alpha_1 x_{1j} + \gamma_{10} z_{1ij} + \gamma_{11} z_{1ij} w_{1j} + (u_{0j} + u_{1j} z_{1ij} + e_{ij}) \tag{2.2.4}$$

Equation 2.2.4 is a *combined model* or *composite model*[2], which is the single equation version of the multilevel model. This model looks similar to an ordinary linear regression model with an interaction term like the *fixed-effect* model shown in eq. 1.4.3. However, unlike the OLS *fixed-*

[1] For fixed level-1 coefficients, their variations at the group level are fixed at zero, and thus no macro models are specified for them.

[2] In econometric terminology, this composite model is called the *reduced model.*

effect model, the *combined model* has a composite error structure. This structure consists of all three terms in parenthesis in eq. 2.2.4; including the level-2 error terms u_{0j} and u_{1j}, level-1 error term e_{ij}, as well as level-1 independent variable z_{1ij}. Notice that the level-2 error terms, u_{0j} and u_{1j}, don't have subscript "*ij* ", but "*j* " only, indicating that their values don't vary within group, but do vary across groups. The observations within each group share common unexplained level-2 random variations. As a result, the within group observations are not independent of each other. In addition, the composite error term $(u_{0j} + u_{1j}z_{1ij} + e_{ij})$ depends on the values of u_{0j}, u_{1j} and z_{1ij}. The term $u_{1j}z_{1ij}$ is an interaction between the groups and variable z_{1j}. Consequently, the composite residual does not have a constant variance, resulting in *heteroscedasticity* in the error term. Because of these unique features of the composite error term, the assumptions for OLS, such as observation independence and equal residual variance or *homescedasticity* are violated. Accordingly, multilevel models must be implemented using special estimation methods. We will discuss estimation methods for multilevel models in Section 2.7.

A two-level model can be expressed in the following generalized form:

$$y_{ij} = \beta_{0j} + \sum_{p=1}^{P} \alpha_p x_{pij} + \sum_{q=1}^{Q} \beta_{qj} z_{qij} + e_{ij} \tag{2.2.5}$$

$$\beta_{0j} = \gamma_{00} + \sum_{m=1}^{M} \gamma_{0m} w_{mj} + u_{0j} \tag{2.2.6}$$

$$\beta_{1j} = \gamma_{10} + \sum_{m=1}^{M} \gamma_{1m} w_{mj} + u_{1j} \tag{2.2.7}$$

$$\vdots$$

$$\beta_{Qj} = \gamma_{Q0} + \sum_{m=1}^{M} \gamma_{Qm} w_{mj} + u_{Qj} \tag{2.2.8}$$

In this example, the level-1 intercept β_{0j} is random, a total of P level-1 explanatory variables x_{pij} have fixed-effect, and a total of Q level-1 explanatory variables z_{qij} have random effects on the dependent variable y_{ij}. Each of the level-1 random effects is specified as a linear function of M level-2 explanatory variables w_{mj} in the $Q + 1$ macro models (eqs. 2.2.6—2.2.8). Conceptually, $J-1$ dummy variables might be used in the macro models to capture all the group features, thus the systematic variation in the random level-1 regression coefficient would be fully explained. This approach, however, is not feasible because it will end up with too many dummy variables in the macro equations and too may cross-level interaction terms in the combined model. Contextual variables that capture group features in different dimensions are usually included in the macro models.

Substituting eqs. 2.2.6—2.2.8 into eq. 2.2.1, we have a combined model:

$$y_{ij} = \gamma_{00} + \sum_{m=1}^{M} \gamma_{0m} w_{mj} + \sum_{p=1}^{P} \alpha_p x_{pij} + \sum_{q=1}^{Q} \gamma_{q0} z_{qij} + \sum_{q=1}^{Q} \sum_{m=1}^{M} \gamma_{qm} w_{mj} z_{qij}$$

$$+ (u_{0j} + \sum_{q=1}^{Q} z_{qij} u_{qj} + e_{ij}) \tag{2.2.9}$$

The combined model consists of two parts: the fixed-effect part—all the regression coefficients (e.g., α and γ coefficients); and the random effects part—all the components in the composite residual term in eq. 2.2.9. The variance/covariance structure of the random components is one of the key issues in growth models or multilevel modeling for longitudinal studies. We will discuss this issue in Chapter 4.

In SAS, multilevel models are specified in matrix format (Littell et al., 1996):

$$Y = X\beta + ZU + e \tag{2.2.10}$$

where Y is the vector of observed outcome measure; X is the design matrix corresponding to the fixed-effect parameter vector β that contains the overall mean and all the fixed-effects (e.g., main effects and interactions); Z is the design matrix corresponding to the vector U that contains all the random effects; and e is a vector of level-1 residuals. Unlike some other multilevel programs (e.g., HLM), macro level or group level models (e.g., Eqs. 2.2.6—2.2.8) are not specified separately in SAS. Instead, the program identifies the variables that have invariant values within groups as group-level explanatory variables, otherwise as level-1 explanatory variables. In the following chapters, we will discuss and demonstrate how to run different multilevel models in SAS with real data.

2.3 Model assumptions

Unlike multiple regression models, multilevel models have micro and macro equations. Each equation has an error term resulting in a complex error structure in the combined model. The error term in the level-1 model (e_{ij} in eq. 2.1.2) represents *within* group variation or level-1 residuals, and the error terms in the level-2 models (u_{0j} and u_{1j} in eqs. 2.2.2 and 2.2.3) represent *between* group variations of the level-1 random regression coefficients (e.g., β_{0j} and β_{1j}), respectively. The level-1 residuals are assumed to follow a normal distribution, level-2 residuals are assumed to follow a multivariate normal distribution, and level-1 residuals are independent of level-2 residuals. Formally, these assumptions are expressed as:

$$e_{ij} \sim N(0, \sigma^2) \tag{2.3.1}$$

$$\begin{pmatrix} u_{0j} \\ u_{1j} \end{pmatrix} \sim N\left[\begin{pmatrix} 0 \\ 0 \end{pmatrix}, \begin{pmatrix} \sigma^2_{u0} & \sigma^2_{u01} \\ \sigma^2_{u01} & \sigma^2_{u1} \end{pmatrix}\right] \tag{2.3.2}$$

$$\mathrm{Cov}(e_{ij}, u_{0j}) = 0, \;\; \mathrm{Cov}(e_{ij}, u_{1j}) = 0 \tag{2.3.3}$$

where the level-2 residuals are not correlated with the level-1 residuals (e.g., $\mathrm{Cov}(e_{ij}, u_{0j}) = 0$ and $\mathrm{Cov}(e_{ij}, u_{1j}) = 0$); but level-2 residuals in different macro equations are allowed to be correlated with each other (e.g., $\mathrm{Cov}(u_{0j}, u_{1j}) = \sigma^2_{u01}$).

The error terms in multilevel models are described in matrix notation in the SAS manual:

$$\mathrm{E}\begin{pmatrix} u \\ e \end{pmatrix} = 0 \tag{2.3.4}$$

$$\mathrm{Var}\begin{pmatrix} u \\ e \end{pmatrix} = \begin{pmatrix} G & 0 \\ 0 & R \end{pmatrix} \tag{2.3.5}$$

where G is the variance/covariance matrix for level-2 residuals, and R is the variance/covariance matrix for level-1 residuals. For a two-level model with a level-1 random intercept and two random slope coefficients, the G matrix is:

$$G = \begin{pmatrix} \sigma^2_{u0} & \sigma^2_{u01} & \sigma^2_{u02} \\ \sigma^2_{u01} & \sigma^2_{u1} & \sigma^2_{u12} \\ \sigma^2_{u02} & \sigma^2_{u12} & \sigma^2_{u2} \end{pmatrix} \tag{2.3.6}$$

The more level-1 random regression coefficients, the more complex is the level-2 variance/covariance structure. In cross-sectional multilevel modeling, level-1 random variation is usually assumed to be a single variance. The R matrix is:

$$R = \sigma^2 I = \begin{pmatrix} \sigma^2 & 0 & \cdots & 0 \\ 0 & \sigma^2 & \cdots & 0 \\ \vdots & \vdots & \ddots & \vdots \\ 0 & 0 & \cdots & \sigma^2 \end{pmatrix} \tag{2.3.7}$$

However, when modeling the longitudinal data, a more complex but appropriate within-subject variance/covariance structure must be specified for each model. This issue will be discussed in Chapter 4.

Finally, as described in Chapter 1, explanatory variables in multilevel models are often assumed to be measured without error, the same as in traditional regression analysis. Measurement errors and multiple sources of variations can be handled simultaneously in multilevel structural equation modeling framework (Muthén & Muthén, 1998—2010). This topic is beyond the scope of this book.

2.4 Fixed and random regression coefficients

In a multilevel model, level-1 regression coefficients may be fixed or random, while all of the group-level regression coefficients in a two-level model are fixed. As in ordinary regression models, a fixed level-1 regression coefficient is constant and does not vary across groups or higher level units; while a random level-1 regression coefficient varies across groups. The following figures illustrate patterns of hypothesized level-1 regression intercepts and slopes for variable x in four groups (i.e., level-2 units). In Figure 2.4.1, within-group regression lines are parallel with the same positive slope but different intercepts. This indicates that this set of within-group regressions have no slope variation, but intercept variation across groups. The bold line represents the average slope or overall relationship between y and x. In Figure 2.4.2, the within-group slopes vary across groups, but the intercept of each regression line is the same. This implies that the group-specific means of y are the same at the point where $x = 0$, but the relationship (i.e., slope) between y and x varies across groups: the relationship is positive in two groups and negative in other two groups. On average, the overall relationship between x and y is close to zero (i.e., the bold overall regression line is almost flat). A more general situation is shown in Figure 2.4.3, in which both within-group intercepts and slopes vary across groups,

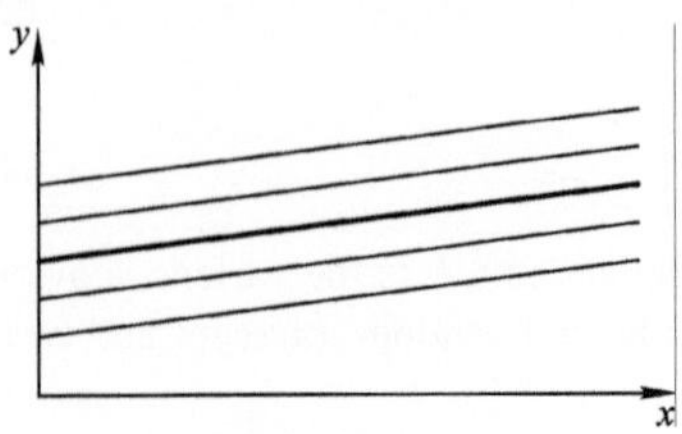

Figure 2.4.1

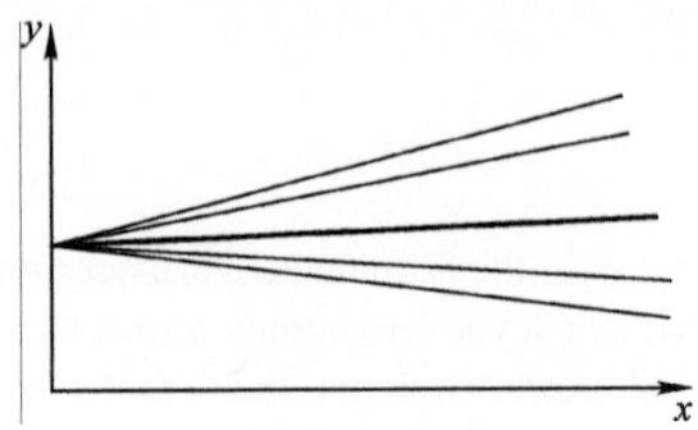

Figure 2.4.2

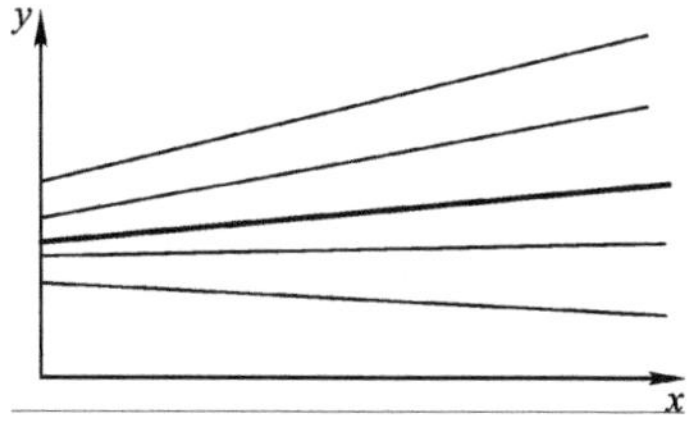

Figure 2.4.3

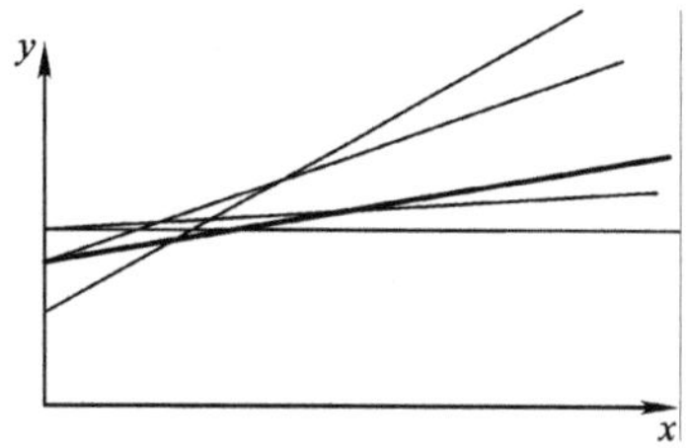

Figure 2.4.4

but on average, the effect of x on y remains positive yet not very strong (i.e., the slope of the bold regression line is not very deep). In Figure 2.4.4 we see a complex interaction between the level-1 independent variable x and group. The within group intercepts do not differ by much, but the variations in the relationships between x and y are profound.

The different patterns of group-specific outcome means and the relationship between level-1 dependent variable and independent variables shown in Figures 2.4.1—2.4.4 illustrate various intercepts and slopes of the within group regression lines. In two-level multilevel modeling, a level-1 random regression coefficient is assumed to vary randomly and be normally distributed in the population. For example, without including covariates in the level-2 equations, the level-1 random intercept and slope coefficients (β_{0j} and β_{1j}) in eq. 2.2.1 can be assumed to vary randomly across groups. This can be shown as below:

$$\beta_{0j} = \gamma_{00} + u_{0j} \tag{2.4.1}$$

$$\beta_{1j} = \gamma_{10} + u_{1j} \tag{2.4.2}$$

$$\begin{pmatrix} \beta_{0j} \\ \beta_{1j} \end{pmatrix} \sim N\left[\begin{pmatrix} \gamma_{00} \\ \gamma_{10} \end{pmatrix}, \begin{pmatrix} \sigma^2_{u0} & \sigma^2_{u01} \\ \sigma^2_{u01} & \sigma^2_{u1} \end{pmatrix}\right] \tag{2.4.3}$$

where the level-1 random intercept β_{0j} represents the jth group-specific mean, which varies across the J level-2 units according to a normal distribution with mean of γ_{00} and variance of σ^2_{u0}. The level-1 slope coefficient β_{1j} represents the jth group-specific effect of z_1 on y_{ij}, which varies across the J level-2 units according to a normal distribution with mean of γ_{10} and variance of σ^2_{u1}. The variances σ^2_{u0} and σ^2_{u1} summarize the degree to which the group-specific intercepts and slopes differ from their overall means.

The variation in each random micro-level coefficient is partially explained by the macro variables (e.g., w_1) included in the macro equations (e.g., eqs. 2.2.2 and 2.2.3). Unexplained variation remains random. R^2 like statistics estimate how much variation macro variables explain in each random coefficient. We will discuss this issue in Section 2.9.

We can conclude that the difference between level-1 random and fixed regression coefficients is that the former include a random term and the latter do not. If the random term in a higher-level equation equals zero, then the corresponding lower-level random coefficient becomes a fixed coefficient. Similarly, testing the hypothesis of zero group level error determines whether a level-1 coefficient is random (Raudenbush & Bryk, 2002). The expected value of a group level error term is normally zero. If the variance of the error term is also zero, then we can conclude the error term is zero. For example, if the test cannot reject the null hypothesis $\sigma^2_{u0} = 0$ (i.e., $\text{Var}(u_{0j}) = 0$), then the level-1 intercept β_{0j} would become a fixed coefficient which is constant across groups.

If all the level-1 regression coefficients are found to be fixed, the composite error term in the combined model would have only one term: e_{ij}. In this case, Model 2.1.4 would become:

$$y_{ij} = \gamma_{00} + \gamma_{01}w_{1j} + \alpha_1 x_{1j} + \gamma_{10}z_{1ij} + \gamma_{11}w_{1j}z_{1ij} + e_{ij} \tag{2.4.4}$$

This model is like the fixed-effect model shown in eq. 1.4.3, and ordinary least squares regression would be appropriate for model estimation. Note that Model 2.4.4 implies the following relations:

$$\beta_{0j} = \gamma_{00} + \gamma_{01}w_{1j} \tag{2.4.5}$$

$$\beta_{1j} = \gamma_{10} + \gamma_{11}w_{1j} \tag{2.4.6}$$

where there are no error terms in the macro models, indicating that the level-1 regression coefficients β_{0j} and β_{1j} vary, but non-randomly, across groups, and the variation in the coefficients are fully explained by the macro variable w_{1j} in the macro models. This situation is unlikely to be true in reality. It is unusual for all level-1 random coefficients to vary only systematically and for their variations to be explained entirely by macro variables included in the macro model. As such, the fixed-effect model shown in eq. 2.4.4 is rarely appropriate for modeling multilevel data.

2.5 Cross-level interactions

An important objective in multilevel modeling is to identify and explain cross-level interactions. If the relationship between a level-1 explanatory variable (e.g., z_{1ij}) and dependent variable (e.g., y_{ij}) varies significantly across groups, we need to identify what contextual variables explain the variation in this relationship. If a contextual variable (e.g., w_{1j}) has a significant effect on random coefficient β_{1j}, then the effect of the level-1 explanatory variable z_{1ij} on y_{ij} may be influenced or moderated by the value of the group-level variable w_{1j}. The interpretation of the main effect of variable z_{1ij} and its interaction effect with variable w_{1j} are the same as those in the ordinary linear regression models. *Cross-level or micro-macro* interactions in multilevel models exist between explanatory variables at different levels. For example, in the combined model shown in eq. 2.4.4, coefficients γ_{10} and γ_{01} are the main effects of the level-1 explanatory variable z_{1ij} and the level-2 contextual variable w_{1j}, respectively; and the coefficient γ_{11} is the interaction effect between z_{1ij} and w_{1j}.

Variation in a micro-level random coefficient can be systematically albeit partially explained by contextual variables at the group level. The unexplained variation is taken into account by the macro error term. When m level-2 explanatory variables (e.g., $w_{1j}, w_{2j}, \cdots, w_{mj}$) are used to predict the random slope coefficient of level-1 explanatory variable z_{ij}, the effect of z_{ij} on y_{ij} would be moderated by the M contextual variables, and there would be M cross-level interactions between the level-1 explanatory variable z_{ij} and the M contextual variables in the combined model. Two points are important to remember in regard to the interactions (Hox, 2002): First, as in multiple regression models, if a cross-level interaction effect is statistically significant, the corresponding main effects, whether they are statistically significant or not, must remain in the model and be included in the effect of the corresponding variable. Second, variables involved in an interaction must have meaningful zero values in their observed measures. Otherwise the main effect of the counterpart variable won't be meaningful. For example, in eq. 2.2.4, the main effect of variable z_{1ij} is interpreted as the effect of z_{1ij}, corresponding to a zero value of variable w_{1j}. If variable w_{1j} does not have a meaningful zero value, we cannot appropriately interpret the main effect of variable z_{1ij}. *Centering* is a process to ensure that variables z_{ij} and w_j would have meaningful zero values (Hox, 1994).

2.6 Measurement centering

In interpreting multiple regression results, analysts usually care less about the intercept than the slope coefficients. Most often the original scale of the explanatory variables is used in modeling. In multilevel modeling, attention is paid to the variations of the level-1 intercepts and slope coefficients that are treated as dependent variables in the macro models. Therefore, it is important for researchers to ensure meaningful interpretation to these level-1 random regression coefficients, and in particular the level-1 random intercept. Otherwise, the level-1 random coefficients are difficult to be interpreted at level 2.

The meaning of the regression intercept is determined by the values and scale of the explanatory variables. If a continuous variable (at either level-1 or level-2) does not have a meaningful 0 value, centering is advisable. Suppose a regression model has a level-1 intercept of 5. We would say that the regression line crosses the *Y*-axis at 5 or the expected value of the dependent variable would be 5 when all explanatory variables in the model have a value of 0. This may not make sense if some explanatory variables don't actually have value 0 in the observed data. For example, when age of adult subjects is included in a regression model, the value of the dependent variable corresponding to age of 0 is meaningless. To make the intercept meaningful, we may use a rescaled or transformed measure of age by centering it; for example subtracting the sample mean age from the observed age for all the cases in the sample. If the sample mean age is 30 and an individual's age is 40, then the value of the centered age for the individual would be $40-30 = 10$, which is a measure of deviation from the mean age. With the centered measure of age, an individual with an average age in the sample would have a value of 0 in the transformed measure of age, and the intercept represents the expected value of the dependent variable for an average individual whose age is the mean age in the sample.

Measurement centering helps ensure appropriate interpretation of main effects of variables that are involved in interactions. When an explanatory variable involved in an interaction has no value of zero in the observed data, it would influence the interpretation of the main effect of the counterpart variable involved in the interaction (see Section 2.5).

In multilevel modeling, centering not only makes regression intercepts more meaningful, but also helps to speed up the model estimation process and reduces convergence problems (Raudenbush & Bryk, 2002; Hox, 2002). For example, in most cases, grand mean centering reduced the correlation between the intercept and slope coefficient estimates across groups. Reduction of covariation between the level-1 random intercepts and slopes can help to alleviate potential level-2 estimation problems due to multicollinearity (Raudenbush, 1989a, 1989b).

There are two different techniques for centering in multilevel modeling: Grand-mean centering and group-mean centering. The former process entails centering the variables on the overall sample mean; and the latter entails centering on the group means. With different centering options, the intercept term takes on different meanings. Grand mean centering yields an intercept equal to the expected value of y_{ij} for an individual with values on all x_{ij}s equal to the mean values of x_{ij}s across all individuals in the sample. Group mean centering yields an intercept equal to the expected value of y_{ij} for an individual in group j whose values on all x_{ij}s are equal to the mean values of the x_{ij}s in that group. For level-1 explanatory variables, both grand-mean centering and group-mean centering can be applied. Centering that can be applied to level-2 explanatory variables is grand-mean centering only.

Grand-mean centering. Grand-mean centering of variable x_{ij} takes the form of $(x_{ij} - \bar{x}_{..})$, where $\bar{x}_{..}$ is the mean of x_{ij} over all individuals in the sample. If both the level-1 explanatory variables x_{ij} and z_{ij} in eq. 2.2.1 are grand-mean centered, we have:

$$y_{ij} = \beta_{0j} + \alpha(x_{ij} - \bar{x}_{..}) + \beta_{1j}(z_{ij} - \bar{z}_{..}) + e_{ij} \tag{2.6.1}$$

where the level-1 intercept β_{0j} represents the group-mean $\bar{y}_{.j}$ for an "average" person who has sample mean values on all the explanatory variables.

The intercept can be expressed as:

$$\beta_{0j} = \bar{y}_{.j} - \alpha(\bar{x}_{.j} - \bar{x}_{..}) - \beta_{1j}(\bar{z}_{.j} - \bar{z}_{..}) \tag{2.6.2}$$

where the level-1 intercept β_{0j} is interpreted as the group-mean $\bar{y}_{.j}$, adjusting for differences in group means, e.g., $(\bar{x}_{.j} - \bar{x}_{..})$ and $(\bar{z}_{.j} - \bar{z}_{..})$.

A grand-mean centered model is statistically *equivalent* to the model with original measures. This is because in grand-mean centering a constant value, which is the grand mean or overall sample mean of a variable is subtracted from the values of the variable of all cases. The model will have the same fit, the same predicted values, and the same residuals, but will change the estimates of the intercept and random slopes by adding a constant to each of the parameter estimates. The intercepts and random slopes of the original and grand-mean centered models can be translated to each other (Kreft & De Leeuw, 1998, pp.108—109).

Group-mean centering. Group-mean centering transforms the variables as deviations from its group mean (e.g., $(x_{ij} - \bar{x}_{.j})$). $\bar{x}_{.j}$ is the mean of x_{ij} over all individuals within group *j*. Since group means are unlikely to be equal among groups, different $\bar{x}_{.j}$ s are subtracted from each x_{ij} by group. As a result, a group-mean centered model is no longer equivalent to the model with the original variables. The level-1 intercept of a group-mean centered model can be interpreted as the within-group expected value of the dependent variable. This can be shown by replacing the x_{ij} and z_{ij} with $(x_{ij} - \bar{x}_{.j})$ and $(z_{ij} - \bar{z}_{.j})$, respectively, in eq. 2.2.1:

$$y_{ij} = \beta_{0j} + \alpha(x_{ij} - \bar{x}_{.j}) + \beta_j(z_{ij} - \bar{z}_{.j}) + e_{ij} \tag{2.6.3}$$

Then, the intercept can be expressed as:

$$\beta_{0j} = \bar{y}_{.j} - \alpha(\bar{x}_{.j} - \bar{x}_{.j}) - \beta_j(\bar{z}_{.j} - \bar{z}_{.j}) = \bar{y}_{.j} \tag{2.6.4}$$

where the level-1 intercept β_{0j} is the unadjusted mean value of the dependent variable for group *j*, and its variance (i.e., σ^2_{u0}) represents between-group variation of the unadjusted group mean $\bar{y}_{.j}$. When using group-mean centered variable (e.g., $(x_{ij} - \bar{x}_{.j})$) in a model, the group mean (i.e., $\bar{x}_{.j}$) is removed from the individual scores. If we posit that the group-level mean value does influence the outcome based on theory or empirical findings, then group mean (e.g., $\bar{x}_{.j}$) should be put back in level-2 models as a group level explanatory variable. If group means are not reintroduced in the macro-level models, important group variation is eliminated and model fit to the data will be weakened.

There are three scaling options for independent variables: 1) raw metric scaling where explanatory variables are used in their original metric; 2) grand-mean centering where the grand mean of a explanatory variables is subtracted from each case (i.e., $(x_{ij} - \bar{x}_{..})$); and 3) group-mean centering where the relevant group mean of an explanatory variable is subtracted from the original score of the variable (i.e., $(x_{ij} - \bar{x}_{.j})$). Less often, a variable may be centered on some other meaningful value, such as population mean. Then the intercept β_{0j} would be interpreted as the mean value of the dependent variable when the variable is equal to the population mean

(Raudenbush & Bryk, 2002). In any case, when an explanatory variable x_{ij} does not have a meaningful value of zero in the data, centering x_{ij} is advised.

Centering a measure creates a relative score, or the deviation of the individual from the group or the full sample. Caution is advised in interpreting the effects of the transformed explanatory variables. When random slopes are involved, the grand-mean centered model is considered statistically equivalent to the raw score model and model parameters can be transformed back to those estimated from the original measures. However, group-mean centering is no longer statistically equivalent to the raw score model. Thus, the choice of a group-mean centered model must be justified by the theory being tested (Kreft et al., 1995; Snijders & Bosker, 2003). Suppose we are assessing the effect of students' high school GPA, z_{ij}, on Scholastic Aptitude Tests (SAT) scores, y_{ij}, among J high schools. A zero GPA (z_{ij}) is impossible. Accordingly, the GPA should be transformed by centering. When group-mean centering, $(z_{ij} - \bar{z}_{.j})$ is used to transform the level-1 explanatory variable z_{ij}, the level-1 intercept, β_{0j}, represents the average SAT score ($\bar{y}_{.j}$) for the jth school, not adjusting for the differences in school mean GPA score (i.e., $\bar{z}_{.j}$); and the level-1 random slope coefficient β_{1j} represents the amount of change in y_{ij} corresponding to a one-unit change in the deviation of the ith individual GPA score (z_{ij}) in the jth school from the school's mean score ($\bar{z}_{.j}$). Including measure $\bar{z}_{.j}$ in the level-2 models provides a measure of the effect of the average level of high school GPA score on the average SAT score; and moderates the effect of individual GPA score on his/her SAT score. In other words, the effect of students' GPA scores on their SAT scores is associated with their relative position in their own school, as well as the school's average GPA. Hofman and Gavin (1998) describe conditions suitable for application of group-mean centering based on the *Frog-pond* theory.

By contrast, under grand-mean centering, $(z_{ij} - \bar{z}_{..})$ represents the relative position of the ith individual of the jth group of the entire sample. The group-mean measures don't need to be included as predictors in level-2 models.

2.7 Model estimation

Multilevel model estimation is complex because of the composite error structure, consisting of both levels 1 and 2 residuals (i.e., ε_{ij}, u_{0j}, and u_{1j}), in the combined model (eq. 2.2.4). Three kinds of parameters, fixed regression coefficients, random regression coefficients and the residual variance/covariance components in matrices G and R need to be estimated simultaneously. Discussion of the statistical theory and estimation algorithms for multilevel modeling is beyond the scope of this book. Readers who are interested in the underlying statistical theory are referred to Mason et al. (1983), Goldstein (1987), Raudenbush & Bryk (2002), and Littell et al. (1996). In this section we limit our discussion to a conceptual, non-technical introduction to model estimation.

If the variance/covariance matrices G and R described in eqs. 2.3.6 and 2.3.7 are known, estimating the combined model with a generalized least square (GLS) model is relatively simple (Goldstein, 1987). Given the known matrixes G and R, the variance of outcome measure y_{ij} can be estimated as:

$$\hat{V} = ZGZ' + R \tag{2.7.1}$$

Using the inverse of matrix $\hat{V}$ as a weight, the model regression coefficients and their standard errors can be estimated via generalized least square. The problem is that the G and R matrices are unknown and must be estimated in the multilevel model. Littell et al. (1996, p.498) state that "the goal thus becomes finding a reasonable estimate of *G* and *R*."

As in many other statistical models, maximum likelihood (ML) estimation is used to estimate the variance-covariance components in matrixes *G* and *R* (e.g., σ^2_{u0}, σ^2_{u1}, σ^2_{u01}, and σ^2) (Littell et al. 1996; Searle et al., 1992). ML estimates generally have very desirable large sample properties: 1) they are consistent (i.e., as the sample size increases, the maximum likelihood estimate tends to approach the true parameter value); 2) they are asymptotically normal (i.e., the maximum likelihood estimate will have an approximate normal distribution centered on the true parameter value), thus significance testing for the parameters becomes possible; and 3) the estimated likelihood function can be used for assessment of the model fit and comparisons among models.

ML estimation is an iterative process. In the beginning of the estimation process starting values for the parameter estimates are generated. The parameter estimates become the starting values of parameters for the next iteration of parameter estimates. The process is repeated many times until the estimates for all the parameters stabilize from one iteration to the next. A default *tolerance* or convergence criterion, which is a very small number (e.g., 1E−8 in SAS *PROC MIXED*), is usually specified. If the difference in parameter estimates between two consecutive iterations reaches the tolerance level, the computation stops, and we say the ML estimation has converged. Sometimes the estimation never converges and the computation process can go on forever. To avoid such an endless computation, a default maximum number of iterations is often set in SAS (e.g., 50 in *PROC MIXED*). If the number of iterations reaches the maximum, the computation process stops. The number of iterations can be modified by users as needed. Lack of convergence generally indicates bad model specification or inadequate sample size.

Different ML-based estimators are available for multilevel modeling. Two frequently used ML-based estimators are full information maximum likelihood (FML) and restricted maximum likelihood (REML) also referred to as residual ML, introduced by Patterson and Thompson (1971). The FML estimator is used in some multilevel model programs (e.g., HLM). SAS documents indicate that ML implemented in *PROC MIXED* is equivalent to FML. The ML and REML likelihood functions used in SAS for estimating matrixes *G* and *R* are described by Littell et al. (1996, p.498) [3]. In SAS *PROC MIXED,* the REML estimator is implemented by default.[4]

The two ML-based estimators are based on different residual terms. ML is based on the ordinary residual term $e = y - X\hat{\alpha} - Z\hat{\beta}$, while REML is based on the residual term $e = y - X\hat{\alpha}$. Note that the variation involved with the level-1 random regression coefficients, $Z\hat{\beta}$, is included in the residual term in REML; thus these residuals contain all of the sources of random variation and are, therefore, called "*full residuals*" to differentiate them from the ordinary residuals (Brown & Prescott, 1999, p. 46).

---

[3] Another non-iterative method called minimum variance quadratic unbiased estimator (MIVQUE) is available in SAS *PROC MIXED* for estimating *G* and *R* matrixes. Compared with ML and REML, MIVQUE does not require a normality assumption and does not involve iterations. MIVQUE is used by default to compute starting values for the iterative ML and REML estimations in *PROC MIXED*. MIVQUE may be used when maximum likelihood iteration does not converge. However, simulation studies have shown that REML and ML are preferable to MIVQUE (Swallow & Monahan, 1984; Littell et al., 1996).

[4] SAS *PROC MIXED* minimizes the ML function using a ridge-stabilized Newton-Raphson algorithm, while the Expectation-Maximum (EM) algorithm is used in HLM. In another multilevel program VARCL (Longford, 1987), the "Fisher scoring" algorithm, which is the classical Newton-Raphson method, is applied to the maximum likelihood estimation.

When the number of groups is not large, ML and REML produce similar level-1 residual variance estimates (i.e., matrix *R*), but REML provides less biased level-2 residual variances/covariances (i.e., matrix *G*) estimates. Therefore, REML is often recommended for estimating models with a smaller number of groups (Kreft, 1996; Raudenbush & Bryk, 2002). However, under the conditions of a small number of groups and/or non-normality, the results of simulation study by Carpenter, Goldstein, and Rasbash (2003) show that the REML estimates may also be biased.

Although REML is considered superior in theory to ML, ML estimation is very useful in multilevel modeling. ML runs much faster than REML and is important for model comparisons. We will discuss this issue in Section 2.8.3.

Other well-known estimators used for multilevel model estimation include the *Iterative Generalized Least Squares* (IGLS) and *Restricted Iterative Generalized Least Squares* (RIGLS) used in the computer program MLwin (Goldstein, 1995). The IGLS iterative estimation procedure starts from an initial OLS model to estimate the fixed regression effects, calculate the OLS residuals and estimate the variance/covariance matrix of the residuals—$\hat{V}$. The inverse of $\hat{V}$ is used in a GLS process to estimate the variances/covariances of the random components (e.g., $\hat{\sigma}^2$, $\hat{\sigma}^2_{u0}$, $\hat{\sigma}^2_{u1}$, and $\hat{\sigma}^2_{u01}$). The estimates obtained from the GLS process are again used to form a weight in a new GLS process in order to obtain new estimates of fixed-effects, residuals, and matrix $\hat{V}$. These two partial estimation processes are alternated, with successive estimates becoming more precise. The iterative estimation process continues until the parameter estimates do not change significantly with successive iterations, at that point the estimation process is said to converge (Goldstein, 1995). IGLS and RIGLS are comparable to ML and REML, respectively (Goldstein, 1989).

Finally, the *empirical Bayes estimator* is very useful for multilevel modeling when some groups have a small sample size (Lindley & Smith, 1972). An empirical Bayes parameter estimate for the jth group is an optimally weighted average of the parameter estimates using the jth group data and the entire sample data. For example, in the two level intercept-only model described in eqs. 2.1.2 and 2.1.3, the empirical Bayes estimate $\beta^*{}_{0j}$ can be expressed as a weighted linear combination of $\hat{\beta}_{0j}$ and $\hat{\gamma}_{00}$:

$$\hat{\beta}^*_{0j} = \rho_{0j}\hat{\beta}_{0j} + (1-\rho_{0j})\hat{\gamma}_{00} \tag{2.7.2}$$

where $\hat{\beta}_{0j}$ is the level-1 OLS estimate of group mean ($\bar{y}_{\cdot j}$) while $\hat{\gamma}_{00}$ is the generalized least square (GLS) estimate of the grand mean or overall mean, of the outcome measure; and the weight ρ_{0j} is the reliability of $\hat{\beta}_{0j}$. If ρ_{0j} for each group approaches 1.0, the multilevel parameter estimates would be equivalent to those of OLS regression. With a high level of reliability of $\hat{\beta}_{0j}$, the empirical Bayes estimate $\hat{\beta}^*_{0j}$ would be weighted more heavily by the group mean $\hat{\beta}_{0j}$; otherwise, $\hat{\beta}^*_{0j}$ would be weighted more by the grand mean $\hat{\gamma}_{00}$. When sub-sample size n_j is small, the weight ρ_{0j} in eq. 2.7.2 would be small, thus the contribution of $\hat{\beta}_{0j}$ to $\hat{\beta}^*_{0j}$ would "*shrink*" and the multilevel model parameter estimate would be pulled toward the grand mean. The empirical Bayes estimator is, therefore, called *a shrinkage estimator* approach (James & Stein, 1961). The degree of shrinkage depends on the precision of the level-1 OLS estimate or its reliability.

The *shrinkage estimator* approach is also called a *borrowing strength* approach; that is, it borrows information from all the groups to support statistical estimation for the groups with insufficient observations (Gelman et al., 2003). This feature of empirical Bayes estimator is very useful when some groups have sparse data (i.e., only a small number of observations available in the groups). As sparseness often characterizes at least some groups in multilevel data, the empirical Bayes estimator is an important method for multilevel data analysis, even though it tends to pull the estimates "too far" toward the sample average (Bryk & Raudenbush, 2002).

Equation 2.7.2 can be readily expanded for conditional shrinkage estimates of group means (i.e., the level-1 random intercept), as well as the level-1 random slopes, in a two-level model with covariates. Derivation of the generalized formulas for conditional shrinkage estimates is available in Bryk & Raudenbush (1992, pp.41—43).

The *empirical Bayes estimator* or *shrinkage estimator* is used for model estimation in computer programs such as HLM. The REML estimator used in SAS *PROC MIXED* has a shrinkage function similar to the *empirical Bayes estimator* (Laird & Ware, 1982; Carlin & Louis, 1996). The *RANDOM statement's SOLUTION* option instructs SAS *PROC MIXED* to employ the *Best Linear Unbiased Prediction* (BLUP), which is a *shrinkage estimator*, to estimate random effects (Littell et al., 1996). In addition, the *PRIOR* statement in *PROC MIXED* will carry out a sampling-based Bayesian analysis (SAS Institute Inc, 2004; Littell et al., 2006).

Reliability of the OLS level-1 coefficients plays an important role in *shrinkage estimation.* Here we briefly discuss how reliability (ρ_{0j}) of $\hat{\beta}_{0j}$ in the *empty model* is estimated. According to classical psychometric true score theory (Lord & Novick, 1968), reliability of a measure is the ratio of the true score's variance to the variance of the observed measure.

$$\rho = \frac{\text{Variance of the true scores}}{\text{Total variance of the observed scores}} \tag{2.7.3}$$

Thus the reliability of the OLS level-1 coefficient $\hat{\beta}_{0j}$ in the *empty model* can be estimated as:

$$\rho_{0j} = \frac{\text{Var}(\beta_{0j})}{\text{Var}(\overline{y}_{.j})} = \frac{\text{Var}(u_{0j})}{\text{Var}(u_{0j} + \overline{e}_{.j})} = \frac{\sigma_{u0}^2}{\sigma_{u0}^2 + (\sigma^2 / n_j)} \tag{2.7.4}$$

where the total variance of the observed group means (i.e., $\text{Var}(\overline{y}_{.j})$) consists of two components: 1) the parameter random variance (i.e., $\text{Var}(u_{0j})$), and 2) the error variance (i.e, $\text{Var}(\overline{e}_{.j})$). The ratio of the parameter random variance to the total variance of the observed group means is reliability of the OLS $\hat{\beta}_{0j}$ or $\overline{y}_{.j}$ in the *empty model.*

From eq. 2.1.7, we have

$$\sigma_{u0}^2 = \frac{\sigma^2 \cdot \text{ICC}}{(1 - \text{ICC})} \tag{2.7.5}$$

Substituting eq. 2.7.5 into eq. 2.7.4, we have

$$\rho_{0j} = \frac{n_j \cdot \text{ICC}}{1 + (n_j - 1)\text{ICC}} \tag{2.7.6}$$

where n_j is the sub-sample size of group j and ICC is the intraclass correlation coefficient in the multilevel data set. The equation shows that reliability ρ_{0j} is a function of n_j. For a given ICC, ρ_{0j} would be larger if n_j is larger. On the other hand, ρ_{0j} is also a function of ICC. For a given n_j, a larger ICC leads to a larger ρ_{0j}.

The generalized formula for estimating reliability of OLS level-1 coefficient $\hat{\beta}_{qj}$ is defined as the following (Bryk & Raudenbush, 1992, p.43):

$$\rho_{qj} = \frac{\sigma^2_{uq}}{\sigma^2_{uq} + \mathrm{Var}(\hat{\beta}_{qj})} \tag{2.7.7}$$

where

$$\mathrm{Var}(\hat{\beta}_{qj}) = \frac{\sigma^2}{\Sigma z^2_{qij}} \tag{2.7.8}$$

where Σz^2_{qij} is the sum of squared values of the corresponding level-1 explanatory variable z_{qij}. Equations 2.7.7 and 2.7.8 show that a larger variation of z_{qij} increases the reliability of $\hat{\beta}_{qj}$.

Reliability of $\hat{\beta}_{qj}$ is very important in multilevel model parameter estimation. Computer programs like HLM provide estimates of reliability of $\hat{\beta}_{qj}$. However, analysts often don't report ρ_{qj} s in multilevel studies, even though ρ_{qj} is used to obtain shrinkage estimates. SAS *PROC MIXED* does not provide estimates of these reliabilities. Accordingly, we will not report reliability in the models demonstrated in this book. In Chapter 3, however, we will demonstrate how to estimate the reliability of $\hat{\beta}_{0j}$ as well as its shrinkage estimate in the *empty model.* The results demonstrate that the BLUP estimates provided by SAS *PROC MIXED* are shrinkage estimates.

We have shown that multilevel models can be applied in studies having sparse data, or a small number of observations within some groups. The application of multilevel model estimation is still a challenge when the number of higher level units is small. In these cases neither ML estimation nor REML estimation work very well. When group size is small or the distribution of outcome measures is non-normal, bootstrap methods can be used for robust multilevel modeling. We will discuss different bootstrapping methods and their applications in Chapter 6.

2.8 Model fit, hypothesis testing, and model comparisons

2.8.1 Model fit

The process of assessing the fit of multilevel models to their data is just like those employed in other statistical procedures such as logistic regression models. SAS *PROC MIXED* produces −2LL or −2 times the log likelihood ratio. A smaller −2LL indicates that a model fits the data better. A fully saturated model fits the data perfectly and has −2LL equal to 0. −2LL of the specified model is a measure of the deviance from the fully saturated model. Because the deviance statistic reflects how much worse the specified model is compared with the saturated model, it is actually a "badness-of-fit" statistic although it is usually considered to be a "goodness-of-fit" statistic. The smaller the deviance, the better the model fits to the data.

In addition, three frequently used information criteria measures[5] are provided by *PROC MIXED* to assess model fit: Akaike's information criterion (AIC) (Akaike, 1974), finite-sample

[5] Two additional criteria measures are available in *PROC MIXED*: Hannan-Quinn Information Criterion (HQIC) (Hannan & Quinn, 1979) and Consistent Akaike s Information Criterion (CAIC) (Bozdogan, 1987). The option IC, standing for information criteria, specified in the *PROC MIXED* statement provides all the information criteria measures, including HQIC and CAIC.

corrected version of AIC (AICC) (Hurvich & Tsai, 1989), and Bayesian information Criterion (BIC) (Schwarz, 1978). Beginning with the SAS release 8.1, *PROC MIXED* provides *the smaller-is-better forms* of the information criteria measures so that AIC, AICC, and BIC all have positive values. A smaller value of those criteria indicates better model fit. The formula for calculating AIC, AICC, and BIC are expressed as (SAS Institute Inc., 2004, p. 2676):

$$\text{AIC} = -2\text{LL} + 2d \tag{2.8.1.1}$$

$$\text{AICC} = -2\text{LL} + 2d \cdot n/(n-d-1) \tag{2.8.1.2}$$

$$\text{BIC} = -2\text{LL} + d \cdot \ln(n) \tag{2.8.1.3}$$

where −2LL is −2 times the log maximum likelihood function, d denotes the number of parameters to estimate in the model, and n is the number of valid observations for maximum likelihood estimation. All three information criteria measures are calculated on the basis of the model likelihood function penalized for the number of model parameters used, while AICC and BIC impose an additional penalty.

While these model fit statistics (e.g., −2LL, AIC, AICC, BIC) provide useful information for model comparison, none of them can tell how well a single model fits the data. Convergence information in iteration history of model estimation can provide useful information on model fit. Rapid convergence suggests that the model fits the data well. On the contrary, failure to converge indicates incorrect model specification. In addition, variation explained by the model can also be used as an indicator of how well a model fits the data. This will be discussed in Section 2.9.

2.8.2 Hypothesis testing

Several hypothesis tests are often needed in multilevel modeling, including hypothesis testing for random effects, fixed-effects[6], and model comparisons.

Testing random effects: Random effects refer to the components of the random variance in the combined model, or the variances/covariances of the macro residuals (e.g., σ^2_{u0}, σ^2_{u1} and σ^2_{u01}). To test significance of the variance/covariance parameters, *PROC MIXED* provides the Wald Z statistic, computed as the ratio of the parameter estimate to its standard error.

A significant level-2 residual variance indicates that the corresponding level-1 regression coefficient is a random-effect coefficient that significantly varies across groups. Thus, significance testing for level-2 residual variances is an approach we use to examine which level-1 regression coefficients (i.e., intercept and slope coefficients) are random or fixed coefficients. Existence of significant variation in a level-1 regression coefficient indicates this coefficient is random and should be treated as a dependent variable in a group-level model. Non-random level-1 regression coefficients detected in exploratory modeling are treated as fixed-effects in the final model.

Several points in regard to significance testing for residual variances/covariances should be noted. First, the Wald Z test for testing residual variance is a one-tailed test of significance (e.g., H_0: $\sigma^2_{u0} = 0$ vs. H_A: $\sigma^2_{u0} > 0$), where the p-value for the Z statistic needs to be divided by 2, since variance cannot be negative by definition.

---

[6] Note that the term "fixed-effects" is often interchangeably used for different concepts. The "fixed-effects" in level-1 model like eqs. 2.2.1 and 2.2.5 refer the level-1 regression coefficients that don't vary across groups. The "fixed-effects" in the combined models like eqs. 2.2.4 and 2.2.9 refer all the regression coefficients in the fixed part of the model.

Second, the likelihood ratio (LR) test that is often used for model comparisons can also be used to test random effects. This can be accomplished by running the model twice with and without specifying the level-1 slope of interest as a random coefficient. The one-tailed LR test gives improved results. Again, the *p*-value for the Chi-square statistic should be divided by 2. The Wald *Z* test and LR test are asymptotically equivalent. When the number of groups is relatively small the Wald *Z* test and LR test may provide different results, and the LR test is preferred (Littell et al., 1996).

Finally, group-level residual covariance (e.g., σ^2_{u01}) may be significant. This would indicate that the effect of the level-1 explanatory variable z_{1ij} on the outcome is related to the group-mean level for the outcome measure. This is an important issue particularly in growth models as applied in multilevel modeling to longitudinal data where the intercept coefficient represents the initial mean level of individual outcome measure and slope coefficient of the time variable represents the rate of change in the outcome. Association between initial level and rate of change in the outcome under study is particularly interesting in outcome change trajectory studies, as described in Chapter 4.

Testing fixed-effects: Fixed-effects are referred to as the fixed components in the combined model, which include the intercept and the main effects of both micro and macro explanatory variables (e.g., γ_{00}, γ_{0k}, α_p, and γ_{q0} in eq. 2.2.9), as well as the cross-level interaction effects between some micro and macro explanatory variables (e.g., γ_{qk} in eq. 2.2.9). *PROC MIXED* provides the Walt *t* statistic for significance testing of fixed-effects. Interpretation of the *t*-tests is similar to multiple regression models. A *t*-ratio greater than 1.96 indicates that the effect is statistically significant at $\alpha = 0.05$, and the null hypothesis of "*no effect*" is rejected.

For a categorical explanatory variable with Q levels, we may generate $(Q-1)$ dummy variables. The results of the Wald *t*-tests for the $(Q-1)$ dummy variables test the difference in outcome means between the specific categories and the reference category (i.e., the omitted category). We can also conduct a global test for significance of a categorical variable as a whole. By default, *PROC MIXED* provides an *F*-test[7] for assessing the fixed-effects of categorical explanatory variables. If the *Chisq* option is specified in the *Model* statement, the program will provide Wald χ^2 statistics for the global test. The χ^2 test is more liberal than the *F*-test because it effectively assumes infinite denominator degrees of freedom (SAS Institute Inc., 2004, p.2752).

Testing a set of coefficients: Researchers may be interested in simultaneously testing the significance of the effects of a set of explanatory variables. This can be done by fitting the model to the same data set with and without the set of variables of interest, then conducting a likelihood ratio (LR) test.

Testing linear contrast of coefficients: Linear contrast of coefficients is often encountered in statistical analysis. For example, suppose we have a three-track treatment measure (e.g., Trt_1, Trt_2, and control) with groups means $\bar{y}_{\text{Trt\_1}}$, $\bar{y}_{\text{Trt\_2}}$, and $\bar{y}_{\text{control}}$. The contrast for comparing the mean of the control group to the mean of the two treatment groups would be $(\bar{y}_{\text{control}} - 0.5\bar{y}_{\text{Trt\_1}} - 0.5\bar{y}_{\text{Trt\_2}})$. Linear contrast allows us to test different hypotheses involving a single variable, as well as two-way or higher interactions in *PROC MIXED*. In Chapter 3, we will discuss how to actually conduct such tests with examples.

[7] The relation between the Wald *t*-test and *F*-test is: $t_{DF} = (F_{1,DF2})^{1/2}$, where DF2 is the denominator degrees of freedom of *F*-test.

2.8.3 Model comparisons

When different models are used to fit the same data, the models may be nested or non-nested. For nested models, i.e., one model is a sub-model of the other, the likelihood ratio (LR) test can be used for model comparison. The difference in deviance statistics or −2LL between two models are used for the LR test. Suppose we have a *null model* which includes K explanatory variables, and the *alternative model* which includes the same K explanatory variables plus additional P explanatory variables. For each model a deviance statistic is computed: $D_A = -2\text{LL}_A$ for the alternative model and $D_0 = -2\text{LL}_0$ for the null model. Thus, the difference in deviance statistics between the two models is:

$$\begin{aligned} D_A - D_0 &= (-2\text{LL}_A) - (-2\text{LL}_0) \\ &= -2\ln\left(\frac{L_A}{L_0}\right) \end{aligned} \tag{2.8.3.1}$$

which is −2 times the natural logarithm of the ratio of the likelihood function of the alternative model to that of the null model. This difference in deviance statistics has an asymptotic chi-square distribution with degrees of freedom equal to the difference in the number of parameters estimated in the two models (i.e., d.f.= P).

The LR test does not work when comparing models that are not nested. Information criteria statistics such as AIC, AICC, and BIC are suitable for comparisons of nested models as well as nonnested models. A model with values of AIC, AICC, and BIC that are closer to zero is preferable. Raftery (1996), based on Jeffreys (1961), suggests some guidelines for the strength of evidence favoring one model against another based on a difference in the value of BIC:

- 0—2: Weak evidence
- 2—6: Positive evidence
- 6—10: Strong evidence, and
- 10 + : Very strong evidence.

Importantly, ML and REML estimators need to be used appropriately for different multilevel model comparisons. When models have the same random effects, but different fixed-effects, model fit statistics used for model comparison should be estimated under ML method. The REML method should be used for models that have the same fixed-effects, but different random effects. Generally speaking, ML is more flexible for model comparisons. When models have both different fixed-effects, as well as different random effects, and the random effects are nested (e.g., one has u_{0j}, u_{1j}, and another model has u_{0j}, u_{1j}, and u_{2j}), ML is recommended. The same principles apply to information criteria measure such as AIC, AICC, and BIC. However, once the best model is determined with ML, we should use REML method to estimate the "final" model and report the REML results.

2.9 Explained level-1 and level-2 variances

In a multiple regression model, the proportion of variation in the outcome measure explained by the explanatory variables is measured by the *coefficient of multiple determination* or R^2. The situation becomes complex in multilevel models as the outcome variance consists of micro and macro components and unexplained variance occurs at both levels. An index of the proportional reduction in variance has been proposed as an analog to R^2 to measure explained outcome variance at level-1 and level-2 (Raudenbush & Bryk, 2002).

Estimating explained level-1 variance: The *empty model* (i.e., the model shown in eqs. 2.1.1—2.1.3) provides the total, within- and between-group variance in the outcome that can be explained by levels-1 and level-2 explanatory variables included in a multilevel model. To illustrate how to estimate explained level-1 variance, we run the *empty model* first; then the model with explanatory variables at level-1, and no level-2 explanatory variables are used in the macro equations:

$$y_{ij} = \beta_{0j} + \alpha_1 x_{1ij} + \beta_{1j} z_{1ij} + e_{ij} \quad (2.9.1)$$

$$\beta_{0j} = \gamma_{00} + u_{0j} \quad (2.9.2)$$

$$\beta_{1j} = \gamma_{10} + u_{1j} \quad (2.9.3)$$

Compare the results of the *empty model* and the specified model. A correctly specified model should reduce the unexplained level-1 variance $\hat{\sigma}^2$ significantly (Raudenbush & Bryk, 2002, p.79):

$$\begin{pmatrix} \text{RB: \% of Explained} \\ \text{level-1 variance} \end{pmatrix} = \frac{\hat{\sigma}^2(\text{empty model}) - \hat{\sigma}^2(\text{specified model})}{\hat{\sigma}^2(\text{empty model})} = 1 - \frac{\hat{\sigma}^2(\text{specified model})}{\hat{\sigma}^2(\text{empty model})} \quad (2.9.4)$$

where RB: % *of Explained level*-1 *variance* stands for estimated proportion of explained level-1 variance using Raudenbush & Bryk's method; $\hat{\sigma}^2$ (*empty model*) and $\hat{\sigma}^2$ (*specified model*) denote the level-1 residual variances in the *empty model* and the specified model, respectively.

*Estimating explained level-*2 *variance*: The model specified in eq. 2.2.1 has two level-1 random regression coefficients: random intercept β_{0j} and slope β_{1j}. Both are specified as a function of group-level explanatory variable w_{1j} in the macro models specified in eqs. 2.2.2 and 2.2.3. By including level-2 variables, we hope to reduce the size of level-2 residual variances, indicating that some of the level-2 residual variances were explained. As such, two indices of explained level-2 variances need to be calculated.

To estimate explained variance in level-1 random intercept, the index of the *proportional reduction in variance* of u_{0j} is defined as (Raudenbush & Bryk, 2002, p.74):

$$\begin{pmatrix} \text{RB: \% of Explained} \\ \text{level-2 variance} \end{pmatrix} = \frac{\hat{\sigma}^2_{u0}(\text{empty model}) - \hat{\sigma}^2_{u0}(\text{specified model})}{\hat{\sigma}^2_{u0}(\text{empty model})} = 1 - \frac{\hat{\sigma}^2_{u0}(\text{specified model})}{\hat{\sigma}^2_{u0}(\text{empty model})} \quad (2.9.5)$$

where RB: % *of Explained level*-2 *variance* stands for the estimated proportion of explained level-2 variance using Raudenbush & Bryk's method; $\hat{\sigma}^2_{u0}$ (*empty model*) is a specific level-2 residual variance in the *intercept-only* model, while and $\hat{\sigma}^2_{u0}$ (*specified model*) is the corresponding level-2 residual variance in the specified model.

To estimate explained variance in the level-1 slope coefficient, the level-2 residual variances $\hat{\sigma}^2_{u1}$ (*empty model*) and $\hat{\sigma}^2_{u1}$ (*specified model*) should be estimated from eqs. 2.9.3 and 2.2.3, respectively, and then the results can be used to calculate the explained variance in level-1 random slope $\hat{\beta}_{1j}$, using eq. 2.9.5.

Note that inclusion of group-level explanatory variables in a model influences level-2 variation, but not level-1 outcome variation because group-level variables have constant values within each group. In contrast, individual-level explanatory variables can influence outcome variations at both individual level and group level because they may have both within-group and between-group variations. As a result, inclusion of a level-1 explanatory variable with no group-variation may increase *variance explained* at level-1, but will not change *variance explained* at level-2. However, inclusion of level-1 explanatory variables with a large between-group variation (e.g., proportion of males greatly varies across groups) would influence *variances explained* at both levels 1 and 2.

Using proportional reduction in estimated residual variances to estimate approximately explained variances may encounter problems. First, adding a level-1 explanatory variable, particularly a group-mean centered variable, into a multilevel model may increase rather than decrease level-1 residual variance and/or level-2 residual variances, resulting in negative values of "R^2" (Snijders & Bosker, 1994, 1999). Second, the estimated residual variances in multilevel models will change with changes in the scales of the explanatory variables when random slope coefficients are involved (Hox, 2002). Snijders and Bosker (1994, 2003) recommend working with the *proportional reduction in mean squared prediction error*, instead of the *proportional reduction in estimated variance*, for estimating unexplained residual variances in multilevel models. Snijders and Bosker's method can be described using the following model:

$$y_{ij} = \beta_{0j} + \alpha_1 x_{1ij} + e_{ij} \tag{2.9.6}$$

$$\beta_{0j} = \gamma_{00} + \gamma_{01} w_{1j} + u_{0j} \tag{2.9.7}$$

$$y_{ij} = \gamma_{00} + \gamma_{01} w_{1j} + \alpha_1 x_{1ij} + u_{0j} + e_{ij} \tag{2.9.8}$$

where the level-1 intercept is a random coefficient and the level-1 explanatory variable has a fixed-effect. For a given value of x_{1ij}, the model prediction error would be:

$$y_{ij} - (\gamma_{00} + \gamma_{01} w_{1j} + \alpha_1 x_{1ij}) = u_{0j} + e_{ij} \tag{2.9.9}$$

The corresponding *mean squared prediction error* is:

$$\begin{aligned}\mathrm{Var}(u_{0j} + e_{ij}) &= \sigma^2_{u0}(\text{specified model}) + \sigma^2(\text{specified model}) \\ &= \sigma^2_{\text{total\_level\_1}}(\text{specified model})\end{aligned} \tag{2.9.10}$$

Using Snijders and Bosker's method, the explained proportion of the level-1 variance can be described as the proportional reduction in the *mean squared prediction error* $\sigma^2_{\text{total\_level\_1}}$, comparing the specified model and the *empty model*:

$$\begin{aligned}\begin{pmatrix}\text{SB: \% of Explained} \\ \text{level -1 variance}\end{pmatrix} &= \frac{\hat{\sigma}^2_{\text{total\_level\_1}}(\text{empty model}) - \hat{\sigma}^2_{\text{total\_level\_1}}(\text{specified model})}{\hat{\sigma}^2_{\text{total\_level\_1}}(\text{empty model})} \\ &= 1 - \frac{\hat{\sigma}^2_{\text{total\_level\_1}}(\text{specified model})}{\hat{\sigma}^2_{\text{total\_level\_1}}(\text{empty model})}\end{aligned} \tag{2.9.11}$$

where SB: % *of Explained level*-1 *variance* denotes for the proportion of explained level-1 variance using Snijders and Bosker's method.

The level-2 model prediction error is described as:

$$\bar{y}_{ij} - (\gamma_{00} + \gamma_{01} w_{1j} + \alpha_1 \bar{x}_{1\cdot j}) = u_{0j} + e_{\cdot j} \tag{2.9.12}$$

The corresponding *mean squared prediction error* is:

$$\begin{aligned}\mathrm{Var}(u_{0j} + e_{\cdot j}) &= \sigma^2_{u0} + \frac{\sigma^2}{n} \\ &= \hat{\sigma}^2_{\text{total\_level\_2}}\end{aligned} \tag{2.9.13}$$

where n is the group size and $\hat{\sigma}^2_{\text{total\_level\_2}}$ is the *mean squared prediction error* for group mean $\bar{y}_{.j}$. With unequal group sizes, n is often replaced with the average group size. Alternatively, n could be replaced with the harmonic mean (Snijders & Bosker, 2003), defined by:

$$\left(\frac{\sum_j 1/n_j}{N} \right)^{-1} = \frac{N}{\sum_j 1/n_j} \tag{2.9.14}$$

The model explained level-2 variance using Snijders and Bosker's method can be described as the proportional reduction in $\hat{\sigma}^2_{\text{total\_level\_2}}$ comparing the specified model with the *empty model*:

$$\begin{pmatrix} \text{SB: \% of Explained} \\ \text{level -2 variance} \end{pmatrix} = \frac{\hat{\sigma}^2_{\text{total\_level\_2}}(\text{empty model}) - \hat{\sigma}^2_{\text{total\_level\_2}}(\text{specified model})}{\hat{\sigma}^2_{\text{total\_level\_2}}(\text{empty model})}$$

$$= 1 - \frac{\hat{\sigma}^2_{\text{total\_level\_2}}(\text{specified model})}{\hat{\sigma}^2_{\text{total\_level\_2}}(\text{empty model})} \tag{2.9.15}$$

In theory, adding additional explanatory variables to the model will decrease the *mean squared prediction errors* for predicting an individual outcome, as well as for predicting a group mean, thus increase the SB: % *of explained levels*-1 *and level*-2 *variances*. However, in empirical studies, it is possible for SB: % *of explained levels*-1 *variance* and/or SB: % *of explained levels*-2 *variance* to decrease when additional explanatory variables are added. A small decrease may be a result of chance effects. However, for a reasonably large sample size, a decrease of 0.05 would suggest a possible misspecification of the model. As such, changes in the wrong direction of SB: % *of explained level*-1 *variance* and/or SB: % *of explained level*-2 *variance* can be used as diagnostics of model misspecification (Snijders & Bosker, 2003).

Once random slopes are involved in a model, calculation for the explained variances becomes complicated (Snijders & Bosker, 1994). Usually researchers can treat all random slopes as fixed and use the model results to calculate the explained within- and between-group outcome variation using eqs. 2.9.11 and 2.9.15. The results are very close to those from the model with random slopes (Snijders & Bosker, 2003, p.105).

Snijders & Bosker's method has some limitations. First, it only allows estimating explained variances of within- and between-group outcome measures. It does not address explained variance in random regression slopes. Second, it is currently applicable only to two-level models (Snijders & Bosker, 1994, 2003). And third, it does not solve the problem that Raudenbush & Bryk's method encounters, that is the residual variances depend on the scale of explanatory variable in a model with random slope coefficients. Further studies are needed to develop measures of explained variances in multilevel modeling.

At present, the Raudenbush & Bryk's *proportional reduction in estimated variance* is most widely applied in multilevel models because it is easy to use, can be used to predict variation not only in mean outcome, but also in random slopes, and it can be readily applied to 3 or higher level models.

2.10 Steps for building multilevel models

In statistical modeling, very often neither theoretical knowledge nor empirical findings provide precise information about what model should be specified for data analysis. Analysts may have a hypothesis about a relationship they would like to model based on theories, literature or previous studies. However, in the beginning of a study, some exploration may be needed to find a satisfactory model that fits the data well, yields meaningful, interpretable results, and is parsimonious too.

In general, model development is an exploratory process that is driven by theory, hypotheses, statistical tests of the hypotheses, and model comparisons. Guidelines and steps for multilevel model development or model specification are discussed in various studies (Hox, 1994; Singer, 1998). This text more closely follows a combination of approaches that Hox (1994) and Singer (1998) recommend. We start by fitting an *empty model*, and then sequentially examine the effects of macro and micro explanatory variables, as well as cross-level interaction effects, on the outcome.

1) *Running the empty model*: The *empty model* (eqs. 2.1.2—2.1.4) is fundamental in multilevel model development. Only when we have identified the existence of significant within-group observation dependence or ICC, then we should build a multilevel model. Otherwise multiple regressions may be appropriate for data analysis. The *empty model* allows us to decompose the total outcome variation into within-group and between-group variations. As discussed in Section 2.1, the micro and macro sources of outcome variation are used to calculate ICC and test the null hypothesis that all groups have the same outcome mean. In addition, the *empty model* provides important information about the grand mean of the outcome measure, reliability of each group's mean outcome measure as an estimate of its population mean, as well as a baseline model with which other complex models can be compared. Running the empty model is always the starting point for multilevel model development.

2) *Adding level-2 explanatory variables into the empty model*: The next step of model development expands the *empty model* by adding level-1 explanatory variables (Hox, 1994) or adding level-2 explanatory variables (Singer, 1998). We prefer Singer's approach. If the *empty model* shows the existence of significant between-group variation in the mean outcome, it also indicates significant unexplained variation in the mean outcome across groups. Thus it is logical to add group-level variables into the model to explain the variation.

In this step, all relevant group-level explanatory variables should be considered. However, the number of group-level variables must not be greater than the number of groups. This rule is similar to the rule that the number of independent variables must not be greater than the number of observations in a regression model. In practice, however a limited number of group level variables are generally used in multilevel model applications. For simplicity, we illustrate a model with only one level-2 explanatory variable, w_{1j}, to predict the level-1 random intercept:

$$y_{ij} = \beta_{0j} + e_{ij} \tag{2.10.1}$$

$$\beta_{0j} = \gamma_{00} + \gamma_{01} w_{1j} + u_{0j} \tag{2.10.2}$$

$$y_{ij} = \gamma_{00} + \gamma_{01} w_{1j} + (u_{0j} + e_{ij}) \tag{2.10.3}$$

This tentative model may be called a random intercept model with main effects of macro explanatory variables only. Compared with the *empty model*, this model has the same random components or random effects ($u_{0j} + e_{ij}$), but different fixed-effects (i.e., γ_{10} in the *empty model* but γ_{10} and γ_{11} in this model). The LR Chi-square test should be estimated under ML instead of REML to compare this model with the *empty model*.

3) *Adding level-1 explanatory variables to the model*: In Step 2, the relationship between group mean outcome and group-level variables is examined without adjusting for the individual characteristics. In this step, level-1 explanatory variables are added to the model, and all the level-1 slope coefficients are treated as fixed. All relevant level-1 variables should be selected into the model on the basis of theory, hypothesis, and empirical findings. For simplicity, we include only one level-1 explanatory variable with fixed-effect in the model:

$$y_{ij} = \beta_{0j} + \alpha_1 x_{1ij} + e_{ij} \tag{2.10.4}$$

$$\beta_{0j} = \gamma_{00} + \gamma_{01} w_{1j} + u_{0j} \tag{2.10.5}$$

$$y_{ij} = \gamma_{00} + \alpha_1 x_{1ij} + \gamma_{01} w_{1j} + (u_{0j} + e_{ij}) \tag{2.10.6}$$

This tentative model is called the *random intercept model* in econometrics, which can be estimated via OLS regression with appropriate transformation of the outcome and explanatory variables (Judge et al., 1982; Wang & Fisher, 1994). In this book, we call this model the *random intercept model with fixed main effects.*

The principle of specifying the level-1 model (eq. 2.10.4) is similar to multiple regression models. All plausible explanatory variables should be included. The rational of testing such a model is that we would like to test the main effects first before considering interaction effects. After this step, we will identify level-1 random slope coefficients and assess cross-level interactions.

The models specified in Steps 2 and 3 have the same random effects (σ^2 and σ^2_{u0}), but different fixed-effects (γ_{00} and γ_{01} in the model specified in Step 2; γ_{00}, γ_{01}, and α_1 in Step 3). Again, the deviances statistics used for LR test should be estimated from ML.

4) *Testing level-*1 *random slope coefficients*: In a multilevel model, not only the group mean outcomes, but also the relationships between level-1 explanatory variables and the outcome measure may vary across groups. That is, each level-1 explanatory variable may have a random slope coefficient, and all the level-1 random coefficients may be correlated with each other. Following Step 3, the next step is to identify random level-1 slope coefficients in the model. In theory, it is possible for the intercept and all the slope coefficients to be random, leading to a *random effect model.* Usually, only some of the slopes in a model are random while those remaining are fixed. Thus, the model is also called a *mixed model* in which fixed and random regression coefficients are mixed together.

In principle, it is advisable to check the randomness of slopes of all level-1 explanatory variables. When the random part of a variable's effect is mistakenly omitted, the estimate of the fixed part effects may be biased. However, specifying too many random slopes in a model may cause estimation problems, often leading to a failure of estimation convergence. This is because the number of level-2 residual variances/covariances increases dramatically when the number of level-1 random coefficients increases. For example, there is only one level-2 residual variance (i.e., σ^2_{u0}) in the *empty model* where random intercept β_{0j} is the only level-1 random coefficient. Adding one level-1 random slope coefficient (e.g., β_{1j}) results in three level-2 residuals variances and covariance, (e.g., σ^2_{u0}, σ^2_{u1}, and σ^2_{u01}); with Q level-1 random slope coefficients plus a random intercept, the number of level-2 residual variances and covariances would be $(Q + 1)[(Q + 1) + 1]/2$. As a result, significantly more information is needed for reasonable estimation of the random part of the model. Thus the data set needs to be very large in order to prevent instability of model estimation with many random coefficients.

Very often, analysts have prior assumptions based on theory or the research design about which level-1 explanatory variables may have random slope coefficients. Significance testing for variances (e.g., $\sigma^2_{u1}, \ldots, \sigma^2_{uQ}$) in level-2 residual variance/covariance matrix G provides information about which level-1 slope coefficients are random. If σ^2_{uq} is not statistically significantly different from zero, it indicates that the corresponding level-1 slope coefficient β_q does not significantly vary across groups; thus it can be treated as a fixed level-1 slope coefficient with its group-level variance fixed at zero (i.e., $\sigma^2_{uq} = 0$), and its corresponding level-1 explanatory variable z_{qij} can be treated as a fixed-effect variable.

Suppose we have a model with one level-1 random intercept and a random slope, and only one group-level explanatory variable so that

$$y_{ij} = \beta_{0j} + \beta_{1j} z_{1ij} + e_{ij} \tag{2.10.7}$$

$$\beta_{0j} = \gamma_{00} + \gamma_{01} w_{1j} + u_{0j} \tag{2.10.8}$$

$$\beta_{1j} = \gamma_{10} + u_{1j} \tag{2.10.9}$$

$$y_{ij} = \gamma_{00} + \gamma_{01} w_{j1} + \gamma_{10} z_{1ij} + (u_{0j} + z_{1ij} \cdot u_{1j} + e_{ij}) \tag{2.10.10}$$

where both the level-1 intercept β_{0j} and slope β_{1j} vary across groups; and the level-2 residual u_{0j} represents how much the jth group outcome mean deviates from the overall mean (γ_{00}), conditional on the value of group-level variable w_{1j}; while u_{1j} represents how much the effect of variable z_{1ij} on y_{ij} in the jth group deviates from the average effect (γ_{10}) of variable z_{1ij}. The covariance between u_{0j} and u_{1j} measures the association between the level-1 random intercept and slope coefficients. The model specified in eqs. 2.10.7—2.10.10 differs only in random components from the model specified in eqs. 2.10.4—2.10.6, thus the models should be estimated under REML for model comparison.

Note that it is not uncommon for model estimation convergence to fail even with a limited number (e.g., two) of level-1 random slope coefficients. If this happens, one way to get around it is to transform the corresponding level-1 explanatory variables by transforming them to have 0 means or making them orthogonal (i.e., have zero correlations) in order to improve convergence (Snijders & Bosker, 2003, p.82).

5) *Including cross-level interactions in the model*: If some level-1 slope coefficients are found to be random, their between-group variations need to be explained at the group-level by treating them as a function of group-level explanatory variables in the macro models. For example, including a group-level explanatory variable w_{1j} in eq. 2.10.9 will result in a multilevel model as the following:

$$y_{ij} = \beta_{0j} + \beta_{1j} z_{1ij} + e_{ij} \tag{2.10.11}$$

$$\beta_{0j} = \gamma_{00} + \gamma_{01} w_{1j} + u_{0j} \tag{2.10.12}$$

$$\beta_{1j} = \gamma_{10} + \gamma_{11} w_{1j} + u_{1j} \tag{2.10.13}$$

$$y_{ij} = \gamma_{00} + \gamma_{01} w_{1j} + \gamma_{10} z_{1ij} + \gamma_{11} w_{1j} z_{1ij} + (u_{0j} + z_{1ij} u_{1j} + e_{ij}) \tag{2.10.14}$$

where the effect (i.e., γ_{01}) of group-level variable w_{1j} on the level-1 random intercept β_{0j} is simply the main effect of w_{1j} on outcome measure y_{ij}. Explaining the variation in level-1 slope coefficient β_{1j} is equivalent to assessing the interaction between the level-1 explanatory variable z_{1ij} and the group-level explanatory variable w_{1j}. ML estimation should be used for comparing this model with the model specified in eqs. 2.10.7—2.10.10 because the two models differ only in fixed-effects.

Ideally, following the above five steps will reveal some ideas about a "final" model for our designed study. A better model fit (e.g., smaller deviance, AIC, and BIC) does not necessarily mean a model is satisfactory. Model selection is both a statistics and theory-driven process. The purpose of model development is to find a parsimonious model that not only fits the data, but also provides interpretable results.

2.11 Higher-level multilevel models

The two-level multilevel model discussed in this chapter can be readily expanded to three or more levels. For example, with three-level hierarchically structured data (e.g., students nested in classes, and classes nested in schools), a three-level model can be constructed in which the outcome variable is still measured at level 1 (e.g., student level) and the level-1 intercept and some slope coefficients may be random. These random level-1 regression coefficients will be treated as dependent variables in level-2 models in which classes are the analytical units. Further, some of the level-2 regression coefficients (e.g., level-2 intercept and slope coefficients) may vary across level-3 units (e.g., schools). Then the level-2 random regression coefficients will be treated as a function of some level-3 contextual variables in level-3 models.

Suppose we have a three-level multilevel data set with outcome variable y_{ijk}, measuring HIV risk behaviors among drug users recruited in multiple research project sites located in different states in the U.S. The subscripts i, j, and k indicate levels 1, 2, and 3 observation units. In this multilevel data set, there are a total of K level-3 units (i.e., states); each level-3 unit has J_k level-2 units (i.e., project sites); and the jth level-2 unit in the kth level-3 unit has n_{ijk} level-1 units (i.e., individual participants). The two-level multilevel model shown in eqs. 2.2.1—2.2.3 can be expanded to a three-level model. The formulation of the three-level model follows:

$$y_{ijk} = \beta_{0jk} + \alpha_1 x_{1ijk} + \beta_{1jk} z_{1ijk} + e_{ijk} \quad (2.11.1)$$

$$\beta_{0jk} = \gamma_{00k} + \gamma_{01} w_{1jk} + u_{0jk} \quad (2.11.2)$$

$$\beta_{1jk} = \gamma_{10} + \gamma_{11k} w_{1jk} + u_{1jk} \quad (2.11.3)$$

$$\gamma_{00k} = \xi_{000} + \xi_{001} \upsilon_{1k} + \upsilon_{00k} \quad (2.11.4)$$

$$\gamma_{11k} = \xi_{110} + \xi_{111} \upsilon_{1k} + \upsilon_{11k} \quad (2.11.5)$$

where the intercept β_{0jk} and the slope coefficient β_{1jk} in eq. 2.11.1 are level-1 random coefficients, which vary across level-2 units (e.g., sites); the intercept γ_{00k} in eq. 2.11.2 and the slope coefficient γ_{11k} in eq. 2.11.3 are level-2 random coefficients, which vary across level-3 units (e.g., states). The two level-3 models (i.e., eqs. 2.11.4—2.11.5) are to explain the variations in the two level-2 random coefficients.

By substituting eqs. 2.11.4—2.11.5 into eqs. 2.11.2—2.11.3, respectively, then substituting the resulting equations into eq. 2.11.1, we will have the combined three-level multilevel model. The composite residuals of this combined model will consist of residuals at three different levels: e_{ijk} at level 1; u_{0jk} and u_{1jk} at level 2; and υ_{00k} and υ_{11k} at level 3.

Formulation of a three or higher level multilevel model is straightforward and the model can be run readily in all the multilevel model computer software and programs noted earlier. However, in real research it is often difficult to find a satisfactory multilevel data set with a sufficient number of level 3 or higher level units. For example, we may not have enough states involved in the research. As such, we may not have satisfactory, hierarchically structured data for three-level modeling. In this case researchers can disaggregate level-3 data to level-2 units by assigning the same level-3 information to each level-2 unit within the same level-3 groups. By reducing three-level data into a two-level data structure, a level-2 multilevel model can then be applied to analyze the data.

Another important reason why three or higher level multilevel models are not often conducted is that models can become very complex with too many model parameters, in particular, when more variables at different levels and more random coefficients are involved. The model results can be very difficult to interpret. In this book, we focus on two-level multilevel models.

Chapter 3
Application of two-level linear multilevel models

In this chapter, we demonstrate how to build two-level multilevel models, using SAS (Statistics Analysis System), a widely-used statistical analysis package. SAS provides a convenient solution for multilevel modeling, including data manipulation, obtaining descriptive statistics, running multilevel models, and reporting results with graphs within the same package. The demonstration proceeds step-by-step using SAS *PROC MIXED* with the actual research data.

3.1 Data

The data used for model demonstrations in this chapter were extracted from a U.S. national database of the National Institute on Drug Abuse's (NIDA) AIDS community-based intervention research program conducted in the early 1990s in U.S. (Needle et al., 1995). The program was designed to monitor and reduce HIV risk behaviors and assess HIV seroprevalence among not-in-treatment injection drug users (IDUs) and crack cocaine users. The program evaluated the efficacy of interventions designed to prevent, eliminate, or reduce HIV risk behaviors. The program engaged 23 project sites between 1990 and 1994. One project site that was located outside U.S. territory and two sites that had too few IDUs or had exclusively Hispanic participants were excluded from our model demonstration. Consequently, only 20 project sites with a total of 9 824 injection drug users (IDUs) were included in our study. All participants included in the study took voluntary and confidential HIV antibody tests during the baseline interview. The percentage of HIV positives among the IDUs was used as the approximate estimate of HIV seropositivity rate among IDUs in the city where the project site was located. All sites were divided into two regions based the level of HIV seropositivity: a Low HIV Prevalence Region (HIV seropositivity rate < = 15%) and a Higher HIV Prevalence Region (HIV seropositivity rate >15%). Overall 9.44% of the IDUs in the sample tested HIV positive. The HIV prevalence rates for the two regions were 5.63% and 19.86%, respectively.

Selected sample statistics for the IDUs are shown in Table 3.1.1 by region and project site. About 73.25% (N = 7 196) of the IDUs were located in the Low HIV Prevalence Region, while 26.75% (N = 2 628) were located in the Higher HIV Prevalence Region. The outcome measure we are going to model is the number of times an IDU injected drugs in the 30 days prior to the baseline interview. The average frequency of drug injection varies between regions and across sites within each region. On average IDUs injected drugs about 56.47 times in the Low HIV Prevalence Region and 84.75 times in the Higher HIV Prevalence Region. Some variations in individual socio-demographic characteristics, such as gender, ethnicity, age, and education, are noticeable across sites and regions. For simplicity, we will only include a limited number of variables in our models:

Table 3.1.1 Sample statistics of injection drug users (IDU) by region and project site

Project Site[1]	N	HIV Sero-positive Rate[2] (%)	Frequency of Drug Injection[3]	Gender (%)		Ethnicity (%)		Mean Age	Education(%)	
				Male	Female	Black	White		<High Sch.	High Sch.+
				Low HIV Prevalence Region[4]						
16	86	1.16	21.22	75.58	24.42	22.09	77.91	33.73	34.88	65.12
24	1036	1.45	56.77	73.94	26.06	70.17	29.83	40.39	38.22	61.78
30	365	1.64	58.02	81.10	18.90	86.03	13.97	43.44	35.07	64.93
21	932	1.82	70.94	69.53	30.47	34.66	65.34	37.50	30.79	69.21
15	279	2.15	52.94	79.93	20.07	10.04	89.96	36.59	21.50	78.49
17	223	3.59	76.36	69.96	30.04	42.15	57.85	39.17	30.94	69.06
32	284	3.87	85.40	60.21	39.79	44.01	55.99	38.84	34.86	65.14
27	328	3.96	29.51	80.49	19.51	67.07	32.93	36.98	36.59	63.41
19	552	3.99	84.86	77.54	22.46	14.86	85.14	37.73	38.95	61.05
23	695	5.61	68.98	69.78	30.22	38.27	61.73	39.41	40.58	59.42
22	469	8.10	51.43	87.21	12.79	87.21	12.79	40.89	58.00	42.00
18	684	10.82	39.26	75.29	24.71	91.96	8.04	41.02	46.20	53.80
10	455	10.99	19.23	69.67	30.33	76.48	23.52	38.47	47.03	52.97
31	254	12.99	35.30	77.95	22.05	78.35	21.65	37.27	38.98	61.02
13	554	13.00	52.70	81.05	18.95	81.95	18.05	39.13	48.74	51.26
Subtotal	**7,196**	**5.63**	**56.47**	**74.90**	**25.10**	**58.88**	**41.12**	**39.15**	**39.70**	**60.30**
				Higher HIV Prevalence Region[4]						
28	910	16.48	86.86	64.51	35.49	99.78	0.22	41.64	42.97	57.03
14	1311	17.93	83.59	67.96	32.04	84.74	15.26	41.90	36.46	63.54
25	89	21.35	96.72	84.27	15.73	66.29	33.71	38.60	41.57	58.43
12	179	29.61	86.55	69.27	30.73	65.36	34.64	40.39	45.81	54.19
11	139	46.76	71.98	68.35	31.65	92.09	7.91	40.50	49.64	50.36
Subtotal	**2,628**	**19.86**	**84.75**	**67.42**	**32.58**	**88.39**	**11.61**	**41.52**	**40.22**	**59.78**
Total	**9,824**	**9.44**	**64.04**	**72.90**	**27.10**	**66.78**	**32.22**	**39.79**	**39.84**	**60.16**

Note:

[1]Data were extracted from a national database of the National Institute on Drug Abuse's (NIDA) AIDS community-based intervention research program conducted in early 1990s in U.S.

[2]HIV seropositivity rate, estimated as the percentage of HIV positives among the IDUs who took voluntary and confidential HIV antibody tests at the baseline interview.

[3]Average frequency of drug injection in 30 days prior to baseline interview.

[4]Sites with an estimated HIV seropositivity rate less/greater than 15.00% are defined as being in a low/higher HIV Prevalence Region (Wang, 2003).

- Outcome measure (dependent variable): Frequency of drug injection.
- Level-1 explanatory variables: Gender, Ethnicity, Age, and Education.
- Level-2 (group) explanatory variable: HIV Prevalence rate by research site.

Some multilevel model computer programs (e.g., HLM) require users to input separate micro and macro level data files that are then linked by group ID. HLM creates a system data file (e.g., SSM file, standing for *sufficient statistics matrices*, for HLM), then separately specifies micro and macro models. SAS also requires creating a multilevel data set using micro and macro data for analysis, but the micro and macro models are not specified separately. In our example, no separate macro data information are available, and the group-level explanatory variables or contextual variables are generated from the original individual level data.

The following SAS program shows how to create a multilevel data from the original sample:

***SAS Program 3.1.1**;

```
LIBNAME Lib 'd:\MLM\data';
Data data1;
  Set Lib.ML_IDUdata;
Proc sort; by site;

*Create grand mean centered level-1 variables;
Proc SQL;
  Create table lib.data_level_1 as select *, log(Inject) as lninj, age-
mean(age) as Gdmc_age from data1;
Quit;

*Create level-2 variables: HIV prevalence region;
Proc SQL;
  Create table data2 as select site, HIV,
  100* (mean(HIV)) as m_HIV from data1
  Group by site;
  Create table lib.data_level_2 as select distinct site,
   m_HIV>15 as HIV_Region from data2;
Quit;

Data lib.ML_data_Chap3;
Merge lib.data_level_1(in = d1) lib.data_level_2(in = d2);
by site;
if d1 = d2;
label HIV_Region = '1-Higher HIV prevalence region; 0-Lowever HIV preva-
lence region';
label Gdmc_age = 'Grand-mean centered age';

Proc means;
Class HIV_Region site;
Var HIV Inject Gender Ethnic Age Highsch;
Run;
```

The program begins by reading the original SAS data into a SAS active file, Data1. Level-1 and Level-2 data sets are created using SAS *PROC SQL*, an implementation of structured query language (SQL) for data manipulation. Then, the micro and macro data sets are merged by site to create the multilevel data set, *ML_data_Chap3*, which is saved as a permanent SAS data file in the SAS library. The variables available in this multilevel data set are:

Variables measured at individual level:

- *Inject*—Outcome measure/dependent variable
- *Gender* —Dummy variable (1-Male; 0-Female)
- *Ethnic*—Dummy variable (1-Black; 0-White)
- *Age*—Original measure of age
- *Gdmc_age*—Grand-mean centered age
- *Highsch*—Dummy variable for level of education (1-High school or above; 0-Less than high school)
- *Edu_gp*—Categorical variable (1-Less than high school; 2-High school; 3-College)
- *Edu_gp*1—Dummy variable (1-Less than high school; 0-Otherwise)
- *Edu_gp*2— Dummy variable (1-High school; 0-Otherwise).

Contextual variables —Variables measured at project site (group) level:

- *M_HIV* —Project site HIV prevalence rate
- *HIV_Region*—Dummy variable (1-Higher HIV prevalence region; 0-Low HIV prevalence region). [1]

3.2 Empty model

The *empty model* is the simplest random effect model referred as one-way random effect ANOVA or *unconditional means model*. The purpose of running this model is to assess within-group homogeneity or between-group heterogeneity. This is Step-1 of model development in Section 2.10 in Chapter 2. Applying the *empty model* specified in eq. 2.1.4, we have:

$$\text{Inject}_{ij} = \beta_{0j} + e_{ij} \tag{3.2.1}$$

$$\beta_{0j} = \gamma_{00} + u_{0j} \tag{3.2.2}$$

$$\text{Inject}_{ij} = \gamma_{00} + u_{0j} + e_{ij} \tag{3.2.3}$$

The SAS program for the *empty model* follows:

***SAS Program 3.2.1**;

```
LIBNAME lib 'D:\MLM\data';
Data Data1;
  Set lib.ML_data_Chap3;
*Empty model;
PROC MIXED METHOD = REML COVTEST;
CLASS Site;
MODEL Inject = /SOLUTION;
RANDOM Intercept /SUBJECT = Site SOLUTION;
Run;
```

[1] Sites with an estimated HIV seropositivity rate less/greater than 15.00% are defined as being in a low/higher HIV Prevalence Region (Wang, 2003). Conceptually, regions where the project sites are located are level-3 observation units. Since there are only two such units, we disaggregate the regiona variable *HIV_Region* to level-2 contextual variable, i.e., the same value of variable *HIV_Region* are assigned to the project sites located in the same region.

where the multilevel data set, *ML_data_Chap*3, generated in Section 3.1 is used for modeling. There are four key SAS statements in *PROC MIXED* for modeling the cross-section multilevel data:

- The *PROC MIXED* statement: This statement calls the *PROC MIXED* procedure for linear multilevel modeling. The estimation method (e.g., ML, REML) is specified with the *METHOD* = option. If no method is specified, REML is the default. The *COVEST* option requests the program to print the standard errors and *z*-tests for the variance/covariance parameter estimates of the random effects.
- The *CLASS* statement: As in many other SAS modeling procedures this statement specifies categorical variables. The model will estimate ($K-1$) parameters for a variable with K levels with the last category of the variable as the reference group. A global test for any differences among the K levels will be produced for the overall effect of the variable. Besides the categorical explanatory variables, the level-2 units identification index (i.e., group ID) is usually specified in the CLASS statement. If the group ID is a numeric measure and is not specified in the CLASS statement, it must be sorted prior to running the *PROC MIXED* statement. Otherwise, *PROC MIXED* would take a record to be from a new group whenever the value of the group ID changes from the previous record. However. Treating group ID as numeric will speed up the computation process when the number of groups is large.
- The *MODEL* statement: The *MODEL* statement specifies the fixed-effect part of a multilevel model (see eq. 2.2.9). All variables specified in the *MODEL* statement that are not specified in the *CLASS* statement will be treated as continuous variables. The intercept is included in SAS *PROC MIXED* by default[2]. Since no variables are specified on the right side of the "=", the model is specified as the *intercept-only model.* The *SOLUTION* (or *S*) option in the *MODEL* statement requests the fixed-effects estimates.
- The *RANDOM* statement: The *RANDOM* statement specifies the random effect part of the model (see eq. 2.2.9). Without a *RANDOM* statement in model specification, the model specified is the same as OLS regression. A level-1 intercept can be specified as random by specifying *INTERCEPT* (or *INT*) in the *RANDOM* statement; while a level-1 slope coefficient can be specified as random by including the name of its corresponding level-1 variable in the *RANDOM* statement. The *SUBJECT* = option identifies the groups or the level-2 units in a two-level multilevel model. The *SOLUTION* (or *S*) option in the *RANDOM* statement instructs the program to print the Best Linear Unbiased Prediction (BLUP) of the random effects.

In our example of the *empty model*, the outcome variable is *Inject.* No predictors or explanatory variables are specified in this model. The model is estimated using REML, the default estimation method in *PROC MIXED.* "*Site*," the level-2 unit, is identified in the *CLASS* statement as a class variable. *INTERCEPT* (or *INT*) specified in the *RANDOM* statement tells *PROC MIXED* to let the level-1 intercept coefficient vary across groups/sites. The *SUBJECT* = *Site* (or *SUB* =) option in the *RANDOM* statement instructs SAS that individuals are nested within sites; thus individuals are treated as level-1 units, and projects sites are treated as level-2 units. Selected SAS output of the model are shown below.

[2] To fit a model without an intercept, the option */NOINT* needs to be specified in the *MODEL* statement.

SAS Output 3.2.1 Selected results of the *empty model*

Iteration History

Iteration	Evaluations	-2 Res Log Like	Criterion
0	1	111038.52376815	
1	2	110256.17252401	0.00000339
2	1	110255.99639867	0.00000017
3	1	110255.98824435	0.00000000

Convergence criteria met.

Covariance Parameter Estimates

Cov Parm	Subject	Estimate	Standard Error	Z Value	Pr Z
Intercept	site	523.57	176.30	2.97	0.0015
Residual		4350.47	62.1381	70.01	<.0001

Fit Statistics

-2 Res Log Likelihood	110256.0
AIC (smaller is better)	110260.0
AICC (smaller is better)	110260.0
BIC (smaller is better)	110262.0

Null Model Likelihood Ratio Test

DF	Chi-Square	Pr > ChiSq
1	782.54	<.0001

Solution for Fixed-effects

Effect	Estimate	Standard Error	DF	t Value	Pr > |t|
Intercept	61.4366	5.1906	19	11.84	<.0001

Solution for Random Effects

Effect	site	Estimate	Std Err Pred	DF	t Value	Pr > |t|
Intercept	10	-41.4489	5.9477	9804	-6.97	<.0001
Intercept	11	9.9472	7.3158	9804	1.36	0.1740
Intercept	12	23.9969	6.9160	9804	3.47	0.0005
Intercept	13	-8.6036	5.8214	9804	-1.48	0.1395

Intercept	14	22.0135	5.4682	9804	4.03	<.0001
Intercept	15	-8.2483	6.3678	9804	-1.30	0.1952
Intercept	16	-36.6724	8.2785	9804	-4.43	<.0001
Intercept	17	14.3904	6.6219	9804	2.17	0.0298
Intercept	18	-21.9073	5.7082	9804	-3.84	0.0001
Intercept	19	23.0730	5.8235	9804	3.96	<.0001
Intercept	21	9.4225	5.5763	9804	1.69	0.0911
Intercept	22	-9.8275	5.9268	9804	-1.66	0.0973
Intercept	23	7.4570	5.7005	9804	1.31	0.1909
Intercept	24	-4.6302	5.5391	9804	-0.84	0.4032
Intercept	25	32.2697	8.2003	9804	3.94	<.0001
Intercept	27	-31.0583	6.2099	9804	-5.00	<.0001
Intercept	28	25.1949	5.5852	9804	4.51	<.0001
Intercept	30	-3.3387	6.1164	9804	-0.55	0.5852
Intercept	31	-25.3133	6.4690	9804	-3.91	<.0001
Intercept	32	23.2836	6.3494	9804	3.67	0.0002

The *Iteration History* panel shows that the model estimation converged successfully after 3 iterations. Quick convergence usually indicates good model fit. The restricted log-likelihood (−2LL) and three information criteria measures (*AIC*, *AICC*, and *BIC*) are reported in the panel of *Fit Statistics*. These statistics don't tell how well the model fits the data. Instead, they are used for comparisons between different models. We will discuss more about this issue later in this chapter.

The variance of the level-1 random intercept coefficients and the level-1 residual variance are shown in panel of *Covariance Parameter Estimates*: $\hat{\sigma}_{u0}^2$ = 523.57 (p = 0.0015) and $\hat{\sigma}^2$ = 4350.47 (p < 0.001). The results suggest that 1) average frequency of drug injection among IDUs varies significantly across sites; and 2) there is a significant, large variation among IDUs within project sites. The within-group variance (i.e., $\hat{\sigma}^2$) is about 8.31 times as large as the between-group variance (i.e., $\hat{\sigma}_{u0}^2$).

$\hat{\sigma}_{u0}^2$ and $\hat{\sigma}^2$ can be used to calculate the measure of within-group homogeneity or between group heterogeneity——the intra-class correlation (ICC) using eq. 2.1.1:

$$\mathrm{ICC} = \frac{\hat{\sigma}_{u0}^2}{\hat{\sigma}_{u0}^2 + \hat{\sigma}^2} = \frac{523.57}{523.57 + 4350.47} = 0.11 \tag{3.2.4}$$

An ICC of 0.11 is moderately large, indicating that about 11% of the total variance in the outcome measure is due to variations between the project sites. In Chapter 1 we indicated that even a small ICC could cause a large Type I error in model estimation. As $\hat{\sigma}_{u0}^2$ is statistically significant, we would conclude that ICC is statistically significant, and that a multilevel model is necessary for data analysis.

The fixed-effect estimates are shown in the *Solution for Fixed-effects* panel. Since no explanatory variable is specified at the micro or macro levels in this model, the only fixed-effect is the intercept of model in eq. 2.1.4, $\hat{\gamma}_{00}$ = 61.44 (SE = 5.19), which is the model estimated overall mean of the outcome measure.

The final panel of the SAS Output 3.2.1 is the *Solution for Random Effects*. For the *empty model*, the random effects are $\hat{u}_{0j}$ only. The *SOLUTION* option in the *RANDOM* statement instructs *PROC MIXED* to uses BLUP, a Bayesian technique, to estimate $\hat{u}_{0j}$ for each group (i.e., site in this study)[3]. $\hat{u}_{0j}$ represents the deviation of the site-specific mean outcome from the overall mean outcome. A positive/negative value of $\hat{u}_{0j}$ indicates how much the mean outcome measure in Site j is larger/smaller than the overall mean outcome. In this example, the within-site mean outcome measure is significantly different from the overall mean outcome for most of the sites, except Sites 11, 13, 15, 23, 24, and 30 (see the p-values in the lower panel of SAS Output 3.2.1).

The *SOLUTION* option in the *RANDOM* statement under REML, produces the BLUP estimates. We will demonstrate that the BLUP estimates $\hat{\beta}_{0j}^{\text{BLUP}}$ are a form of *shrinkage estimates* or *empirical Bayes estimates* using our *empty model* as an example. First, we calculate the $\hat{\beta}_{0j}^{\text{BLUP}}$ as:

$$\hat{\beta}_{0j}^{\text{BLUP}} = \hat{\gamma}_{00} + \hat{u}_{0j} \tag{3.2.5}$$

where $\hat{\gamma}_{00}$ and $\hat{u}_{0j}$ are the BLUP estimates of the grand mean γ_{00} and random effect u_{0j}, respectively. Second, we use the formula shown in eq. 2.7.4 to estimate reliability ($\hat{\lambda}_{0j}$) of the OLS estimated group means (i.e., $\bar{y}_{.j}$):

$$\hat{\lambda}_{0j} = \frac{\hat{\sigma}_{u0}^2}{\hat{\sigma}_{u0}^2 + (\hat{\sigma}^2 / n_j)} \tag{3.2.6}$$

Third, we use the following equation to estimate the *empirical Bayes estimate* or the *shrinkage estimate* of β_{0j}:

$$\hat{\beta}_{0j}^{\text{EB}} = \hat{\lambda}_{0j}\bar{y}_{\cdot j} + (1 - \hat{\lambda}_{0j})\hat{\gamma}_{00} \tag{3.2.7}$$

where $\bar{y}_{.j}$ and $\hat{\gamma}_{00}$ are group mean and grand mean estimates, respectively. The following SAS program is used for this comparison:

***SAS Program 3.2.2**;

```
...
PROC MIXED Data = Data1 METHOD = REML noclprint COVTEST;
CLASS site;
MODEL Inject = /SOLUTION;
RANDOM INT /SUBJECT = site SOLUTION;
ODS output solution = Ran_effect;
Run;

Data Data2 (keep = site u0j);
 Set Ran_effect;
rename Estimate = u0j;

Proc SQL;
  Create table data3 as select site, mean(Inject) as y_bar_j,
  count(site) as n_j from data1
  group by site;
Quit;

Data Data4;
```

[3] Note that specifying *SOLUTION* option in the *RANDOM* statement can produce the BLUP estimates of the random effects under either REML or ML, but the results would be slightly different. We recommend that REML should be used for random effects estimation.

```
Merge data2(in = d1) data3(in = d2);
By site;
Grand_mean = 61.4366; *estimated grand mean in the empty model;
Var_U0j = 523.57; *variance of random effect u0j;
Var_eij = 4350.47; *variance of level-1 residuals eij;
BLUP_Beta0j = u0j + Grand_mean;
Lambda_0j = Var_U0j/(Var_U0j + (Var_eij/n_j));
EB_Beta0j = Lambda_0j*y_bar_j + (1-Lambda_0j)*Grand_mean;

Proc print Data = Data4 noobs;
var site n_j Grand_mean y_bar_j u0j Lambda_0j BLUP_Beta0j EB_Beta0j;
Run;
```

where the random effects estimated from the *empty model* are saved in the active SAS data file *Ran_effect*, using the *ODS output SOLUTIONR* = option. In the output data file the random effect $\hat{u}_{0j}$ is internally given a variable name *estimate*. This variable is renamed to *u0j* when the data file, *Ran_effect,* is read into Data file *Data2*. New variables *y_bar_j* (i.e., $\bar{y}_{.j}$) and *n_j* (i.e., sub-sample size of Site *j*) are created using *PROC SQL*; then the BLUP estimate—*BLUP_Beta0j* (i.e., $\hat{\beta}_{0j}^{\text{BLUP}}$), the reliability—*Lambda_0j* (i.e., $\hat{\lambda}_{0j}$), and the *empirical Bayes estimate*—*EB_Beta0j* (i.e., $\hat{\beta}_{0j}^{\text{EB}}$) are calculated using eqs. 3.2.5—3.2.7, respectively, given the *empty model* results ($\hat{\gamma}_{00}$ = 61.44, σ^2_{u0} = 523.57, and σ^2 = 4350.47). The results are shown in the following:

SAS Output 3.2.2 Shrinkage estimates of random intercept and reliability of observed sample mean outcome by project site

site	n_j	grand_ mean	y_bar_j	u0j	Lambda_0j	BLUP_ Beta0j	EB_Beta0j
10	455	61.4366	19.2308	-41.4489	0.98207	19.9877	20.0344
11	139	61.4366	71.9784	9.9472	0.94359	71.3838	71.5305
12	179	61.4366	86.5475	23.9969	0.95564	85.4335	85.5489
13	554	61.4366	52.7040	-8.6036	0.98522	52.8330	52.8714
14	1311	61.4366	83.5896	22.0135	0.99370	83.4501	83.4665
15	279	61.4366	52.9427	-8.2483	0.97108	53.1883	53.2635
16	86	61.4366	21.2209	-36.6724	0.91189	24.7642	24.9934
17	223	61.4366	76.3632	14.3904	0.96408	75.8270	75.9205
18	684	61.4366	39.2632	-21.9073	0.98800	39.5293	39.5605
19	552	61.4366	84.8569	23.0730	0.98517	84.5096	84.5481
21	932	61.4366	70.9431	9.4225	0.99116	70.8591	70.8821
22	469	61.4366	51.4350	-9.8275	0.98259	51.6091	51.6544
23	695	61.4366	68.9827	7.4570	0.98819	68.8936	68.9243
24	1036	61.4366	56.7693	-4.6302	0.99204	56.8064	56.8271
25	89	61.4366	96.7191	32.2697	0.91461	93.7063	93.9284
27	328	61.4366	29.5915	-31.0583	0.97529	30.3783	30.4425
28	910	61.4366	86.8615	25.1949	0.99095	86.6315	86.6550
30	365	61.4366	58.0219	-3.3387	0.97774	58.0979	58.1558
31	254	61.4366	35.2953	-25.3133	0.96832	36.1233	36.2058
32	284	61.4366	85.4014	23.2836	0.97157	84.7202	84.7941

Because the sub-sample sizes of all the project sites are large (the smallest one is $n_j = 86$), all of the reliability estimates (i.e., $\hat{\lambda}_{0j}$ s) are above 0.90, indicating that the sample outcome means are reliable estimates of their population means. In addition, we can see that the reliability estimates ($\hat{\lambda}_j$) are positively related to n_j: the larger/smaller the n_j is, the larger/smaller the $\hat{\lambda}_j$ would be. For sites with smaller/larger n_j, the parameter estimates using the entire sample (i.e., the grand-mean estimate $\hat{\gamma}_{00}$ here) had larger/smaller weights in estimating their $\hat{\beta}_{0j}^{\text{EB}}$. For example, the sub-sample size for Sites 16 (n_{16} = 86) and 25 (n_{25} = 89) are smaller, compared with other project sites; thus $\hat{\beta}_{016}^{\text{EB}}$ and $\hat{\beta}_{025}^{\text{EB}}$ were pulled toward the parameter estimate of the overall mean estimate ($\hat{\gamma}_{00}$) from their observed within-site mean outcome measures ($\bar{y}_{.j}$). Site 24 has a large group size (n_{24} = 1 036), thus $\hat{\beta}_{024}^{\text{EB}}$ remains the same as its within-site mean outcome measure.

The SAS Output 3.2.2 shows that the $\hat{\beta}_{0j}^{\text{EB}}$'s are almost identical to the $\hat{\beta}_{0j}^{\text{BLUP}}$'s. This confirms that BLUP estimates in SAS *PROC MIXED* provides shrinkage estimates.

3.3 Predicting between-group variation

The results of the *empty model* show that there is significant between-group variation in the mean outcome measure ($\hat{\sigma}_{u0}^2$ = 523.57, p = 0.0015). In other words, there exists significant unexplained variation in the mean outcome across project sites. As such, in the next step (Step-2) of model development, we need to explore what contextual variables or site-level variables explain the between-site variation. We tested the two contextual variables (*M_HIV* and *HIV_Region*) in the exploratory modeling. Only *HIV_Region* shows a significant effect on between-group variation in drug injection frequency; thus this group-level variable is included in the following model to explain between-group variation. The model follows:

$$\text{Inject}_{ij} = \beta_{0j} + e_{ij} \tag{3.3.1}$$

$$\beta_{0j} = \gamma_{00} + \gamma_{01}\text{HIV\_Region}_{1j} + u_{0j} \tag{3.3.2}$$

$$\text{Inject}_{ij} = \gamma_{00} + \gamma_{01}\text{HIV\_Region}_{1j} + (u_{0j} + e_{ij}) \tag{3.3.3}$$

We may call this tentative model the *random intercept model with macro explanatory variables only*. As no level-1 explanatory variables are included, within-group variation is not explained and no cross-level interaction is involved in the model. The corresponding SAS program is:

```
*SAS Program 3.3.1;
...
Data Data1;
  Set lib.ML_data_Chap3;
PROC MIXED METHOD = REML COVTEST;
CLASS Site;
MODEL Inject = HIV_Region/SOLUTION DDFM = BW;
RANDOM INT/SUBJECT = Site;
Run;
```

All micro and macro explanatory variables are specified in the same *MODEL* statement, and macro or group-level variables will be internally identified by *PROC MIXED*. A variable that has constant values within each group will be treated as a group-level explanatory variable. The data set, therefore, must be formatted as shown in Table 1.3.3 for two-level multilevel modeling.

For the model discussed in this section, only one variable, *HIV_Region*, is specified on the right side of "=" in the *MODEL* statement in PROC MIXED, and it is treated as a group-level explanatory variable because it has constant values within each group (i.e., project site). Selected model output is shown below:

SAS Output 3.3.1 Selected results of the random intercept model with a group-level explanatory variable *HIV_Regrion*

Iteration History

Iteration	Evaluations	-2 Res Log Like	Criterion
0	1	110705.91097375	
1	2	110241.79764849	0.00000927
2	1	110241.29349196	0.00000110
3	1	110241.23841823	0.00000002
4	1	110241.23741776	0.00000000

Convergence criteria met.

Covariance Parameter Estimates

Cov Parm	Subject	Estimate	Standard Error	Z Value	Pr Z
Intercept	site	353.17	123.53	2.86	0.0021
Residual		4350.43	62.1371	70.01	<.0001

Fit Statistics

-2 Res Log Likelihood	110241.2
AIC (smaller is better)	110245.2
AICC (smaller is better)	110245.2
BIC (smaller is better)	110247.2

Solution for Fixed-effects

Effect	Estimate	Standard Error	DF	t Value	Pr > |t|
Intercept	53.7510	4.9407	18	10.88	<.0001
HIV_Region	31.2957	9.9697	18	3.14	0.0057

Type 3 Tests of Fixed-effects

```
                        Num      Den
Effect                   DF       DF    F Value    Pr > F

HIV_Region                1       18       9.85    0.0057
```

The output shows that the model converged quickly in just four iterations, indicating appropriate model fit. Two fixed-effects are reported in the panel of *Solution for Fixed-effects* of the output. The estimated overall mean outcome, corresponding to the low HIV prevalence region (i.e., *HIV_Region* = 0) is $\hat{\gamma}_{00}$ = 53.75 (p < 0.0001). The main effect of *HIV_Region* is $\hat{\gamma}_{01}$ = 31.30 (p = 0.0057), indicating that the frequency of drug injection among IDUs in the higher HIV prevalence region ($\hat{\gamma}_{00} + \hat{\gamma}_{01}$ = 53.75 + 31.30 = 85.05) is significantly higher than that in the low HIV prevalence region ($\hat{\gamma}_{00}$ = 53.75).

An additional panel— *Type* 3 *Tests of Fixed-effects*—for fixed-effects is reported in the SAS output. The *F*-test in the *Type* 3 *Tests of Fixed-effects* panel is useful for testing the global effects of categorical variables that are specified in the *CLASS* statement. This panel may not be necessary in models without categorical variables. The option *NOTEST* can be used in the *MODEL* statement to suppress this additional panel of output.

In SAS program 3.3.1, a new option *DDFM* = is included in the *MODEL* statement that specifies the method used to calculate degrees of freedom (d.f.) for fixed-effects. Attention needs to be paid to obtaining correct d.f. for the fixed-effects in multilevel models. Because the denominator d.f. of Wald *t*-tests and *F*-tests are substantially affected by multiple sources of errors (i.e., micro and macro errors) in multilevel models, calculation of the correct d.f. is not straightforward. In Section 3.6, we will discuss different methods of d.f. calculation for fixed effects in linear multilevel models.

The *empty model* and the current model are nested, differing only by one fixed-effect (i.e., γ_{10}, in the *empty model* but two, γ_{10} and γ_{11}, in the current model). The random effects part (i.e., $u_{0j} + e_{ij}$) is the same in both models. To compare such models, fit statistics (e.g., −2LL and information criterion measures) should be estimated under ML instead of REML by (by replacing METHOD = REML with METHOD = ML in the PROC MIXED statement in SAS Program 3.3-1). Option *IC*, standing for information criteria, instructs *PROC MIXED* to display various information criteria measures, including *HQIC* and *CAIC* in addition to *ACI*, *AICC*, and *BIC*, as well as the −2LL and the number of parameters estimated in the model, in the *Information Criteria* panel of SAS output. The following are the ML estimated model fit statistics for the *empty model* and the model with one contextual variable *HIV_Region.*

The −2LL or deviance statistic for the current model and the *empty model* are 110252.3 and 110261.1, respectively. Thus the LR test Chi-square statistic is (110261.1−110252.3) = 8.8 with d.f. = 4−3 = 1 (p = 0.0030). [4] The p-value for the Chi-square statistic can be found from the χ^2 Distribution Table in statistics text books or it can be readily calculated using SAS *Probchi*() function. The p-values for χ^2 = 8.8 with d.f. = 1 can also be calculated in the following SAS program[5]:

[4] Under the ML, *IC* option prints 3 model parameters (1 fixed-effect (i.e., γ_{10}) and 2 random effects (i.e., u_{0j} and e_{ij})) in SAS output for the empty model, and 4 model parameters (2 fixed-effects (i.e., γ_{10} and γ_{11}) and 2 random effects (i.e., u_{0j} and e_{ij})) for the current model. Under REML, the model parameters would be 2 for both the empty model and the current model because REML is for comparing models with the same fixed-effects, but different random effects.

[5] Note that the estimated p-value is printed in the SAS Log file instead of the SAS Output file.

***SAS Program 3.3.2;**

```
Data _NULL_;
P_value = 1-Probchi(8.8, 1);
Put P_value = ;
Run;
```

Significant reduction in −2LL means that the model with a random intercept and the group-level explanatory variable *HIV_Region* fits the data better than the *empty model*. This conclusion is supported by reductions in all the information criteria measures (*AIC*, *AICC*, and *BIC*) (see SAS Output 3.3.2).

SAS Output 3.3.2 ML estimated model fit statistics for comparison between the empty model and the model specified in eqs. 3.3.1—3.3.3

Empty Model:

```
                     Covariance Parameter Estimates

                                            Standard            Z
Cov Parm       Subject      Estimate          Error         Value          Pr Z

Intercept      site           494.87         162.38          3.05        0.0012
Residual                     4350.50        62.1387         70.01        <.0001


                  Fit Statistics

   -2 Log Likelihood              110261.1
   AIC (smaller is better)        110267.1
   AICC (smaller is better)       110267.1
   BIC (smaller is better)        110270.1


                              Information Criteria

  Neg2LogLike   Parms    AIC       AICC      HQIC      BIC      CAIC

    110261        3     110267    110267    110268    110270    110273
```

Model Specified in eqs. 3.3.1—3.3.3:

```
                       Covariance Parameter Estimates

                                Standard            Z
Cov Parm       Subject          Estimate          Error         Value         Pr Z

Intercept      site               315.03         104.78          3.01       0.0013
Residual                         4350.46        62.1377         70.01       <.0001
```

```
                 Fit Statistics

-2 Log Likelihood              110252.3
AIC (smaller is better)        110260.3
AICC (smaller is better)       110260.3
BIC (smaller is better)        110264.3

                              Information Criteria

Neg2LogLike  Parms     AIC      AICC      HQIC       BIC      CAIC

110252         4     110260    110260    110261    110264    110268
```

The within-group variance estimated from the current model is $\hat{\sigma}^2 = 4350.43$ is almost identical to that ($\hat{\sigma}^2 = 4350.47$) estimated from the *empty model*. However, the between-group variance, $\hat{\sigma}^2_{u0}$, has decreased substantially from 523.57 to 353.17. The results indicate that inclusion of the group-level explanatory variable, *HIV_Region*, does not explain within-group variation, but does explain between-group variation.

The intraclass correlation coefficient (ICC) after including covariates in the model is called the *conditional ICC*, which is a measure of the degree of within-group homogeneity or between-group heterogeneity after controlling for contextual variables. For the current model, the *conditional ICC* is:

$$\text{ICC} = \frac{\hat{\sigma}^2_{u0}}{\hat{\sigma}^2_{u0} + \hat{\sigma}^2} = \frac{353.17}{353.17 + 4350.43} = 0.08 \tag{3.3.4}$$

The proportion of explained between-group variation in the average frequency of drug injection can be estimated using both Raudenbush & Bryk's method (eq. 2.9.5) and Snijders & Bosker's method (eq. 2.9.15).

Using Raudenbush & Bryk's method, the proportion of explained between-group outcome variation is estimated from the proportion reduction in level-2 variance $\hat{\sigma}^2_{u0}$—the random intercept variance, using eq. 2.9.5:

$$\begin{pmatrix} \text{RB: \% of explained} \\ \text{between-group} \\ \text{outcome variation} \end{pmatrix} = 1 - (353.17/523.57) = 0.325 \tag{3.3.5}$$

That is, about 33% of the variation in frequency of drug injection among IDUs is explained by HIV prevalence region.

Using Snijders & Bosker's method, the proportion of explained between-group outcome variation is estimated as the proportional reduction in the variance of level-2 mean squared prediction error $\hat{\sigma}^2_{\text{total level2}}$, comparing the current model with the *empty model*. First, we calculate the total level-2 variance for the *empty model* and the currently specified model, using eq. 2.9.13):

$$\begin{pmatrix} \hat{\sigma}^2_{\text{total level2}} \text{ of} \\ \text{empty model} \end{pmatrix} = 523.57 + (4350.47/491) = 532.43 \tag{3.3.6}$$

$$\begin{pmatrix} \hat{\sigma}^2_{\text{total level2}} \text{ of} \\ \text{specified model} \end{pmatrix} = 353.17 + (4350.43/491) = 362.03 \tag{3.3.7}$$

where the average number of cases ($n = 491$) among the project sites is used to estimate the total level-2 variance. The proportion of explained between-group outcome variation is (see eq. 2.9.15):

$$\begin{pmatrix}\text{SB: \% of explained}\\ \text{between-group}\\ \text{outcome variation}\end{pmatrix} = 1-(362.03/532.43) = 0.320 \tag{3.3.8}$$

The results of Raudenbush & Bryk's and Snijders & Bosker's methods are very close in this example.

3.4 Predicting within-group variation

From the results of model development discussed in the last section, we have gained some knowledge about how the contextual variables (e.g., *HIV_Region* in our example) explain variation of group mean outcomes without adjusting for the individual characteristics. In this section, we test a tentative model by adding level-1 explanatory variables into the model discussed in Section 3.3, and the effects of all the level-1 explanatory variables will be treated as fixed. The model is expressed as:

$$\text{Inject}_{ij} = \beta_{0j} + \beta_1\text{Ethnic}_{ij} + \beta_2\text{Gender}_{ij} + \beta_3\text{Gdmc\_age}_{ij} + \beta_4\text{Highsch}_{ij} + e_{ij} \tag{3.4.1}$$

$$\beta_{0j} = \gamma_{00} + \gamma_{01}\text{HIV\_Region}_j + u_{0j} \tag{3.4.2}$$

$$\text{Inject}_{ij} = \gamma_{00} + \gamma_{01}\text{HIV\_Region}_j + \beta_1\text{Ethnic}_{ij} + \beta_2\text{Gender}_{ij} + \beta_3\text{Gdmc\_age}_{ij} + \beta_4\text{Highsch}_{ij} + (u_{0j} + e_{ij}) \tag{3.4.3}$$

where four individual socio-demographic measures, gender, ethnicity, age (grand-mean centered age), and education, are included into the micro model (i.e., eq. 3.4.1), while only one group-level variable *HIV_Region* is included in the macro model (eq. 3.4.2) to predict variation in the level-1 random intercept coefficient β_{0j}. The corresponding SAS program follows:

***SAS Program 3.4.1**;

```
...
PROC MIXED METHOD = REML COVTEST;
CLASS site;
MODEL inject = HIV_Region Gender Ethnic Gdmc_age Highsch
 /SOLUTION DDFM = BW NOTEST;
RANDOM INT /SUBJECT = site;
Run;
```

where *Gender* (1-Male; 0-Female), *Ethnic* (1-Black; 0-White), *Gdmc_age* (Grand-mean centered age), and *Highsch* (1-High school or above; 0-Less than high school) are level-1 explanatory variables. All levels 1 and 2 variables are specified in the same *MODEL* statement in PROC MIXED. The program identifies macro and micro variables by checking the values of the variables. To illustrate, let's look at the selected records of the data used for our models:

SAS Output 3.4.1 Selected Data Records

Site	Pid	Inject	HIV_ Region	Gender	Ethnic	Gdmc_Age	Highsch
10	101234	4	0	0	1	-7.7865	1
10	101235	45	0	1	1	-4.7865	1
10	101236	60	0	1	0	-8.7865	1

```
...
 11     116191     180     1     0     1     -3.7865     0
 11     116192      56     1     0     1     -5.7865     1
 11     116194     150     1     0     1      0.2135     0
...
 12     120080     112     1     0     1     10.2135     0
 12     120102      45     1     1     0      2.2135     1
 12     120105      10     1     0     1     -2.7865     0
...
 32     320002     150     0     0     0     -4.7865     1
 32     320003       7     0     0     0    -12.7865     0
 32     320022     300     0     0     0    -16.7865     0
```

The outcome measures and all explanatory variables are sorted by the site identification index (i.e., Site) and the participant identification number (i.e., PID), displaying only the first three cases in each of the selected project sites. We can see from SAS Output 3.4.1 that the values of variable *HIV_Region* vary across project sites, but remain constant (either 0 or 1) within each site. As such, *HIV_Region* is treated by *PROC MIXED* as a level-2 variable. All other variables' values vary both within and between sites, thus are treated as level-1 variables. Selected model results are shown in SAS Output 3.4.2.

SAS Output 3.4.2 Selected results of the model specified in eqs. 3.4.1—3.4.3

```
                     Iteration History

Iteration    Evaluations    -2 Res Log Like       Criterion

        0              1    110513.49981830
        1              2    110178.97953202      0.00002665
        2              1    110177.46349547      0.00000647
        3              1    110177.11763157      0.00000058
        4              1    110177.08906592      0.00000001

                 Convergence criteria met.


              Covariance Parameter Estimates

                                         Standard         Z
Cov Parm       Subject     Estimate         Error     Value          Pr Z

Intercept      site          300.30        106.01      2.83        0.0023
Residual                    4327.30       61.8202     70.00        <.0001


              Fit Statistics

-2 Res Log Likelihood          110177.1
AIC (smaller is better)        110181.1
AICC (smaller is better)       110181.1
```

```
BIC (smaller is better)        110183.1
```

Solution for Fixed-effects

Effect	Estimate	Standard Error	DF	t Value	Pr > |t|
Intercept	64.0393	4.8764	18	13.13	<.0001
HIV_Region	33.3228	9.2439	18	3.60	0.0020
gender	-5.7093	1.5371	9800	-3.71	0.0002
ethnic	-11.1986	1.7299	9800	-6.47	<.0001
gdmc_age	0.2904	0.09223	9800	3.15	0.0016
highsch	1.1269	1.3729	9800	0.82	0.4118

where the panel of *Type 3 Tests of Fixed-effects* in the SAS output was suppressed by specifying the option *NOTEST* in the *MODEL* statement in *PROC MIXED.* The model estimation converged quickly within only 4 iterations. Three of the level-1 explanatory variables (gender, ethnicity, and age) have significant effects on the frequency of drug injection among IDUs, net of HIV prevalence region. Female and White IDUs have lower frequency of drug injection on average compared with male and Black IDUs, controlling for other individual characteristics and region. In addition, age has a significant positive effect, while education has no significant effect, on drug injection frequency.

The intercept estimate is 64.04, unlike the estimate in previous models. This is because it represents a different overall outcome mean. Here, $\hat{\gamma}_{00}$ = 64.04 represents the model estimated overall mean outcome, corresponding to the situation where all the explanatory variables take a zero value: *Gender* = 0, *Ethnic* = 0, *Highsch* = 0, and *Gdmc_age* = 0, as well as *HIV_Region* = 0. That is, the model estimated frequency of drug injection among female White IDUs with average age (i.e., *Gdmc_age* = 0 or *age* = 39.79) of the entire sample in the low HIV prevalence region. The corresponding figure for the group in the Higher HIV prevalence region (*HIV_Region* = 1) is 64.04 + 33.32 = 97.36.

We will now compare the current model with the model specified in Section 3.3 in order to measure the model improvement realized by adding level-1 explanatory variables. All level-1 explanatory variables in the current model have fixed-effects. Thus, the current model and the model specified in Section 3.3 have the same random effects ($u_{0j} + e_{ij}$), but different fixed effects (γ_{00} and γ_{01} in the model specified in Section 3.3; γ_{00}, γ_{01}, and β_1, β_2, β_3, and β_4 in the current model). Therefore, the deviance statistics used for model comparison should be estimated using ML instead of REML.

SAS Output 3.4.3 ML estimated model fit statistics for comparison between models specified in eqs. 3.3.1—3.3.3 and 3.4.1—3.4.3

Model specified in eqs. 3.3.1—3.3.3:

```
                Fit Statistics

-2 Log Likelihood              110252.3
AIC (smaller is better)        110260.3
AICC (smaller is better)       110260.3
```

```
              BIC (smaller is better)         110264.3

                 Information Criteria

Neg2LogLike  Parms     AIC      AICC     HQIC     BIC      CAIC

  110252       4      110260   110260   110261   110264   110268
```

Model specified in eqs. 3.4.1 — 3.4.3:

```
                       Fit Statistics

              -2 Log Likelihood               110193.0
              AIC (smaller is better)         110209.0
              AICC (smaller is better)        110209.0
              BIC (smaller is better)         110216.9


                 Information Criteria

Neg2LogLike  Parms     AIC      AICC     HQIC     BIC      CAIC

  110193       8      110209   110209   110211   110217   110225
```

Adding the four level-1 explanatory variables (*Gender*, *Ethnic*, *Gdmc_age*, and *Highsch*) into the model has reduced −2LL from 110252.3 to 1101193.0, resulting in a LR χ^2 = 58.7 with d.f. = 8−4 = 4, which is statistically significant ($p < 0.0001$). In addition, all the information criteria measures have also decreased substantially (see SAS Output 3.4.3). The model fit statistics provide evidence that including the level-1 explanatory variables has significantly improved the model.

Now, let's see how much of the levels 1 and 2 variances have been explained by the model. To calculate the proportion of explained levels 1 and 2 variances, we need to compare the current model with the *empty model* in which no variation was explained. The proportions of explained levels 1 and 2 variances are calculated below, using both Raudenbush & Bryk's and Snijders & Bosker's methods.

The estimated proportions of explained levels 1 and 2 variance, using Raudenbush & Bryk's method, are shown below:

$$\begin{pmatrix}\text{RB: \%of Explained}\\ \text{Level 1 variance}\end{pmatrix} = 1 - \frac{\hat{\sigma}^2(\text{specified model})}{\hat{\sigma}^2(\text{empty model})} = 1 - \frac{4327.30}{4350.47} = 0.005 \tag{3.4.4}$$

$$\begin{pmatrix}\text{RB: \%of Explained}\\ \text{Level 2 variance}\end{pmatrix} = 1 - \frac{\hat{\sigma}^2_{u0}(\text{specified model})}{\hat{\sigma}^2_{u0}(\text{empty model})} = 1 - \frac{300.30}{523.57} \approx 0.426 \tag{3.4.5}$$

The estimated proportions of explained level 1 and level 2 variance, using Snijders & Bosker's method, are shown below:

$$\begin{pmatrix}\text{SB: \%of Explained}\\ \text{Level 1 variance}\end{pmatrix}=1-\frac{\hat{\sigma}^2_{\text{total level1}}(\text{specified model})}{\hat{\sigma}^2_{\text{total level1}}(\text{empty model})}$$

$$=1-\frac{(300.30+4327.30)}{(523.57+4350.47)}=1-\frac{4627.6}{4874.04}=0.051 \tag{3.4.6}$$

$$\begin{pmatrix}\text{SB: \%of Explained}\\ \text{Level 2 variance}\end{pmatrix}=1-\frac{\hat{\sigma}^2_{\text{total level2}}(\text{specified model})}{\hat{\sigma}^2_{\text{total level2}}(\text{empty model})}$$

$$1-\frac{300.30+(4327.30/491)}{523.57+(4250.47/491)}=1-\frac{309.11}{532.43}=0.419 \tag{3.4.7}$$

The estimated proportions of explained level-1 variance differ by method. Raudenbush & Bryk's method suggest that less than 1% of the level-1 variance is explained by the model. In contrast, about 5% of the level 1 variance is explained according to Snijders & Bosker's method. It seems that Snijders & Bosker's method may provide more reasonable results for the proportion of explained level-1 variance. The two methods provide similar results for level 2 variance explained (42.6% for For Raudenbush & Bryk's method and 41.9% for Snijders & Bosker's method) in this example.

The proportion of explained level-2 variance (42%) in the current model is about 10% greater than the *random intercept model with contextual variables only* as specified in Section 3.3. The inclusion of level-1 explanatory variables into the model has not only explained level-1 variance (level-1 residual variance σ^2 reduced from 4350.43 to 4327.30), but also contributed to explaining more level-2 variance (level-2 variance σ^2_{u0} reduced from 353.17 to 300.30). This suggests that level-1 explanatory variables may have both within-group and between-group variations and may influence outcome variations at both the individual level and group level (see Section 2.9 of Chapter 2).

3.5 Testing level-1 random

In previous sections, we have focused on variation of mean outcome (i.e., the level-1 intercept) across groups and how this variation is explained by macro and micro explanatory variables. The aim of this section is to test whether the effects of level-1 explanatory variables on the outcome measure vary across groups, i.e., to test whether any level-1 slope coefficient varies across groups. The testing results will help us correctly identify which level-1 slopes should be specified as fixed, and which are random in our final model under study. The test hypothesis is H_0: σ^2_{uq} = 0. If the H_0 is rejected, the corresponding level-1 slope should be specified as a random coefficient. Once the level-1 random slopes are identified, their variations need to be explained by including contextual variables or group-level explanatory variables in the macro models.

Each level-1 explanatory variable may have a random slope coefficient, and all the level-1 random coefficients may be correlated with each other. In this example we have neither theoretical support nor empirical findings to guide us to decide which level-1 explanatory variables should be specified as random and which ones should be fixed. Thus, we need to test for randomness of the slope coefficient for each of the level-1 explanatory variables (i.e., *Gender*, *Ethnic*, *Gdmc_age*, and *Highsch*).

The following SAS program is employed to conduct in exploratory modeling for preliminary assessment of the randomness of level-1 slopes:

***SAS Program 3.5.1**;

```
...
PROC MIXED METHOD = REML COVTEST;
CLASS site;
MODEL Inject = HIV_Region Ethnic Gender Gdmc_age Highsch
  /SOLUTION DDFM = BW NOTEST;
RANDOM Int Ethnic Gender Gdmc_age Highsch /SUBJECT = site G TYPE = VC;
Run;
```

In the program, all level-1 slope are treated as random by specifying the corresponding level-1 explanatory variables (i.e., *Ethnic*, *Gender*, *Gdmc_age*, and *Hishsch*) in the *RANDOM* statement. The option *TYPE = VC*, the default option, requests SAS to estimate only *variance components* of the random coefficients (i.e., the diagonal values of the *G* matrix, see eq. 2.3.6). In the upper panel of SAS Output 3.5.1, we see that the different variances on the diagonal, and zeros off the diagonal, of the *G* matrix. The lower panel of the SAS Output shows that in addition to the level-1 intercept, the slope coefficient of variable *Ethnic* also varies significantly across project sites ($z = 1.76$, $p = 0.0388$),[6] and none of the remaining level-1 explanatory variables has a random slope.

SAS Output 3.5.1 Preliminary results of randomness of level-1 slopes

Estimated G Matrix

Row	Effect	site	Col1	Col2	Col3	Col4	Col5
1	Intercept	10	329.28				
2	ethnic	10		79.3635			
3	gender	10			8.5772		
4	gdmc_age	10				0.05286	
5	highsch	10					11.3281

Covariance Parameter Estimates

Cov Parm	Subject	Estimate	Standard Error	Z Value	Pr Z
Intercept	site	329.28	124.76	2.64	0.0042
Ethnic	site	79.3635	44.9790	1.76	0.0388
Gender	site	8.5772	15.2283	0.56	0.2866
Gdmc_age	site	0.05286	0.07735	0.68	0.2472
Highsch	site	11.3281	16.6029	0.68	0.2475
Residual		4306.65	61.7358	69.76	<.0001

The *Type* option in the *RANDOM* statement can specify various variance/covariance structures of the *G* matrix. The most often used option is *Type = UN*, which specifies an unstructured variances and covariances in the *G* matrix. That is, it allows the variance/covariance structure of

[6] Recall that the significance testing for variance is a one-tailed test and the p-value is divided in half.

the random effects to be determined by the data. However, when too many random slopes are specified with option *Type* = *UN*, the computation becomes very time consuming and may not converge. In our example, when all the four level-1 slopes are specified as random with option *Type* = *UN*, the program does not converge. As such, we also decided to test the randomness of slope one variable at a time. The results were consistent with SAS Output 3.5.1: only the level-1 intercept ($p = 0.0042$) and the slope of variable *Ethnic* ($p = 0.0388$) are random coefficients. We define the current tentative model as:

$$\text{Inject}_{ij} = \beta_{0j} + \beta_{1j}\text{Ethnic}_{ij} + \beta_2\text{Gender}_{ij} + \beta_3\text{Gdmc\_age}_{ij} + \beta_4\text{Highsch}_{ij} + \text{e}_{ij} \quad (3.5.1)$$

$$\beta_{0j} = \gamma_{00} + \gamma_{01}\text{HIV\_Region}_j + u_{0j} \quad (3.5.2)$$

$$\beta_{1j} = \gamma_{10} + u_{1j} \quad (3.5.3)$$

$$\text{Inject}_{ij} = \gamma_{00} + \gamma_{01}\text{HIV\_Region}_j + \gamma_{10}\text{Ethnic}_{ij} + \beta_2\text{Gender}_{ij} + \beta_3\text{Gdmc\_age}_{ij} + \beta_4\text{Highsch}_{ij} + (u_{0j} + \text{Ethnic}_{ij} * u_{1j} + \text{e}_{ij}) \quad (3.5.4)$$

In this model, the effects of all level-1 explanatory variables (i.e., *Gender*, *Gdmc_age*, and *Highsch*), except *Ethnic*, on the outcome measure *Inject* are fixed. Thus, this is a mixed model with random level-1 intercept (i.e., β_{0j}), one level-1 random slope (i.e., β_{1j}) and three fixed level-1 slopes (β_{2j}, β_{3j}, and β_{4j}). The SAS program for the model follows:

***SAS Program 3.5.2**;

```
...
PROC MIXED METHOD = REML COVTEST IC;
CLASS site;
MODEL Inject = HIV_Region Ethnic Gender Gdmc_age Highsch
  /SOLUTION DDFM = BW NOTEST;
RANDOM Int Ethnic /SUBJECT = site G TYPE = UN;
Run;
```

where option *TYPE* = *UN* is specified in the *RANDOM* statement to let the data determine the variance/covariance structure of the random effects. With two random level-1 coefficients, the model has four variance/covariance estimates: *UN(1,1)*—$\hat{\sigma}^2_{u0}$ for β_{0j}; *UN(2,2)*—$\hat{\sigma}^2_{u1}$ for β_{1j}; *UN(1,2)*—$\hat{\sigma}^2_{u01}$; and *Residual* —$\hat{\sigma}^2$ for e_{ij}.[7] The option *TYPE* = *UN* also provides a χ^2 statistic for a global significance test for variance/covariance components. Option *IC* requests all the information criterion measures. Note that the number of parameter option *IC* requests is the number of variances/covariances when the REML estimator is used, but the number of total parameters estimated, including the fixed-effects, when the ML estimator is used. The model results are shown in SAS Output 3.5.2.

SAS Output 3.5.2 Selected results of the model specified in eqs. 3.5.1—3.5.4

```
                    Estimated G Matrix

Row     Effect          site          Col1          Col2

  1     Intercept       10          404.92       -135.38
```

[7] When option *TYPE*=*UN* is specified, the parameter estimates reported in the panel of the *Covariance Parameter Estimates* in the SAS output are named as UN (1,1) for the first random effect ,UN(2,2) for the second random effect specified in the *RANDOM* statement, respectively; and UN (1,2) as the covariance between the two random effects.

2	ethnic	10	-135.38	118.16

Covariance Parameter Estimates

Cov Parm	Subject	Estimate	Standard Error	Z Value	Pr Z
UN(1,1)	site	404.92	153.89	2.63	0.0043
UN(2,1)	site	-135.38	87.9365	-1.54	0.1237
UN(2,2)	site	118.16	65.9876	1.79	0.0367
Residual		4311.70	61.6506	69.94	<.0001

Fit Statistics

-2 Res Log Likelihood	110157.8
AIC (smaller is better)	110165.8
AICC (smaller is better)	110165.8
BIC (smaller is better)	110169.8

Null Model Likelihood Ratio Test

DF	Chi-Square	Pr > ChiSq
3	355.72	<.0001

Information Criteria

Neg2LogLike	Parms	AIC	AICC	HQIC	BIC	CAIC
110158	4	110166	110166	110167	110170	110174

Solution for Fixed-effects

Effect	Estimate	Standard Error	DF	t Value	Pr > |t|
Intercept	64.2019	5.4435	18	11.79	<.0001
HIV_Region	35.3035	8.6236	18	4.09	0.0007
Ethnic	-13.2971	3.2121	9800	-4.14	<.0001
Gender	-5.6524	1.5364	9800	-3.68	0.0002
Gdmc_age	0.2890	0.09229	9800	3.13	0.0017
Highsch	1.0366	1.3721	9800	0.76	0.4500

In the SAS Output the global test for the variance/covariance components is highly statistically significant (χ^2 = 355.72, d.f. = 3, p < 0.0001), and both $\hat{\sigma}^2_{u0}$ = 404.92 (Z = 2.63, p = 0.0043) and $\hat{\sigma}^2_{u1}$ = 118.16 (Z = 1.79, p = 0.0367) are statistically significant. The data lead us to reject null hypotheses H$_0$: σ^2_{u0} = 0 and H$_0$: σ^2_{u1} = 0; that is, the The level-1 intercept β_{0j} and slope β_{1j} are random coefficients.

The covariance between the two random effects u_{0j} and u_{1j} is not statistically significant ($\hat{\sigma}^2_{u01} = -135$, $p = 0.1237$), indicating that the level-1 intercept β_{0j} and the slope β_{1j} are not significantly correlated with each other. In other words, the effect of the variable *Ethnic* on the outcome *Inject* is not significantly related to the average level of the outcome measure in project sites. In the application of multilevel models in longitudinal studies, the covariance between the intercept and slope has a special meaning, which will be discussed in Chapter 4.

By letting the slope coefficient of the level-1 explanatory variable *Ethnic* vary across groups, the model fit has also significantly improved. This can be tested by comparing the current model and the model specified in eqs. 3.4.1—3.4.3 in Section 3.4. The difference in −2LL between the two models is: LR = (110177.1−110157.8) = 19.3, d.f. = 2, $p <= 0.0001$. In addition, all the information criterion measures are smaller for the current model (see SAS Output 3.4.2 and 3.5.2).

Attention needs to be paid to two issues. First, appropriate model estimation methods (e.g., ML or REML) should be used when models are compared. To compare nested models with the same fixed-effects, but different random effects, REML should be used for model estimation (see Section 2.8.3 in Chapter 2). The current model and the model specified in eqs. 3.4.1—3.4.3 in Section 3.4 have the same fixed-effects (i.e., regression coefficients in the combined models eq. 3.4.3 and eq. 3.5.4 for explanatory variables *HIV_Region*, *Ethnic*, *Gender*, *Gdmc_age*, and *Highsch*), but different random effects (i.e., two residual variances ($\hat{\sigma}^2$ and $\hat{\sigma}^2_{u0}$) in the model specified in eq. 3.4.3; three residual variances ($\hat{\sigma}^2$, $\hat{\sigma}^2_{u0}$, $\hat{\sigma}^2_{u1}$) and one residual covariance ($\hat{\sigma}^2_{u01}$) in eq. 3.5.4). Therefore, the −2LLs used for this model comparison were estimated under REML, and the d.f. for the LR test was 2.

Second, the Wald Z test results may not be reliable when the number of groups or level-2 units is not large enough. In such a circumstance, the LR test is often preferred (Littell et al., 1996; 2006). This can be done by running the model twice, with and without specifying the corresponding level-1 slope as a random coefficient, and then assess model improvement. A better way to deal with a small number of groups in multilevel modeling is to apply the bootstrap technique. We will discuss this issue in Chapter 6.

In the current example, the P-value of the Wald Z test ($p = 0.0367$) of testing the randomness of the slope coefficient of the variable *Ethnic* is much larger than that of the LR test ($p < 0.0001$). The LR test indicates greater likelihood of randomness of the slope coefficient β_{1j}. The process of testing randomness was repeated for each of the level-1 slopes in the model specified in eqs. 3.4.1—3.4.3. The results of Wald Z and LR Tests are shown in Table 3.5.1.

Table 3.5.1 Results of significance testing for variances of level-1 slopes[1]

Variable	Wald Z Test[2]			LR Test[2]		
	Z	d.f.	P-value	χ^2	d.f.	P-value
Ethnic	1.79	1	0.0367	19.3	2	<0.0001
Gender	1.01	1	0.1569	5.2	2	0.0371
Gdmc_age	0.41	1	0.3411	3.3	2	0.0960
Highsch	1.19	1	0.1170	0.4	2	0.4094

Note:

[1] - The model was run four times; each time the level-1 intercept and one of the slopes were specified as random coefficients.

[2] - One-tailed significance test for the variance of level-1 random slope.

When the slope of variable *Gender* was specified as random the Wald Z score was 1.01 (d.f. = 1, $p = 0.1569$), which is not statistically significant. However, the LR test produced a different result (i.e., $\chi^2 = 5.2$, d.f. = 2, $p = 0.0371$), indicating that the variable *Gender* has a random slope coefficient or the effect of gender on frequency of drug injection among the IDUs varied significantly across project sites. As mentioned earlier, when the results of Wald *Z* test and the LR tests are not consistent, we prefer the LR test.

A logical step for further model development is to set the slope coefficients of both variables *Ethnic* and *Gender* as random in the model. Having taken a long time to converge, the model results show that the slope variance is still statistically significant for the variable *Ethnic* ($Z = 2.52$, $p = 0.0058$), but not for the variable *Gender* ($Z = 0.80$, $p = 0.2105$); and the corresponding LR test is not statistically significant either for *Gender* (LR $\chi^2 = 4.0$, d.f. = 3, $p = 0.1308$). Based on the results, we decided that only one level-1 slope coefficient (i.e., the slope of variable *Ethnic*) should be specified as random in our final model.

When random slope coefficients are involved in a model, the explained variance at either micro or macro level (R_1^2 or R_2^2) can be estimated using the same concept of proportional reduction in the prediction error for y_{ij} and $\bar{y}_{ij}$, but the calculation is complicated (Snijder & Bosker, 1994). Treating slope as fixed or random usually does not change R_1^2 and R_2^2 much. Thus, R_1^2 and R_2^2 of a model with random slopes are usually estimated using the results of a random intercept model with the same set of slopes as fixed coefficients. The estimated R_1^2 and R_2^2 values would be very close to those for the random slope model (Snijder & Bosker, 2003).

3.6 Across-level interactions

The model specified in eqs. 3.5.1—3.5.4 tests the main effects of the levels 1 and 2 explanatory variables. In this section, we address one of the most important components in multilevel modeling—the *cross-level interactions*—which tells how the effects of level-1 variables on the outcome measure are moderated by the group-level variables. Testing the cross-level interactions is equivalent to setting level-1 random slope coefficients as dependent variables at the group level:

$$\text{Inject}_{ij} = \beta_{0j} + \beta_{1j}\text{Ethnic}_{ij} + \beta_2\text{Gender}_{ij} + \beta_3\text{Gdmc\_age}_{ij} + \beta_4\text{Highsch}_{ij} + \text{e}_{ij} \qquad (3.6.1)$$

$$\beta_{0j} = \gamma_{00} + \gamma_{01}\text{HIV\_Region}_j + u_{0j} \qquad (3.6.2)$$

$$\beta_{1j} = \gamma_{10} + \gamma_{11}\text{HIV\_Region}_j + u_{1j} \qquad (3.6.3)$$

$$\begin{aligned}\text{Inject}_{ij} = {} & \gamma_{00} + \gamma_{01}\text{HIV\_Region}_j + \gamma_{10}\text{Ethnic}_{ij} + \gamma_{11}\text{HIV\_Region}_j * \text{Ethnic}_{ij} \\ & + \beta_2\text{Gender}_{ij} + \beta_3\text{Gdmc\_age}_{ij} + \beta_4\text{Highsch}_{ij} + (u_{0j} + \text{Ethnic}_{ij} * u_{1j} + \text{e}_{ij}) \qquad (3.6.4)\end{aligned}$$

where the two level-1 random regression coefficients β_{0j} and β_{1j} are regressed on the group-level explanatory variable *HIV_Region* in macro eqs. 3.6.2 and 3.6.3. In the combined model eq. 3.6.4, the term *HIV_Region*Ethnic* is a cross-level interaction. The slope coefficient γ_{11} measures the moderating effect of *HIV_Region* on the effect of *Ethnic*. In other words, the relationship of the level-1 explanatory variable *Ethnic* with the outcome measure *Inject* varies depending upon the values of the group-level variable *HIV_Region*. The SAS program for the model follows:

```
*SAS Program 3.6.1;
...
PROC MIXED METHOD = REML COVTEST IC;
CLASS site;
MODEL Inject = HIV_Region|Ethnic Gender Gdmc_age Highsch
  /SOLUTION DDFM = BW NOTEST;
RANDOM INT Ethnic /SUBJECT = site G TYPE = UN;
Run;
```

where the interaction between variables *HIV_Region* and *Ethnic* is specified by *HIV_Region|ethnic*. SAS Output 3.6.1 shows selected results of the model. The interaction between *HIV_Region* and *Ethnic* is not statistically significant ($p = 0.2269$), indicating that the effect of variable *Ethnic* on the outcome measure does not depend upon the values of the group-level variable *HIV_Region*. That is, the difference in drug injection frequency between White and Non-White IDUs was not significantly different irrespective of whether they were located in the high or low HIV prevalence regions.

SAS Output 3.6.1 Selected results of the model specified in eqs. 3.6.1—3.6.4

Fit Statistics

-2 Res Log Likelihood	110150.2
AIC (smaller is better)	110158.2
AICC (smaller is better)	110158.2
BIC (smaller is better)	110162.2

...

Solution for Fixed-effects

Effect	Estimate	Standard Error	DF	t Value	Pr > |t|
Intercept	65.8454	5.5964	18	11.77	<.0001
HIV_Region	25.4727	11.8461	18	2.15	0.0454
Ethnic	-15.0578	3.5124	9799	-4.29	<.0001
HIV_Region*ethnic	10.2047	8.4447	9799	1.21	0.2269
Gender	-5.6520	1.5363	9799	-3.68	0.0002
Gdmc_age	0.2891	0.09228	9799	3.13	0.0017
Highsch	1.0439	1.3721	9799	0.76	0.4468

Thus far, all the level 1 and 2 explanatory variables, as well as the cross-level interactions, have been included in the model. The method used to calculate d.f. for all the fixed-effects (i.e., the fixed component in eq. 2.2.9) is *DDFM = BETWITHIN or DDFM = BW* (see SAS Program 3.6.1). In SAS Output 3.6.1, Fixed-effects have different d.f.. Here, we briefly introduce the *DDFM* option in the SAS *PROC MIXED* procedure that provides different methods to calculate d.f. for the Wald *t*-tests and *F*-tests for fixed-effects.

- The option *DDFM = RESIDUAL* uses the d.f. of residual variance for each fixed-effect. As such, it often overestimates the optimal denominator d.f.. This method could be useful only if the asymptotic statistical distributions are well approximated in a large sample. Among all the *DDFM* options, this method is least appealing. *DDFM = RESIDUAL* was the default in very early releases of the *PROC MIXED* procedure, but it is no longer so.

- The option *DDFM = CONTAIN* uses the containment method to determine the d.f. when the fixed-effect of interest is contained in a random effect (SAS Institute Inc., 2004). *DDFM = CONTAIN* requests balanced data (e.g., in an experimental design, the same number of observations is taken for each combination of experiment factors). When the *RANDOM* statement is specified, *DDFM = CONTAIN* is the default in the current version of *PROC MIXED*. This method is primarily useful in experimental design situations. For multilevel models discussed in this book, the containment method is not very useful, and the default *DDFM = CONTAIN* should be overridden.
- The option *DDFM = BETWITHIN or DDFM = BW* is the default when the *REPEATED* statement is used to specify the variance-covariance structure in longitudinal data and when the *RANDOM* statement is not used. It decomposes the residual d.f. by assigning the between-subject d.f. to the between-subject variables/factors and within-subject d.f. to the within-subject variables/factors. However, when the *TYPE = UN* option is used together with *REPEATED* statement, but no *RANDOM* statement is specified, all fixed-effects are assigned the between-subject degrees of freedom.
- The option *DDFM = SATTERTHWAITE or DDFM = SATTERTH* denotes the wellknown Satterthwaite's approximation (Satterthwaite, 1946), which is generalized and used in SAS *PROC MIXED* for estimating denominator d.f. (Fai & Cornelius, 1996). Because the estimated denominator d.f. is a complicated weighted average of the between and within d.f., which is an approximation of the true d.f., some d.f. provided by Satterthwaite's method may have decimals.
- The option *DDFM = KENWARDROGER or DDFM = KR* denotes another useful denominator d.f. approximation method developed by Kenward-Roger (Kenward & Roger, 1997; Freund et al., 2002). This method provides Satterthwaite-type d.f. For unbalanced longitudinal data or when *TYPE = UN* option is used, *DDFM = KR* is recommended (SAS Institute Inc., 2004). Like *DDFM = SATTERTH*, the option *DDFM = KR* also provides d.f. with decimals.

The first three options (i.e., *DDFM = RESIDUAL, DDFM = CONTAIN*, and *DDFM = BW*) are "*exact methods*" used in situations where the design is balanced and the variance-covariance structure is simple. When data are unbalanced and/or the model involves complicated covariance structures, the test statistics for fixed-effects follow unknown distributions. In such a case, the "*approximate*" methods such as Satterthwaite's method or Kenward-Roger's method should be used to approximate d.f. for the fixed-effects in a multilevel model.

The following SAS output provides the solutions of different d.f. estimation methods, other than the *Between- Within* (*BW*) method that is used in SAS Program 3.6.1, for the fixed-effects in our current model.

SAS Output 3.6.2 Solutions for fixed-effects by DDFM options (for the results of DDFM = BW, see SAS Output 3.6.1)

DDFM = RESIDUAL

Solution for Fixed-effects

Effect	Estimate	Standard Error	DF	t Value	Pr > |t|
Intercept	65.8454	5.5964	9817	11.77	<.0001
HIV_Region	25.4727	11.8461	9817	2.15	0.0316
Ethnic	-15.0578	3.5124	9817	-4.29	<.0001
HIV_Region*ethnic	10.2047	8.4447	9817	1.21	0.2269

Gender	-5.6520	1.5363	9817	-3.68	0.0002
Gdmc_age	0.2891	0.09228	9817	3.13	0.0017
Highsch	1.0439	1.3721	9817	0.76	0.4468

DDFM = CONTAIN

Solution for Fixed-effects

Effect	Estimate	Standard Error	DF	t Value	Pr > |t|
Intercept	65.8454	5.5964	18	11.77	<.0001
HIV_Region	25.4727	11.8461	9781	2.15	0.0316
Ethnic	-15.0578	3.5124	18	-4.29	0.0004
HIV_Region*ethnic	10.2047	8.4447	9781	1.21	0.2269
Gender	-5.6520	1.5363	9781	-3.68	0.0002
Gdmc_age	0.2891	0.09228	9781	3.13	0.0017

DDFM = SATTERTH

Effect	Estimate	Standard Error	DF	t Value	Pr > |t|
Intercept	66.1456	5.1230	20.7	12.91	<.0001
HIV_Region	25.5478	10.7335	23.4	2.38	0.0258
Ethnic	-15.6693	3.1034	17.6	-5.05	<.0001
HIV_Region*ethnic	10.5327	7.4122	23.1	1.42	0.1687
Gender	-5.5693	1.5366	9809	-3.62	0.0003
Gdmc_age	0.2825	0.09230	9797	3.06	0.0022
Highsch	1.0307	1.3724	9807	0.75	0.4526

DDFM = KR

Solution for Fixed-effects

Effect	Estimate	Standard Error	DF	t Value	Pr > |t|
Intercept	66.1456	5.1307	20.7	12.89	<.0001
HIV_Region	25.5478	10.8108	23.4	2.36	0.0268
Ethnic	-15.6693	3.1280	17.6	-5.01	<.0001
HIV_Region*ethnic	10.5327	7.5678	23.1	1.39	0.1773
Gender	-5.5693	1.5370	9809	-3.62	0.0003
Gdmc_age	0.2825	0.09235	9797	3.06	0.0022
Highsch	1.0307	1.3728	9807	0.75	0.4528

When different d.f. methods are used, parameter estimates, standard errors, *t*-ratio and *F*-value are basically the same, but the denominator d.f. and *p*-values are different for fixed effects. In general, Satterthwaite and Kenward-Roger' methods, particularly the latter, are recommended by statisticians. However, both *SATTERTH* and *KR* methods are computationally intensive, particularly when the model is complex and sample size is large. In comparison, the option *DDFM = BW* provides very similar results to *DDFM = SATTERTH* and *DDFM = KR* with much less computation time. Thus we have used the *DDFM* = *BW* option to demonstrate our models discussed in this Chapter. However, as we pointed out earlier, in general, *DDFM* = *KR* or *DDFM* = *SATTERTH* are preferred d.f. methods for reporting final model results.

Calculation of d.f. in multilevel models is complicated. For a theoretical discussion of this subject, readers are referred to Elston (1998). The SAS Manual provides application details about the *DDFM* = option (SAS Institute Inc., 1999).

3.7 Other issues in model development

In the previous sections of this chapter, we have demonstrated the process of multilevel model development, step by step. Original observed measures were used to illustrate models under the assumption of normality. In real-life research, however, observed measures, especially the outcome measures, often have non-normality problems (e.g., high skewness and/or kurtosis), resulting in violations of regression assumptions. As such, variable distributions should be checked before proceeding to complex modeling.

To test normality of the measures a variety of approaches are useful such as visual inspection of data plots, quantile-quantile plots (or Q-Q plots), and statistical testing for skew and kurtosis. The following SAS program produces distribution and Q-Q plot of the standardized the outcome measure *Inject*:

***SAS Program 3.7.1**;

```
LIBNAME Lib 'D:\MLM\data';

PROC STANDARD Data = Lib.ML_data_Chap3 MEAN = 0 STD = 1 OUT = New_data;
  VAR Inject;
Run;

PROC UNIVARIATE Data = New_data Noprint;
VAR Inject;
HISTOGRAM / Normal;
QQPLOT / Normal;
TITLE1 'Figure 3.7.1. Distribution of standardized outcome measure';
Run;
```

In this program, the *PROC STANDARD* procedure is used to standardize the outcome measure *Inject*, saving the standardized variable with the same name in the SAS active data file *New_data*. *PROC UNIVARIATE* produces a histogram of the standardized outcome variable. If the observed outcome measure *Inject* is normally distributed with mean μ_{Injet} and standard deviation σ_{Injct}, the standardized variable *Inject* should be normally distributed with mean 0 and standard deviation 1. The *NORMAL* option in the SAS program superimposes a normal distribution curve on the histogram so that the distribution of the observed outcome measure can be compared with the theoretical normal distribution.

The SAS *QQPLOT* command and the *NORMAL* option create a quantile-quantile (Q-Q) plot (see Figure 3.7.2) that graphically compares the quantiles of the standardized variable *Inject* against the quantiles of the theoretical standardized normal scores. By a quantile, we mean the percentage of points below the given value. For example, the 0.25 (or 25%) quantile is the point below which 25% percent of the data fall and above which 75% fall. If the observed outcome *Inject* has a normal distribution, the points in the Q-Q plot in Figure 3.7.2 should fall approximately along a 45° line. The greater is the departure from this line, the greater the support for the conclusion that the observed outcome variable is not normally distributed. From both Figures 3.7.1 and 3.7.2, we can tell that the outcome measure *Inject* is not normally distributed.

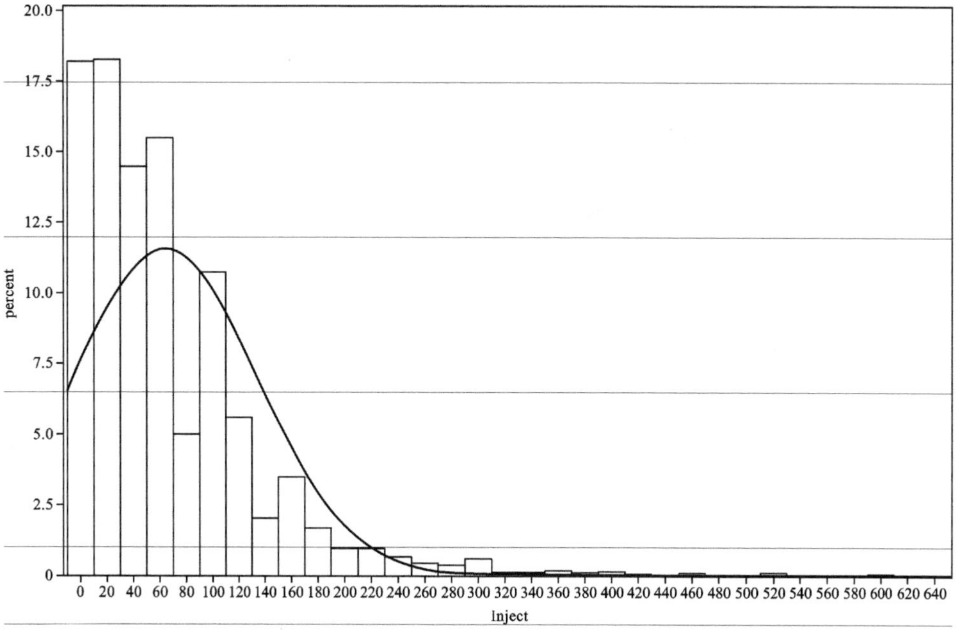

Figure 3.7.1 Distribution of standardized outcome measure

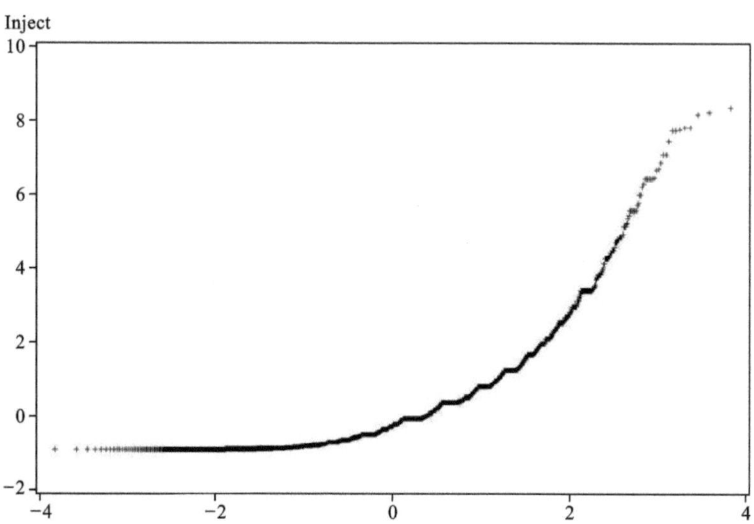

Figure 3.7.2 Q-Q plot of standardized outcome measure

In statistical analysis, when an outcome measure appears to have a non-normal distribution, a log-transformation of this variable may create an approximation of normality. We have, therefore, taken a log-transformation of our outcome measure *Inject*, resulting in a new outcome measure— *Lninj*. In this case, the log-transformation has made the distribution of the outcome measure approximately normal (see Figure 3.7.3). Accordingly, the outcome measure *Inject* in the model shown in eqs. 3.6.1—3.6.4 was replaced with the log-transformed variable *Lninj*, and the model was re-estimated.

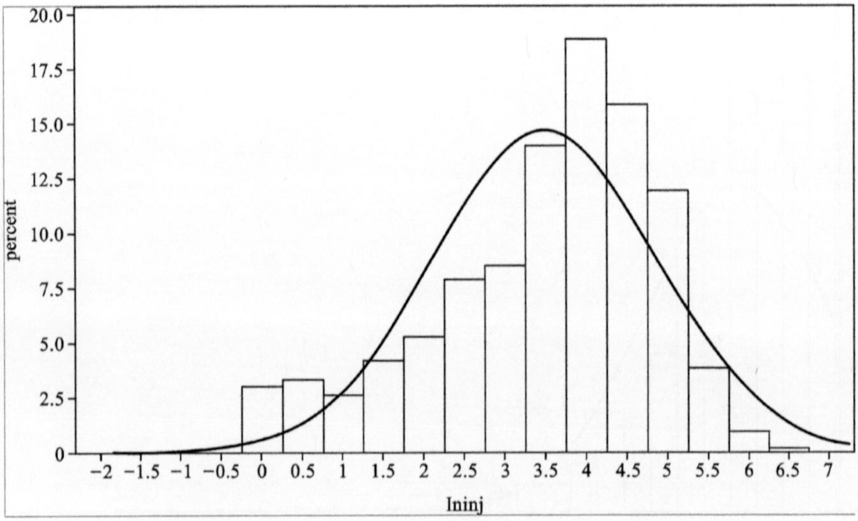

Figure 3.7.3 Distribution of log-transformed outcome measure

The model results are shown in Table 3.7.1, in which the fixed-effects are reported first, followed by the random component in the middle, and finally the model fit statistics in the lower part, of the table. Among the fixed-effects, the main effect of the group-level explanatory variable *HIV_Region* is only marginally significant (γ_{01} = 0.53, p = 0.0681), but its interaction with level-1 explanatory variable *Ethnic* is statistically significant (γ_{11} = 0.33, p = 0.0489). Because the interaction effect is statistically significant, the corresponding main effect, whether it is statistically significant or not, must be reported together with the interaction.

We also need to address the issue of coding categorical variables. In all of the models we have demonstrated so far, no categorical variables are included. In fact, most of the level-1 explanatory variables (e.g., Gender, Ethnic, Highsch) and the group-level explanatory variable HIV_Region are categorical variables with two categories. These variables are coded as dummy variables (i.e., 1 vs. 0) and they don't have to be specified as categorical variables in SAS programs. By default, the "0" category of a dummy variable is treated as the reference group. Dummy variables can be specified as categorical variables by putting them in the CLASS statement in *PROC MIXED*. SAS treats the last category of a categorical variable as the reference group. If we put a dummy variable Gender (1-Male; 0-Female) in the CLASS statement, we would have a slope coefficient estimate with the same absolute value but opposite sign because the "1" category, instead of the "0" category would be treated as the reference group. As such, caution is necessary when coding categorical variables in SAS.

Table 3.7.1 Results of the "final" model with log-transformed outcome measure

Parameter	Estimate	S.E.	*P*-Value
	Fixed-effect		
γ_{00}	3.51	0.1318	<0.0001
γ_{01}	0.53	0.2738	0.0681
γ_{10}	−0.55	0.0706	<0.0001
γ_{11}	0.33	0.1668	0.0489
β_2	−0.05	0.0292	0.1124
β_3	0.01	0.0018	<0.001
β_4	0.05	0.0261	0.0522
	Variance Component[1]		
Var(e_{ij}): σ^2	1.55	0.0222	<0.0001
Var(u_{0j}): τ_{00}	0.24	0.0883	0.0037
Var(u_{1j}): τ_{11}	0.05	0.0287	0.0443
Cov(u_{0j},u_{1j}): τ_{01}	−0.01	0.0415	0.7521

Mode fit statistics:
−2LL = 32331.5; AIC = 32339.5; AICC = 32339.5; BIC = 32343.5

Notes:
γ_{00} —conditional grand mean; γ_{01} —main effect of *HIV_Region*; γ_{10} —main effect of *Ethnic* ; γ_{11} —interaction between *Ethnic* and *HIV_Region*; β_2 —main effect of *Gender*; β_3 —main effect of *Gdmc_age*; β_4 —main effect of *Highsch*. [1]—Significance testing for variance is a one-tailed test.

For some variables like education, rather than using a two-category dummy variable (i.e., 1-High school or above education; 0-less than high school education), we may specify more categories. One of the advantages of using categorical variables in a model is that various cross-category comparisons can be conducted, providing in-depth information about the effect of the categorical variable. In the "final" model we previously demonstrated, the dummy variable *Highsch* has only a marginally significant effect (β_4 = 0.05, p = 0.0522) (see Table 3.7.1). This may be due to the way education was categorized. We can further assess the effect of education by using a 3-level categorical variable E*du_gp* (1-Less than high school; 2-High school; 3-College/above). The SAS program follows:

***SAS Program 3.7.2**;

```
...
PROC MIXED METHOD = REML noclprint COVTEST;
CLASS site Edu_gp;
MODEL Lninj =  HIV_Region|Ethnic Gender Gdmc_age Edu_gp
  /SOLUTION DDFM = BW;
RANDOM INT Ethnic/SUBJECT = site TYPE = UN;
LSMEANS Edu_gp/DIFF;
ESTIMATE 'Edu_gp 2 & 3 vs. Edu_gp 1' Edu_gp -1 0.6873 0.3127;
```

```
ESTIMATE 'Edu_gp 3 vs. Edu_gp 1 & 2' Edu_gp -0.4907 -0.5093 1;
CONTRAST 'Edu_gp 2 & 3 vs. Edu_gp 1' Edu_gp -1 0.6873 0.3127;
CONTRAST 'Edu_gp 3 vs. Edu_gp 1 & 2' Edu_gp -0.4907 -0.5093 1;
Run;
```

where the new variable *Edu_gp* is specified in the *CLASS* statement, and the last category of the variable (3-College/above) is treated as the reference group. The three new statements (i.e., *LSMEANS*, *ESTIMATE*, and *CONTRAST*) used after the *MODEL* statement function the same as in SAS *PROC GLM*. Working together with the *CLASS* and *MODEL* statement, these statements estimate and compare the fixed-effects of specific categories of the variable *Edu_gp*. Corresponding model results are shown in SAS Output 3.7.2.

The global test results show that the categorical variable *Edu_gp* has no significant overall effect ($p = 0.0685$) (see the Section of *Type 3 Tests of Fixed-effects* in SAS Output 3.7.2). However, the mean outcome among IDUs with less than high school education (i.e., *Edu_gp* = 1) was significantly smaller than that among those with college or higher education (i.e., *Edu_gp* = 3) (see the upper panel of SAS Output 3.7.2). Further studies are needed to explain why IDUs with higher education were more likely to inject drugs than those with lower education, controlling for other factors.

The *LSMEANS* statement computes least-squares means (LS-mean) for each category of the variable, *Edu_gp*. With option *DIFF*, pair-wise differences in LS-means between any two categories of the categorical variable were tested. The pair-wise comparisons provided by the *LSMEANS* statement (see the lower panel of SAS Output 3.7.2) are identical to the two fixed effects of variable *Edu_gp*; in addition, it also shows that IDUs with less than high school education (i.e., *Edu_gp* = 1) were not significantly different from high school graduates ($p = 0.2162$), with respect to frequency of drug injection.

Alternatives to pair-wise comparisons, both *ESTIMATE* and *CONTRAST* statements allow testing specific hypotheses of differences between various category combinations of a categorical variable. The *ESTIMATE* statement produces an estimated mean difference, standard error, *t*-test, and associated probability value, while the *CONTRAST* statement only provides *F*-test. The results of the statements are shown in the panels of *Estimates* and *Contrasts*, respectively, in SAS Output 3.7.1. The *ESTIMATE* and *CONTRAST* statements provide additional comparisons in mean outcome measure between the following groups: Categories 2 & 3 vs. Category 1; and Category 3 vs. Categories 1 & 2.

Attention needs to be paid to the *weight* in the *ESTIMATE* and *CONTRAST* statements. The three categories of the categorical variable *Edu_gp* in our example have different frequencies: $n_1 = 3914$ for Category 1; $n_2 = 4062$ for Category 2; and $n_3 = 1848$ for Category 3. To compare Categories 2 and 3 with Category 1, the corresponding weights in the *ESTIMATE* and *CONTRAST* statements should be: -1.0 for Category 1; $4062/(4062 + 1848) = 0.6873$ for Category 2 , and $1848/(4062 + 1848) = 0.3127$ for Category 3. Accordingly, to compare Category 3 with Categories 1 and 2, the corresponding weights are: $-(3914/(3914 + 4062)) = -0.4907$ for Category 1; $-(4062/(3914 + 4062)) = -0.5093$ for Category 2; and 1.0 for Category 3, respectively. We may use weights like 1, 1, -2 or 0.5, 0.5, -1 in the *ESTIMATE* and *CONTRAST* statements with balanced group size. Otherwise, comparisons would be biased.

SAS Output 3.7.2 Selected results of the model with categorical measure of education

```
Solution for Fixed-effects
```

Effect	Education: 1-<High school; 2-High schooe; 3-College +	Estimate	Standard Error	DF	t Value	Pr > |t|
Intercept		3.5955	0.1335	18	26.94	<.0001
HIV_Region		0.5310	0.2742	18	1.94	0.0687
Ethnic		-0.5476	0.07072	9798	-7.74	<.0001
HIV_Region*ethnic		0.3311	0.1671	9798	1.98	0.0476
Gender		-0.04522	0.02918	9798	-1.55	0.1213
Gdmc_age		0.01395	0.001756	9798	7.95	<.0001
Edu_gp	1	-0.08517	0.03578	38	-2.38	0.0224
Edu_gp	2	-0.04973	0.03527	38	-1.41	0.1667
Edu_gp	3	0	.	.	.	.

Type 3 Tests of Fixed-effects

Effect	Num DF	Den DF	F Value	Pr > F
HIV_Region	1	18	3.75	0.0687
Ethnic	1	9798	59.97	<.0001
HIV_Region*ethnic	1	9798	3.92	0.0476
Gender	1	9798	2.40	0.1213
Gdmc_age	1	9798	63.16	<.0001
Edu_gp	2	38	2.88	0.0685

Estimates

Label	Estimate	Standard Error	DF	t Value	Pr > |t|
Edu_gp 2 & 3 vs. Edu_gp 1	0.05099	0.02605	38	1.96	0.0577
Edu_gp 3 vs. Edu_gp 1 & 2	0.06712	0.03261	38	2.06	0.0465

Contrasts

Label	Num DF	Den DF	F Value	Pr > F
Edu_gp 2 & 3 vs. Edu_gp 1	1	38	3.83	0.0577
Edu_gp 3 vs. Edu_gp 1 & 2	1	38	4.24	0.0465

Least Squares Means

Effect	Education: 1-<High school; 2-High school; 3-College +	Estimate	Standard Error	DF	t Value	Pr > |t|
Edu_gp	1	3.3320	0.1127	38	29.57	<.0001
Edu_gp	2	3.3675	0.1126	38	29.91	<.0001
Edu_gp	3	3.4172	0.1147	38	29.80	<.0001

Differences of Least Squares Means

Effect	Education: 1-<High school; 2-High School; 3-College +	Education: 1-<High school; 2-High school; 3-College +	Estimate	Standard Error	DF	t Value	Pr > |t|
Edu_gp	1	2	-0.03544	0.02818	38	-1.26	0.2162
Edu_gp	1	3	-0.08517	0.03578	38	-2.38	0.0224
Edu_gp	2	3	-0.04973	0.03527	38	-1.41	0.1667

The new model results show a significant difference in the mean outcome between those who had high school or lower education and those who have college or higher education. Further studies are needed to explain why IDUs with higher education were more likely to inject drugs than those with lower education after controlling for other factors.

In this chapter we have presented the application of multilevel modeling to cross-sectional data. Details in model development and corresponding SAS programming were discussed. In next chapter, we will discuss applications of multilevel modeling to longitudinal data.

Chapter 4
Application of multilevel modeling to longitudinal data

In Chapter 3, we demonstrated how to run multilevel models using cross-sectional data. In this chapter we will expand the application of multilevel modeling to longitudinal data that are collected repeatedly from the same subjects. Longitudinal data are also considered as hierarchically structured data in which repeated measures are nested within subjects.

4.1 Features of longitudinal data

There are two major longitudinal study designs: cohort and panel. The former samples cohorts, such as birth groups, and follows them over time. Panel studies begin with a cross-sectional sample and follow the subjects over time.

Longitudinal data have several distinctive features. First, there are two sources of variations: within-subject or intra-subject variation and between-subject or inter-subject variations. Second, the within-subject observations are usually not independent of each other. Third, the between-subject variation may not be constant over time. And finally, longitudinal data are often incomplete or unbalanced as subjects may drop out of a study at any time for various reasons. Even those who stay in the study may miss some of the follow-up interviews or observations. As such, the actual number of follow-ups and the time intervals between the follow-ups also vary by subjects.

Figure 4.1.1 plots the trajectories of individual *General Life Satisfaction Scores* with an interval of 6 months during a 3-year observation period[1]. From the plots we can see that subjects are different at the beginning of the study and the rate of change in their general life satisfaction varies across individuals. Individuals that started with higher general life satisfaction scores tended to remain higher over the course of the study. Most subjects showed an upward trend over time at different rates. Some subjects showed no significant change. Between-subject variability appears to increase as individual growth lines diverge over time. From Figure 4.1.1, we also see that not every person completed every follow-up interview.

These features make the analysis of longitudinal data more challenging than cross-sectional data. Statistical methods for longitudinal studies must account for within-subject observation dependence, between-subject variation, non-constant variance, and unbalanced data simultaneously. Multilevel models provide appropriate techniques for managing such challenges.

---

[1] Data displayed in Figure 4.1.1 are a sub-set of subjects selected from a New Hampshire Dual Disorder Study. Measurement of the *General Life Satisfaction Score*, as well as the data set, will be described in detail later.

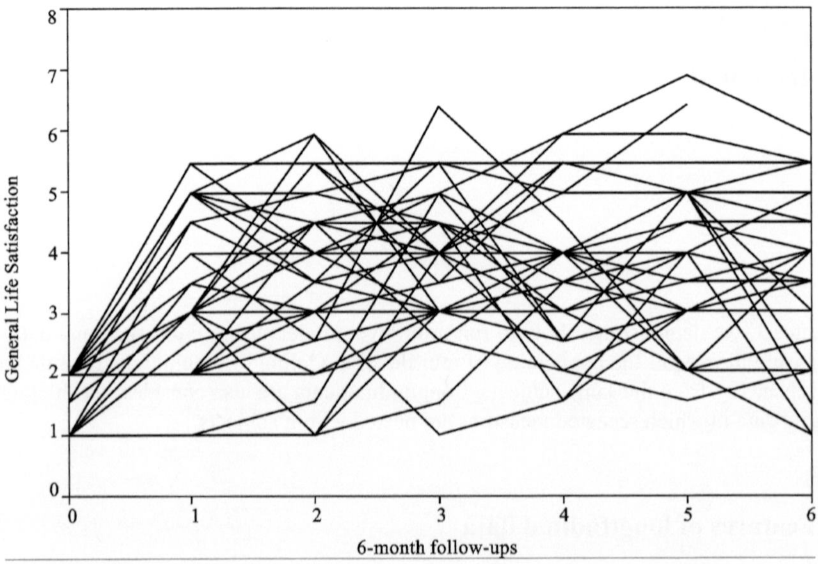

Figure 4.1.1 Observed individual general life satisfaction score over three years

4.2 Limitations of traditional approaches for modeling longitudinal data

The ordinary least square (OLS) is inappropriate for analyzing longitudinal data because: 1) it assumes normality, independence and common variance in the measures, all unrealistic in longitudinal data; and 2) it analyzes only the average growth trajectory for all individuals since the models parameters (intercept and slope) are not allowed to vary by individuals. Traditionally, two approaches have been applied to longitudinal analyses: univariate repeated measure (URM) ANOVA, and multivariate repeated measures (MRM) ANOVA. the URM ANOVA approaches assume that person-specific effects are random, and the effects of time and other independent variables are fixed. In this way URM ANOVA takes into account some of the subject-specific effects, thus reducing the residual variance. However, a particular residual variance/covariance structure, called "compound symmetry (CS)", is assumed for this model. This structure assumes that residual variances are the same at different time points and residual covariances are the same regardless of time lag between measurements (Girden, 1992). This assumption is very restrictive and unlikely to be valid in reality.

An alternative assumption proposed by Huyhn & Feldt (1970) and Rouant & Lépine (1970) is called "*sphericity*" or "*circularity.*" This requires equal variance difference between any two adjacent time points. The sphericity assumption is less restrictive than compound symmetry assumption. Still, for repeated measured data, it is less likely to be the case that the variance difference between two adjacent time points is identical to the variance difference at any other two adjacent points in time. In addition, to deal with the implausibility of the equal covariance assumptions, adjustments such as the Greenhouse and Geisser modifications have been proposed to reduce the degrees of freedom for statistical tests by a correction factor. These adjusted tests can be very conservative (Gibbons et al., 1993).

MRM ANOVA takes a very general approach by making no assumptions about the within-subject variance/covariance structure; that is, the residual variance/covariance structure is unstructured. However, with MRM ANOVA, all subjects would be measured at all time points and with the same length of time intervals. Finally, other than the unstructured form, MRM ANOVA does not allow any other forms of the residual variance/covariance structure. When the number of repeated observations is large, it is not desirable to have unstructured residual variances and covariances structure as it may result in over-parameterization.

4.3 Advantages of multilevel modeling for longitudinal data

Longitudinal data are hierarchically structured with repeated measures nested within individuals. We can treat measurement occasions as level-1 units and individuals as level-2 units. As such, the two-level multilevel model can be applied.

Multilevel models have several advantages over the traditional models for longitudinal data analysis. First, multilevel models are capable of dealing with unbalanced or incomplete data under the assumption that observations are "missing at random" (MAR). The estimation based on either maximum likelihood (ML) or restricted maximum likelihood (REML) techniques use all available data. Thus one neither has to throw away a subject that has a missing observation nor to impute the missing observations.

Second, multilevel models do not require within-subject observations to be independent of each other, nor to be bounded by restrictive assumption such as the compound symmetry. Multilevel modeling manages the within-subject or intra-class correlation from either a between-subject variability perspective or a within-subject variability perspective or both.

From a between-subject variability perspective, the model assumes that the within-subject observation dependence is due to the heterogeneity of subjects caused by unmeasured factors. The model accommodates this by incorporating individual-specific effects in the model, and thus allowing model parameter estimates to vary from subject to subject. These subject-specific effects are assumed to be randomly distributed in the population.

From the within-subject variability perspective, the observation dependence can be modeled by imposing an appropriate residual variance/covariance structure. If modeling variances from either "between-subject"with random effects or from "within-subject" with a structured variance/covariance matrix is still inadequate, one may formulate a full mixed-effects model with both random effects and structured error variance/covariance components.

Finally, multilevel modeling can easily accommodate both time-invariant and time-varying covariates. For repeated measures data, subject-level covariates such as gender and ethnicity are time-invariant because they don't change over time; but observation-level covariates such as time measure, health status, cultural beliefs/practices, and social behaviors, etc. could be time-varying because their values could be different at different time points. Subject-level covariates and observation-level covariates can be easily incorporated in the model.

4.4 Formulation of growth models

Applications of multilevel models to longitudinal data are also called growth models (GM) or individual growth models. The multilevel model formulation for longitudinal data is the same as for cross-sectional clustered data. For a two level growth model, in which level-1 units are repeated measures and level-2 units are individual subjects, the level-1 model (within-subject models) can be described as the following:

$$y_{ij} = \beta_{0j} + \beta_{1j} t_{ij} + e_{ij} \tag{4.4.1}$$

where y_{ij} is the ith occasion (level-1 unit) of outcome measure for the jth individual (level-2 unit), β_{0j} is the intercept coefficient, t_{ij} is time score, β_{1j} is the slope coefficient of t_{ij}, and e_{ij} is the error term assuming to be normally distributed with mean zero and variance σ^2. Note that both β_{0j} and β_{1j} are random regression coefficients indicating that both the initial level of outcome measures and the rate of outcome change vary across individuals.

The level 2 models (between-subject models) take the form:

$$\beta_{0j} = \beta_{00} + \beta_{01} x_j + u_{0j} \tag{4.4.2}$$

$$\beta_{1j} = \beta_{10} + \beta_{11} x_j + u_{1j} \tag{4.4.3}$$

where β_{00} and β_{10} are the overall mean intercept and overall mean slope adjusting for the subject-level (level-2) predictor or covariate x_j. The regression coefficients β_{01} and β_{11} are fixed-effects of the subject-level (level-2) predictor x_j, while u_{0j} and u_{1j} are random effects of the jth person, on the level-1 random intercept and slope coefficients, respectively. The distribution of these random effects is assumed to be normal with means of zero and variances of σ_{u0}^2 and σ_{u1}^2, respectively.

Substituting eqs. 4.4.2 and 4.4.3 into eq. 4.4.1 yields the combined model:

$$y_{ij} = (\beta_{00} + \beta_{01} x_j + \beta_{10} t_{ij} + \beta_{11} x_j t_{ij}) + (u_{0j} + u_{1j} t_{ij} + e_{ij}) \tag{4.4.4}$$

where $(\beta_{00} + \beta_{01} x_j + \beta_{10} t_{ij} + \beta_{11} x_j t_{ij})$ is the fixed component of the model, and $(u_{0j} + u_{1j} t_{ij} + e_{ij})$ is the random component of the model with u_{0j} and u_{1j} as level-2 errors and e_{ij} as level-1 residuals.

The general format of the multilevel model for longitudinal data analysis can be expressed in matrix notation as shown in eq. 2.2.10:

$$Y = X\beta + ZU + e \tag{4.4.5}$$

where Y denotes the vector of observed outcome, β is an unknown vector of fixed-effects with known design matrix X, and U is an unknown vector of random effects with known design matrix Z, and e is a vector of residuals. Random effects and residuals are assumed to be normally distributed with expected values and variances indicated below:

$$\mathrm{E}\begin{pmatrix} u \\ e \end{pmatrix} = 0 \ \textit{and} \ \mathrm{Var}\begin{pmatrix} u \\ e \end{pmatrix} = \begin{pmatrix} G & 0 \\ 0 & R \end{pmatrix} \tag{4.4.6}$$

where G is the variance/covariance matrix of level 2 errors, and R is the variance/covariance matrix of level 1 residuals. The total variance of Y is therefore

$$\mathrm{Var}(Y) = ZGZ' + R \tag{4.4.7}$$

The models formulated above in either non-matrix form or matrix-form are full expression of the multilevel model with both fixed and random effects. As in linear models, the mean value or expected value of Y is modeled through fixed-effects, β, while the variance of Y is modeled through random components, G and R. In this "full" model we can specify appropriate variance/covariance structures for both G and R.

Depending on the nature of the data, it is not always necessary to specify a full model. In some cases, a partial model may be adequate. Two examples are: 1) when a random coefficient approach is taken as a main modeling strategy, the R matrix can be set $R = \sigma^2 I$. 2) If the focus is on modeling within-subject variance/covariance rather than between-subject variability, the residual variance (R matrix) can be modeled by imposing appropriate variance/covariance structures, and the personal level random components U may be set to zero. In the latter case, the equation is reduced to a fixed-effects model with structured residual variances-covariance matrix:

$$Y = X\beta + e \tag{4.4.8}$$

with $e \sim N(0, R)$. The R-matrix can be specified in various forms. If R is also assumed to be $\sigma^2 I$, then the model is further reduced to the traditional multiple regression. It is often a good practice to begin with analysis of partial models before considering a full model.

The rest of the chapter will elaborate the longitudinal model in a step-by-step fashion. We will start with the simplest model and gradually increase the level of complexity by adding more components to the base model. All models will be illustrated using real data.

4.5 Data and variable description

The data used in this chapter were collected in the New Hampshire Dual Disorder (NHDD) Study on mental illness and substance abuse disorders (NHDD) (Drake et al., 1998). Data ($N = 203$) used in this analysis were drawn from the first three years (1998—2001) of the study. Participants were randomly assigned into two treatment groups: assertive community treatment (ACT) ($n = 98$) and standard case management (SCM) ($n = 105$). The former was considered the *experimental treatment group* and the latter the *control group*. Participants were interviewed at baseline and every six months thereafter for three years. Consequently there were a total of 7 observation time-points: the baseline interview and six follow-up interviews. The following are the variables used in the models discussed in this chapter:

Outcome variable:

QOL: General life satisfaction or general quality of life. The question "how do you feel about your life as a whole" was asked at the beginning and at the end of each interview and *QOL* was computed as the average of the two measures. This variable is treated as a continuous measure with values ranging from 1 to 7, where 1 stands for "terrible" and 7 stands for "delighted."

Level 1 explanatory variables:

Time: time was coded as 0, 1, 2, 3, 4, 5 and 6 representing measurement time points with a 6-month time interval from baseline to 36 months after baseline.

SATS: Substance Abuse Treatment Scale is a within-subject time-varying covariate. *SATS* is an 8-point scale measure. A higher score represents a more progress toward recovery from a substance use disorder.

Level 2 explanatory variables:

Trt: dummy variable (1-treatment group; 0-control group).

Age: continuous measure, participant's age at baseline interview.

Gender: dummy variable (1-male; 0-female).

Edu: education, dummy variable (1-high school or above; 0-less than high school).

Diagnosis: primary psychiatric diagnosis, dummy variable (1-schizophrenia/schizoaffective disorder; 0-bipolar disorder).

Our goal is to illustrate the concepts and features of the growth model, rather than mental health research. Demographic and background variables are included as covariates in the model for illustrative purposes.

The SAS *PROC MIXED* procedure is used to fit the 2-level growth model. The person-period data format is required for the modeling; that is, each person may have a multiple number of records (i.e., rows in the data matrix), and the number of records may vary across individuals if there are missing observations or dropouts. As such, the first step before using SAS *PROC MIXED* procedure is to construct a data set with correct format.

The original data are stored in the following format where each individual has a single record that contains: *ID*, *QOL* measured at the baseline and every 6 months for 3 years (*Qol0*, *Qol1*, *Qol2*, *Qol3*, *Qol4*, *Qol5*, *Qol6*), *SATS* measured at every 6 months for 3 years (*Sats0*, *Sats1*, *Sats2*, *Sats3*, *Sats4*, *Sats5*, *Sats6*), while *Trt*, *Age*, *Gender*, *Diagnosis*, and *Edu* are time-invariant measures. The data structure for 4 selected individuals is displayed below:

ID	t1	t2	t3	t4	t5	t6	t7	QOL0	QOL1	QOL2	QOL3	QOL4	QOL5	QOL6
150	0	1	2	3	4	5	6	4.5	4.0	4.0	4.0	4.5	5.0	4.0
151	0	.	2	3	4	5	.	4.5	.	3.0	4.0	4.0	3.5	.
152	0	1	2	3	4	5	6	5.0	4.0	5.0	5.0	5.0	5.0	5.5
153	0	1	2	3	4	.	6	1.0	1.0	1.0	3.0	3.0	.	1.0

ID	SATS0	SATS1	SATS2	SATS3	SATS4	SATS5	SATS6	AGE	Trt	gender	EDU	DIAGNOSIS
150	3.0	3	3	3	3	3	3	31	1	1	1	1
151	3.0	.	6	5	6	7	6	27	1	1	1	1
152	3.0	3	3	3	5	5	4	37	1	1	1	1
153	3.0	3	3	5	6	6	7	46	1	1	1	1

The following SAS program can be used to convert the data set into a person-period format.

***SAS Program 4.5.1**;

```
LIBNAME lib 'd:\MLM\data';
Data lib.ML_data_Chap4;
Keep id Time QOL SATS Trt Age Gender Edu Diagnosis;
 Set lib.Chap4_rawdata;
Array T(7) t1—t7;
Array A(7) qol0—qol6;
Array B(7) sats0—sats6;
Do i = 1 to 7;
 Time=T(i);
 QOL=A(i);
 SATS=B(i);
Output;
End;
Run;
```

This program converted the original longitudinal data set *Chap*4*_rawdata* into "stacked" person-period data, in which each individual has multiple records corresponding to different time points. In the original data, outcome measure (*QOL*), time score (*T*1—*T*7), and time-varying covariate (*SATS*) have a set of 7 observations corresponding to each of the 7 time points. The SAS program converts these variables into three new variables: *QOL*, *time*, and *SATS*. The level-2 (between-subject) variables, such as *Trt*, *Age*, *Gender*, *Edu*, and *Diagnosis* are time-invariant. The newly created SAS data set, *ML_data_Chap*4, contains the variables specified in *KEEP* statement. The following data are from the same 4 individuals after conversion:

ID	TIME	QOL	SATS	TRT	AGE	GENDER	EDU	DIAGNOSIS
150	0	4.5	3	1	31	1	1	1
150	1	4.0	3	1	31	1	1	1
150	2	4.0	3	1	31	1	1	1
150	3	4.0	3	1	31	1	1	1
150	4	4.5	3	1	31	1	1	1
150	5	5.0	3	1	31	1	1	1
150	6	4.0	3	1	31	1	1	1
151	0	4.5	3	1	27	1	1	1
151	.	.	.	1	27	1	1	1
151	2	3.0	6	1	27	1	1	1
151	3	4.0	5	1	27	1	1	1
151	4	4.0	6	1	27	1	1	1
151	5	3.5	7	1	27	1	1	1
151	.	.	.	1	27	1	1	1
152	0	5.0	3	1	37	1	1	1
152	1	4.0	3	1	37	1	1	1
152	2	5.0	3	1	37	1	1	1
152	3	5.0	3	1	37	1	1	1
152	4	5.0	5	1	37	1	1	1
152	5	5.0	4	1	37	1	1	1
152	6	5.5	4	1	37	1	1	1
153	0	1.0	3	1	46	1	1	1
153	1	1.0	3	1	46	1	1	1
153	2	1.0	3	1	46	1	1	1
153	3	3.0	6	1	46	1	1	1
153	4	3.0	6	1	46	1	1	1
153	.	.	.	1	46	1	1	1
153	6	1.0	7	1	46	1	1	1

where a missing value in variable *Time* indicates that the participant missed a corresponding follow-up interview. For example, the participants with ID numbers 150 and 152 did not have any missing values; participant with ID 151 missed the first and the last follow-up interviews; and another one with ID 153 missed the fifth follow-up interview. In the sample of 203 participants, the follow-up rate was very high, with over 90% of the interviews completed in each of the six-month follow-up waves over the 36 month period.

4.6 Linear growth models

We will start with a simple model testing linear growth with one fixed level-1 slope coefficient (i.e., slope of the variable *Time*), one level-1 random coefficient (i.e., random intercept), and no other covariates. We will then augment this base model by adding more variables and other modeling components.

4.6.1 The shape of average outcome change over time

To begin model development, it will be helpful to plot an average growth line for the outcome variable over time. The plot for our outcome variable, *QOL*, shows that the average general life satisfaction score for study sample tends to go slightly up over 3 years, but not in exactly linear fashion (Figure 4.6.1.1). To determine whether the growth trajectory of the outcome is linear or curvilinear, we fitted two fixed-effect models, quadratic and cubic, with an unspecified variance/covariance structure (model results are not reported here). Neither quadratic nor cubic terms are significant. The overall model does not improve significantly (the difference in –2LL between the two models is 2.3, $p = 0.129$). Therefore a linear time trend model is determined to be an appropriate base model.

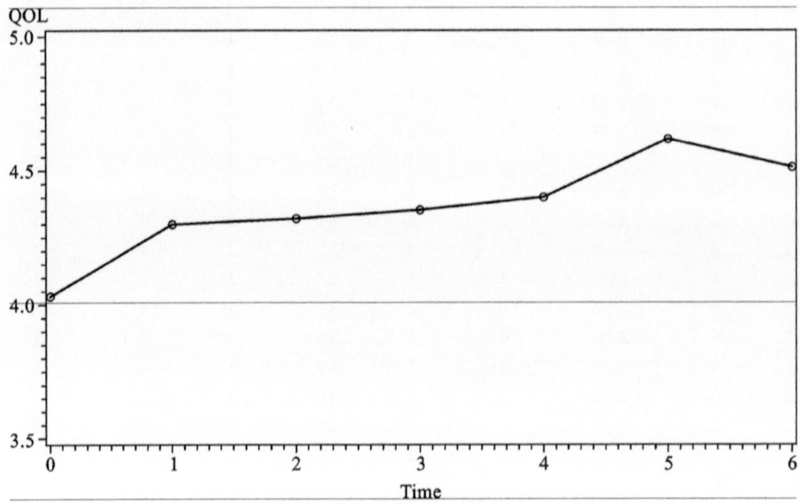

Figure 4.6.1.1 Average general life satisfaction score over three years

4.6.2 Random intercept growth models

The simple extension of classic linear regression models is to replace the fixed intercept with a random intercept leading to the *random intercept model*. This model can be represented as the within-subject (level-1) model,

$$\text{QOL}_{ij} = \beta_{0j} + \beta_{1j}\text{Time}_{ij} + e_{ij} \tag{4.6.2.1}$$

the between-subject (level-2) model,

$$\beta_{0j} = \gamma_{00} + u_{0j} \tag{4.6.2.2}$$

$$\beta_{1j} = \gamma_{10} \tag{4.6.2.3}$$

and the combined model,

$$\text{QOL}_{ij} = \gamma_{00} + \gamma_{10}\text{Time}_{ij} + (u_{0j} + e_{ij}) \tag{4.6.2.4}$$

In the within-subject model individual *j*'s response or outcome value at i^{th} measurement is determined by his/her initial outcome level, β_{0j}, and the rate of outcome change, β_{1j}. The between-subject model decomposes the initial outcome measure into two components: the overall average outcome γ_{10} and the individual specific effect u_{0j}. By adding a random error for individual uniqueness, the initial levels of the outcome measure are individual-specific. In this model, the slope coefficient of *TIME* is restricted to be the fixed; that is, the rate of outcome change over time is assumed to be the same for everyone, thus the individual growth lines would be parallel to the overall average growth line.

The level-2 random errors, u_{0j}, represent individual specific effects that are assumed to be a random sample from a normally distributed population with mean 0 and variance σ_{u0}^2 .The error term e_{ij} in the model is assumed to be normally and conditionally independently distributed in the population with zero mean and variance σ^2 as noise due to between-subject variation is assumed to have been removed from the unexplained variance by including level-2 random effects. The two random error terms, u_{0j} and e_{ij}, however, are assumed not to be correlated.

This *random-intercept model* takes into account the variation in the initial level of outcome. The level-2 error term, u_{0j}, represents the deviation of each individual regression lines from the overall average growth line, and the variance of these deviations, σ_{u0}^2, represents the spread of all individual-specific growth lines around the overall average growth line. If σ_{u0}^2 equals zero, the model would only estimate the overall average growth or trajectory line as in the traditional linear regression model. The SAS program for the random intercept model follows:

***SAS Program 4.6.2.1**;

```
Libname lib 'd:\MLM\data';
Data data1;
  Set lib.ML_data_Chap4;

Proc Mixed covtest noclprint;
Class ID;
ID ID Time QOL; *the 1st ID is a SAS statement & the 2d ID is a variable
name in the data;
Model QOL = Time/S DDFM = KR OUTP=PQOL(keep=ID TIME QOL PRED);
Random Int /subject = ID;
Run;

Proc Gplot data = PQOL;
Symbol1 v=none repeat=203 i=join color=cyan;
Symbol2 v=none i=sm50s color=red w=4;
Plot  PRED*Time=id  /haxis=0 to 6 nolegend;
Plot2 PRED*Time;
Label Time = Month;
Title 'Fitted individual and average lines for random intercept model';
Run;
Quit;
```

where the option *INT* (or *INTERCEPT*) specified in the *RANDOM* statement requests that random intercept model be fitted. The *NOCLPRINT* option specified in the *PROC MIXED* statement suppresses printing classification information. The *ID* statement specifies which variables from the input data set are to be included in the output data set specified in *OUTP* option of the *MODEL* statement. Removing the *ID* statement would cause all variables in the original data to be included in the new output data set. The *KEEP* option causes the predicted value along with observed values of *ID*, *Time* and *QOL* to be included in the new data set *PQOL*, which is used for plotting.

The model was estimated using the default REML method. The default method of calculating denominator d.f. for fixed-effects was replaced with *DDFM* = *KR*. As discussed in Section 3.6 of Chapter 3, for longitudinal data, especially unbalanced longitudinal data and model with complex variance/covariance structures, the Kenward-Roger (*KR*) method is recommended (SAS Institute Inc., 2004).

The fitted growth lines predicted from the random intercept model are demonstrated graphically in Figure 4.6.2.1. The solid line in the middle of the figure is estimated overall average growth line, and the lines around this average line are individual-specific growth lines.

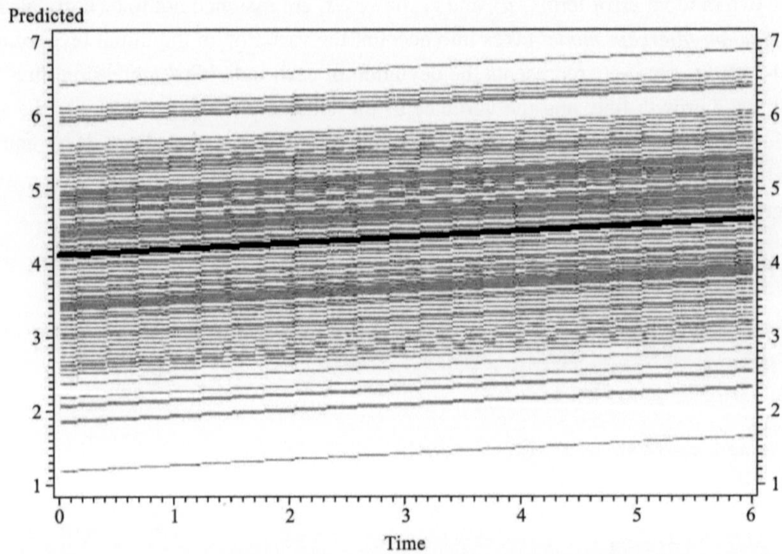

Figure 4.6.2.1 Individual and average growth lines fitted from the random intercept model

The selected output of the random-intercept model is presented in SAS Output 4.6.2.1. Model estimation converged in 3 iterations. The *Covariance Parameter Estimates* panel is the output of variance/covariance parameter estimates: $\hat{\sigma}^2_{u0} = 0.9659$ ($p < 0.0001$) and $\sigma^2 = 1.0001$ ($p < 0.0001$), Both are statistically significant. The significance of $\hat{\sigma}^2_{u0}$ suggests that individuals do differ in their initial level of *QOL*, and the significance of the residual variance σ^2 implies that there was significant within-subject variation after random intercepts are specified in the model. The estimated intra-class correlation (ICC) is 0.9659/(0.9659 + 1.0001) = 0.49, indicating that almost a half of the total variation is due to between-subject heterogeneity.

As REML was used for model estimation (the default estimator), the restricted−2LL can be used to conduct LR test to compare nested models with the same fixed-effects but different random effects. The information criteria statistics *AIC* (Akaike's information criterion), *AICC* (a finite-sample corrected version of *AIC* that is preferred over the *AIC* for small samples), and *BIC* (Schwarz's Bayesian information criterion) can be used to compare non-nested models. Note that these statistics do not tell how a single model fits the data, but are used only for model comparisons to test which model has a relatively better fit.

The *Solution for Fixed-Effects* shows the parameter estimates for fixed-effects. The intercept, representing the estimated average initial *QOL* score, was about 4.13. The linear time effect $\hat{\gamma}_{10} = 0.08$ ($p < 0.0001$), representing the average rate of change in *QOL*, was about 0.08 points per 6-month interval. The *Type* 3 *Tests of Fixed-Effects* section provides results of testing the global effects of the categorical variables using the *F*-test. If no categorical variables are specified in the *CLASS* statement, it is not necessary to request the type 3 tests. The option *NOTEST* can be used in the *MODEL* statement to suppress this additional panel of output.

SAS Output 4.6.2.1 Selected results of the random intercept model

```
                    Iteration History

Iteration   Evaluations   -2 Res Log Like    Criterion

        0             1     4714.88136693
        1             2     4209.29235952   0.00000027
        2             1     4209.29212297   0.00000000
       Convergence criteria met.

           Covariance Parameter Estimates

Cov Parm    Subject  Estimate  Standard Error  Z Value     Pr Z

Intercept   ID         0.9659          0.1113     8.68   <.0001
Residual               1.0001         0.04198    23.82   <.0001

                    Fit Statistics

           -2 Res Log Likelihood        4209.3
           AIC (smaller is better)      4213.3
           AICC (smaller is better)     4213.3
           BIC (smaller is better)      4219.9

             Solution for Fixed-Effects

                         Standard
  Effect      Estimate      Error      DF    t Value    Pr > |t|

  Intercept     4.1314    0.08474     338      48.75      <.0001
  Time         0.07865    0.01375    1140       5.72      <.0001

                 Type 3 Tests of Fixed-Effects

                       Num      Den

         Effect         DF       DF   F Value   Pr > F

         Time            1     1140     32.71   <.0001
```

4.6.3 Random intercept-slope growth models

A random intercept model assumes that individual growth or trajectory lines start at different levels, but grow at the same rate over time. In reality, however, individuals are likely to have different growth trajectories with different initial levels and rates of outcome change over time. Accordingly, a random slope coefficient of *Time* should be added to the model:

$$\text{QOL}_{ij} = \beta_{0j} + \beta_{1j}\text{Time}_{ij} + e_{ij} \quad (4.6.3.1)$$

$$\beta_{0j} = \gamma_{00} + u_{0j} \quad (4.6.3.2)$$

$$\beta_{1j} = \gamma_{10} + u_{1j} \quad (4.6.3.3)$$

$$\text{QOL}_{ij} = \gamma_{00} + \gamma_{10}\text{Time}_{ij} + (u_{0j} + u_{1j}\text{time}_{ij} + e_{ij}) \quad (4.6.3.4)$$

where u_{0j} and u_{1j} represent how the jth individual's initial level and rate of change in outcome deviate from the average initial level and average rate of outcome change. It is assumed that u_{0j} and u_{1j} have a bivariate normal distribution $N\sim(0, G)$, with a variance/covariance matrix:

$$G = \begin{pmatrix} \sigma_{u0}^2 & \sigma_{u01}^2 \\ \sigma_{u01}^2 & \sigma_{u1}^2 \end{pmatrix} \quad (4.6.3.5)$$

where σ_{u0}^2 and σ_{u1}^2 are variances of the random intercept and slope coefficients, respectively, and σ_{u01}^2 is the covariance between the initial level and rate of outcome change.

This model is commonly called the *random intercept-slope* (RIS) *model*. The model is demonstrated graphically in Figure 4.6.3.1, in which the overall average growth line (the bold line) shows slight upward trend over 36 months. The individual growth lines vary with different intercepts and slopes. Some of the individual growth lines slope up, some slope down and others are virtually flat. The following SAS Program 4.6.3.1 was used to fit this model and generate the plot:

***SAS Program 4.6.3.1**;

```
...
Proc Mixed covtest noclprint;
Class ID;
ID ID Time QOL;
Model QOL = Time
 /S DDFM=KR notest OUTP = PQOL(keep = ID TIME QOL PRED);
Random Int Time/subject = ID G TYPE = UN;
Run;
Proc Gplot data = PQOL;
Symbol1 v = none repeat=203 I = join color = cyan;
Symbol2 v = none I = sm50s color = red w = 4;
Plot  PRED*ime=id  /haxis=0 to 6 nolegend;
Plot2 PRED*Time;
Label Time = Month;
Title 'Fitted individual and average lines for random intercept-slope
model';
Run;
Quit;
```

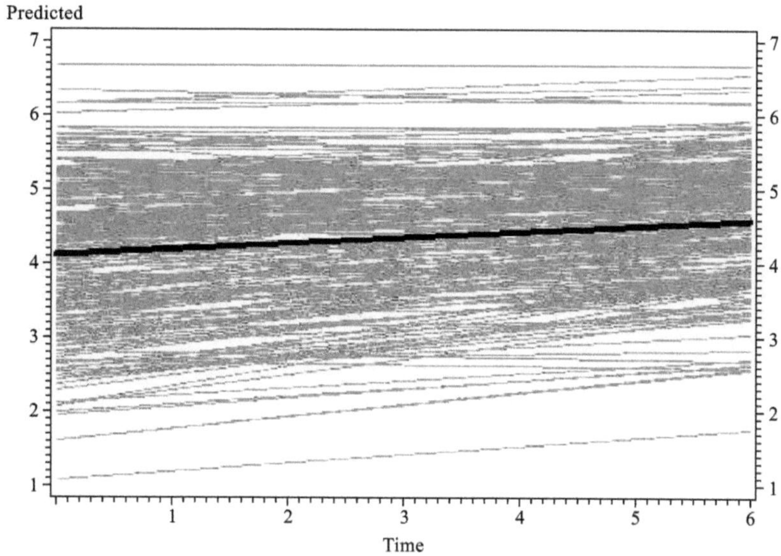

Figure 4.6.3.1 Individual and average growth lines for random intercept-slope model

where the option "*TYPE* = *UN*" in the *RANDOM* statement requests that the variance/covariance *G* matrix to be unstructured. The selected results of the model are shown in SAS Output 4.6.3.1.

SAS Output 4.6.3.1 Selected Results of the Random Intercept-Slope Model

Covariance Parameter Estimates

Cov Parm	Subject	Estimate	Standard Error	Z Value	Pr Z
UN(1,1)	ID	1.3457	0.1793	7.50	<.0001
UN(2,1)	ID	-0.08983	0.02610	-3.44	0.0006
UN(2,2)	ID	0.01887	0.005562	3.39	0.0003
Residual		0.9124	0.04211	21.67	<.0001

Fit Statistics

-2 Res Log Likelihood	4188.0
AIC (smaller is better)	4196.0
AICC (smaller is better)	4196.0
BIC (smaller is better)	4209.3

Null Model Likelihood Ratio Test

DF	Chi-Square	Pr > ChiSq
3	526.86	<.0001

```
                         Solution for Fixed-Effects

Effect        Estimate     Standard Error    DF    t Value    Pr > |t|

Intercept       4.1341           0.09410    203      43.93      <.0001
Time           0.07779           0.01635    201       4.76      <.0001
```

As in the random intercept model, the current model has two fixed-effects: the overall mean outcome measure or overall average intercept $\hat{\gamma}_{00}$ = 4.13 ($p < 0.0001$), and the average rate of outcome change over time or average slope $\hat{\gamma}_{10}$ = 0.08 ($p < 0.0001$). The model also has three parameter estimates of the random effects: the variance of the intercept $\hat{\sigma}^2_{u0}$ = 1.35 ($p < 0.0001$), the variance of the slope for *Time* $\hat{\sigma}^2_{u1}$ = 0.02 ($p = 0.0003$), and the covariance between the two random coefficients $\hat{\sigma}^2_{u01}$ = −0.09 ($p = 0.0006$). The significant variance of the intercept indicates that the participants' *QOL* scores significantly differ at baseline; the significant variance of the slope indicates that the rate of change in *QOL* varied significantly among the participants. The negative covariance indicates that individuals who started with higher *QOL* scores tend to have a lower rate of change over time. By adding random slope into the model, the level-1 residual variance was reduced from 1.0001 to .9124 (about 8.7%). The model has a better fit with the data than the random intercept model: −2LL has changed from 4209.3 to 4188.0. The difference is 21.3 and the improvement is significant ($p < 0.0001$ with d.f. = 2).

4.6.4 Intercept and slope as outcomes

To explore whether the initial outcome level and the rate of outcome change are functions of subject-level (level-2) covariates, and how much variation in the random intercept and slope can be explained by such covariates, *the intercept and slope as outcome model* is considered.

For our example, the *intercept and slope as outcomes model* can be formulated by including the subject-level covariate trt (1-treatment group; 0-control group) in the between-subject equations (i.e., level-2 equations):

$$\text{QOL}_{ij} = \beta_{0j} + \beta_{1j}\text{Time}_{ij} + e_{ij} \tag{4.6.4.1}$$

$$\beta_{0j} = \gamma_{00} + \gamma_{01}\text{Trt}_j + u_{0j} \tag{4.6.4.2}$$

$$\beta_{1j} = \gamma_{10} + \gamma_{11}\text{Trt}_j + u_{1j} \tag{4.6.4.3}$$

$$\text{QOL}_{ij} = \gamma_{00} + \gamma_{01}\text{Trt}_j + \gamma_{10}\text{Time}_{ij} + \gamma_{11}\text{Trt}_j * \text{Time}_{ij} + (u_{0j} + u_{1j}\text{Time}_{ij} + e_{ij}) \tag{4.6.4.4}$$

where γ_{00} is the overall average initial level of *QOL* for the individuals in the control group (i.e., Trt = 0), and γ_{01} is the difference in the average initial level of *QOL* between treatment group and control group; Similarly, γ_{10} denotes the average rate of change for the control group (i.e., Trt = 0), and γ_{11} is the difference in rate of change between the treatment group (i.e., Trt = 1) and the control group. The SAS program for the model follows:

***SAS Program 4.6.4.1;**

```
...
Proc Mixed METHOD=REML covtest noclprint IC;
```

```
Class ID;
Model QOL =Trt|Time / S DDFM=KR notest;
Random Int Time/subject=ID G TYPE=UN;
Run;
```

where the main effects of variable *Time* and *Trt*, as well as their interaction, is specified by *Tret | Time* in the MODEL statement; this statement is identical to statement *Trt Time Trt*Time*. Selected model results are displayed below:

SAS Output 4.6.4.1 Selected results of the random intercept/ slope as outcome model

Covariance Parameter Estimates

Cov Parm	Subject	Estimate	Standard Error	Z Value	Pr Z
UN(1,1)	ID	1.3464	0.1798	7.49	<.0001
UN(2,1)	ID	-0.08877	0.02610	-3.40	0.0007
UN(2,2)	ID	0.01863	0.005561	3.35	0.0004
Residual		0.9127	0.04213	21.66	<.0001

Information Criteria

Neg2LogLike	Parms	AIC	AICC	HQIC	BIC	CAIC
4193.4	4	4201.4	4201.4	4206.8	4214.6	4218.6

Solution for Fixed-Effects

Effect	Estimate	Standard Error	DF	t Value	Pr > |t|
Intercept	4.2262	0.1358	203	31.12	<.0001
Trt	-0.1771	0.1884	202	-0.94	0.3483
Time	0.05655	0.02346	199	2.41	0.0168
Trt*Time	0.04111	0.03264	199	1.26	0.2094

The intercept of the model ($\hat{\gamma}_{10}$ = 4.23) represents the overall mean *QOL* score for the control group at the baseline. The main effect of *Trt* ($\hat{\gamma}_{01}$ = −0.18) shows that the baseline mean *QOL* score was a little lower for the treatment group, compared with the control group. However, this difference is not statistically significant (p = 0.3483) as participants were randomly assigned into the two groups. The main effect of *Time* ($\hat{\gamma}_{10}$ = 0.06, p = 0.0168) represents the rate of change in *QOL*, or average increase in *QOL* score per 6 month after the baseline for the control group (*Trt* = 0). The positive interaction ($\hat{\gamma}_{11}$ = 0.04) between *Trt* and *Time* stands for additional increase in *QOL* score for the treatment group. People in the treatment group may have been more likely to have improved *QOL* scores over time than the control group, but the difference was not statistically significant (p = 0.2094).

The overall fit for the current model was not significantly improved by adding the variable *Trt*. The difference in the ML estimated −2LL between the current model and the random intercept-slope model is 4178.2 −4176.6 = 1.6, d.f. = 2, *p*-value = 0.4493,[2] which is not statistically significant. From a model fit perspective it is not necessary to add *Trt* into the model. However, from a mental health research perspective it is meaningful to include *Trt* in the model. One goal of the NHDD project is to test the effect of the treatment. For this reason, we kept this variable in the model.

The *Covariance Parameter Estimates* section in SAS Output 4.6.4.1 show that the variances of the random intercept and slope coefficients remained essentially unchanged. This indicates that the subject-level covariate *Trt* did not explain much the variations in the level-1 random intercept and slope coefficients.

4.6.5 Controlling for individual background variables in models

In real research settings, in order to separate the effect of the intervention from other "noise", we should control for the theoretically relevant individual background variables in the model as covariates. For our data, some demographic variables such as *Age*, *Gender*, *Edu*, and *Diagnosis*, are likely to be associated with the outcome, and might confound intervention effects. Therefore, we included these covariates into the model:

$$\text{QOL}_{ij} = \beta_{0j} + \beta_{1j}\text{Time}_{ij} + e_{ij} \tag{4.6.5.1}$$

$$\beta_{0j} = \gamma_{00} + \gamma_{01}\text{Trt}_j + \gamma_{02}\text{Age}_j + \gamma_{03}\text{Gender}_j + \gamma_{04}\text{Edu}_j + \gamma_{05}\text{Diagnosis}_j + u_{0j} \tag{4.6.5.2}$$

$$\beta_{1j} = \gamma_{10} + \gamma_{11}\text{Trt}_j + u_{1j} \tag{4.6.5.3}$$

$$\begin{aligned}\text{QOL}_{ij} = {} & \gamma_{00} + \gamma_{01}\text{Trt}_j + \gamma_{10}\text{Time}_{ij} + \gamma_{11}\text{Trt}_j * \text{Time}_{ij} \\ & + \gamma_{02}\text{Age}_j + \gamma_{03}\text{Gender}_j + \gamma_{04}\text{Edu}_j + \gamma_{05}\text{Diagnosis}_j + (u_{0j} + u_{1j}\text{Time}_{ij} + e_{ij})\end{aligned} \tag{4.6.5.4}$$

In this model we assume that the individual background variables (*Age*, *Gender*, *Edu*, and *Diagnosis*) may influence the outcome, but that their effects don't change over time. That is, these variables are assumed to have no interactions with the time variable[3], and therefore, they were not included in eq. 4.6.5.3. The corresponding SAS program follows:

***SAS Program 4.6.5.1;**

```
...
Proc Mixed covtest noclprint IC;
Class id;
ID ID Time QOL;
Model QOL = Trt|Time Age Gender Edu Diagnosis/ S DDFM = KR notest;
Random int Time/subject = ID G TYPE = UN;
Run;
```

[2] Note that SAS Output 4.6.3.1 and Output 4.6.4.1 were model results estimated using REML. As these two models have the same random effects, but different fixed-effects, REML results should not be used to compare the model fit. Instead, −2LL estimated with ML should be used for such model comparisons (see Section 3.3 in Chapter 3).

[3] Our exploratory modeling results show that this assumption holds.

SAS Output 4.6.5.1 contains selected model results. *Age* had a significant negative effect on the outcome indicating that younger people were likely to report higher *QOL*. None of the other individual background variables had significant effects on outcome. The estimated intercept represents the mean *QOL* score of the reference group at *Time* = 0, *Trt* = 0, *Gender* = 0, *Edu* = 0, *Diagnosis* = 0, and *Age* = 0. Since it is not meaningful to have an age of 0 in the sample under study, the intercept estimate is not meaningful, though the slope is interpretable as the change in *QOL* for each additional year of age of the respondent. As discussed in Chapter 3, a grand-mean centered measure of age should be used. We leave this variable recoding and remodeling to the readers as an exercise.

SAS Output 4.6.5.1 Selected results of random intercept-slope as outcome model controlling for background covariates

Covariance Parameter Estimates

Cov Parm	Subject	Estimate	Standard Error	Z Value	Pr Z
UN(1,1)	ID	1.2895	0.1770	7.28	<.0001
UN(2,1)	ID	-0.08855	0.02614	-3.39	0.0007
UN(2,2)	ID	0.01875	0.005572	3.37	0.0004
Residual		0.9124	0.04211	21.67	<.0001

Information Criteria

Neg2LogLike	Parms	AIC	AICC	HQIC	BIC	CAIC
4192.4	4	4200.4	4200.5	4205.8	4213.7	4217.7

Solution for Fixed-Effects

Effect	Estimate	Standard Error	DF	t Value	Pr > |t|
Intercept	5.1171	0.4178	214	12.25	<.0001
Trt	-0.1091	0.1884	200	-0.58	0.5631
Time	0.05648	0.02348	199	2.41	0.0171
Trt*time	0.04090	0.03268	199	1.25	0.2122
AGE	-0.03189	0.009432	198	-3.38	0.0009
Gender	0.1631	0.1690	197	0.97	0.3356
EDU	0.08749	0.1504	199	0.58	0.5613
DIAGNOSIS	-0.03580	0.1745	196	-0.21	0.8377

4.6.6 Coding time score

An important issue in growth modeling is selecting an appropriate measure for time, or specifying appropriate coding for variable *Time*. Different coding schemes may require different interpretations of model estimates. In our previous models we coded *Time* as linear scores with 0 at the baseline (e.g., 0, 1, 2, 3, 4, 5, 6), thus we define the first time point or baseline as the centering point or reference point. With this coding scheme in *the intercept-slope as outcome model*, the overall mean intercept coefficient represents overall average value of the outcome measure in the control group at the baseline; while the main effect of *Trt* refers to the difference between the treatment and control groups at the baseline.

Depending on researcher's interests, different time points can be picked as the centering point. Horizontal shifts of the time axis result in different intercepts for the growth curves. For example, we may treat the mid-point of the observation period, in this case the 18th month of the study, as the reference point by coding time variable as: *Time_mid* = −3, −2, −1, 0, 1, 2, 3.

In randomized studies outcome measures are expected to be approximately equal between the treatment and control groups at the baseline, and researchers are usually interested in the outcome difference between the groups at the end of study period. Thus, it may be more interesting to treat the endpoint as the reference time so that treatment effect would represent the outcome difference between the treatment and control groups at the end of the study. To do this we need to define the time variable as: *Time_end* = −6, −5, −4, −3, −2, −1, 0.[4] By treating the 36th month, i.e., the endpoint of the observation period in this example, as the centering time point, the SAS program follows:

***SAS Program 4.6.6.1**;

```
...
Time_end=Time-6; *recode time score;
Proc Mixed covtest noclprint IC;
Class ID;
ID ID Time QOL;
Model QOL = Trt|Time_end / S DDFM=KR notest;
Random Int Time_end/subject=ID G TYPE=UN;
Run;
```

SAS Output 4.6.6.1 Selected results of model with rescaled time-score

Covariance Parameter Estimates

Cov Parm	Subject	Estimate	Standard Error	Z Value	Pr Z
UN(1,1)	ID	0.9519	0.1415	6.73	<.0001
UN(2,1)	ID	0.02302	0.02184	1.05	0.2920
UN(2,2)	ID	0.01863	0.005561	3.35	0.0004
Residual		0.9127	0.04213	21.66	<.0001

Information Criteria

Neg2LogLike	Parms	AIC	AICC	HQIC	BIC	CAIC
4193.4	4	4201.4	4201.4	4206.8	4214.6	4218.6

[4] By horizontally shifting the centering time point to different locations of time measurement, the intercept estimate will change, and at the same time, the variance components of the model, specifically the intercept variance, and the covariance between intercept and slope will also change.

```
                    Solution for Fixed-Effects
                              Standard
Effect           Estimate        Error      DF    t Value    Pr> |t|

Intercept          4.5655       0.1200     201      38.03     <.0001
Trt               0.06958       0.1674     202       0.42     0.6781
Time_end          0.05655      0.02346     199       2.41     0.0168
Trt*Time_end      0.04111      0.03264     199       1.26     0.2094
```

From the output we can see that the main effect of variable *Time*, as well as its interaction with *Trt* (i.e., *Time* * *Trt*), are identical to those in SAS Output 4.6.4.1. However, the intercept and main effect of treatment are different from previous estimates. The model intercept now reflects the mean outcome of the control group at the end of the observation period, and the main effect of *Trt* represents the mean outcome difference between the treatment and control groups at that point. SAS Output 4.6.4.1 showed that the control group had a slightly higher, 0.18 point, average *QOL* score at the baseline. In this model the experimental treatment group has a slightly higher, 0.07 point, *QOL* mean score at the end of the observation period. The positive main effect of treatment in the current model (0.07, $p = 0.6781$) indicates that participants in the experimental group may have improved faster in their general *QOL* scores than those in the control group, but the difference was not statistically significant.

4.6.7 Residual variance/covariance structures

Up to this point we have focused on modeling variation in outcome between subjects by including random intercept and slope coefficients in the models. This approach is referred to as a *random coefficient model* (Littell et al., 2006). Alternatively, we can model variance from a within-subject perspective by imposing an appropriate residual variance/covariance structure in the *R*-matrix. This approach is sometimes called a *covariance pattern model* (Fitzmaurice et al., 2004; Hedeker & Gibbons, 2006). In fact, the *random coefficient* may be regarded as a type of covariance structure (Hedeker & Gibbons, 2006; Jennrich & Schluchter, 1986).

In order to identify the appropriate within-subject covariance structure, it is helpful to inspect the unstructured residual (or within-subject) variance-covariance and correlation matrix by running the following SAS program:

***SAS Program 4.6.7.1;**

```
...
Timec = Time;
Proc Mixed covtest noclprint ic;
Class ID Trt Timec;
Model QOL = Trt|Time/S DDFM = KR;
Repeated Timec/subject = ID R RCORR TYPE = UN;
Run;
```

where the new variable "*Timec*" is copied from the variable *Time*. *Timec* is treated as a categorical or classification variable specified in the *CLASS* statement. Putting this variable in the

REPEATED statement in *PROC MIXED* procedure, SAS created a set of dummy measures of time to ensure data are lined up by measurement time points in case there are any missing values. The *REPEATED* statement is used instead of the *RANDOM* statement to fit the covariance structure model. The *REPEATED* statement instructs *PROC MIXED* to impose certain structures for the residual variances/covariance matrix (*R* matrix). The *RANDOM* statement models between-subject variation while the *REPEATED* statement models within-subject variation. The options *R* and *RCORR* in the *REPEATED* statement request SAS to print out the residual variance/covariance matrix and residual correlation matrix, respectively. Table 4.6.7.1 exhibits the variance/covariance (upper right off-diagonal) and correlation matrix (lower left off-diagonal) output from this program with *TYPE* = *UN* option. This table shows that variance on the diagonal is not constant, and that correlations in repeated outcome measures decrease as the time lag increases, but the pattern is not consistent.

Table 4.6.7.1 Unstructured residual variance/covariance and correlation matrix

Time	0	1	2	3	4	5	6
0	2.2627	1.1825	1.1348	0.9363	0.9669	0.8117	0.7597
1	0.5224	2.2649	1.2785	0.8989	1.0656	0.7988	0.8744
2	0.5366	0.6042	1.9770	1.0938	1.0886	0.9796	0.8677
3	0.4520	0.4338	0.5650	1.8960	1.0685	0.8573	0.8070
4	0.4532	0.4992	0.5458	0.5470	2.0121	1.0540	0.9610
5	0.4023	0.3958	0.5195	0.4642	0.5540	1.7986	0.9061
6	0.4032	0.4638	0.4927	0.4678	0.5408	0.5394	1.5692

Note:
The figures on the diagonal are residual variances, those in the upper right off-diagonal are residual covariances, and those in the lower left off-diagonal are residual correlations.

Table 4.6.7.1 provides useful information for specification of residual variance/covariance structure. A rigorous way to choose an appropriate residual variance/covariance structure is to conduct a formal statistical test. The SAS *PROC MIXED* procedure provides over twenty optional residual variance/covariance structures. Here we fit models with five frequently applied residual variance/covariance structures: unstructured (UN), compound symmetry (CS), the first order auto-regressive (AR(1)), Toeplitz (TOEP), and Huynh-Feldt (HF). We could fit the model with different residual variance/covariance structures, one at a time. The following SAS macro allows us to fit several models with different variance/covariance structures sequentially with less redundant coding.

***SAS Program 4.6.7.2;**

```
...
Timec=Time;
%Macro Fit(COV);
ODS exclude ClassLevels IterHistory;
Proc Mixed covtest noclprint ic;
  Class ID Timec;
  Model QOL= Trt|Time / S DDFM=KR;
  Repeated Timec/subject=ID R TYPE=&COV;
Run;
%Mend Fit;
%Fit(UN)
%Fit(CS)
%Fit(AR(1))
```

```
%Fit(TOEP)
%Fit(HF)
```

where the *TYPE* option defined a parameter (i.e., *COV*) for the SAS macro *%macro Fit(COV)*. The *UN* option requests an unrestricted variance/covariance structure in which all variance/covariance parameters are estimated. The *CS* option assumes equal variances and covariances, in which only two parameters—common variance and common covariance—are to be estimated. The *AR* (1) option specifies a first-order auto-regressive structure, frequently used in time series data. *AR* (1) assumes equal residual variance and exponentially declining correlations between time lag residuals over time. In *AR* (1) two parameters—the common variance and serial correlation coefficient—are estimated. The *TOEP* option for time series data specifies common variance but arbitrary correlations for each time lag. *TOEP* is less restrictive than *AR* (1) because no assumption is made about serial correlation declining exponentially over time. The *HF* option is used to specify heterogeneous compound-symmetry residual structure in which variances on diagonal are different, and the covariance between any two elements is equal to the average of their variances minus a constant. We may say that *AR*(1) is a special case of the *TOEP* structure, and *CS* is a special case of the *HF* structure.

The *random intercept-slope model* (*RIS*) discussed in early sections of this chapter appears to be different from the residual variance/covariance pattern models discussed here. However, from a "variance component" perspective, all the models including the *RIS* model are alternative ways to model total outcome variance. The difference is that in a random coefficient model we partition the overall variance into two components: the random effects or between-subject variation component (i.e., ZGZ' in eq. 4.4.7), and the residual variance or measurement error component (i.e., R in eq. 4.4.7). Analysts often assume an identity residual matrix ($R = \sigma^2 I$) for the random intercept-slope model.

The *random coefficient model* and *residual covariance pattern model* are considered partial or "submodels". Partial models may not always be adequate. For example, the *RIS* model assumes an identity matrix (i.e., $R = \sigma^2 I$) for the "within-subject" residual variance/covariance structure, conditional on random regression coefficients. For some data, however, there may still be some remaining residual variation to explain. As such, the two sub-models can be combined to fit a "full model" in which outcome total variance would be modeled from both between-subject and within-subject perspectives. Accordingly, in addition to fitting five models with different covariance structures we also fitted a series of "full models" with different combinations of random effects and residual variance/covariance structures. In this case the model with random intercept-slope plus *AR*(1) structure (*RIS-AR*(1)) was the best choice among the full models.

Model fit information for six partial models (five covariance pattern models and the *RIS* model) and the full model (*RIS-AR*(1) model) is summarized in Table 4.6.7.2. Among the partial models, the *RIS* model has better fit than other models as they have smaller information criteria indexes (*AIC*, *AICC* and *SBC*). However, comparing the *RIS* model with the full model, the latter appears to have a better fit. Since *RIS-AR* (1) and *RIS* models are nested, the difference in −2LL was also used to compare these models. If there is no statistically significant difference between the two models it would not be necessary to use the more complex model with combinations of *RIS* and *AR* (1), and a more parsimonious *RIS* model should be selected as the final model.

In this example, the likelihood ratio test turned out to be significant ($\chi^2 = 4193.4 - 4187.2 = 6.2$, d.f. = 1, $p = 0.0128$), indicating that the *RIS-AR*(1) model is preferred over the *RIS* model. Therefore, *RIS-AR* (1) was selected as the final model. The SAS program for the *RIS-AR*(1) model follows:

Table 4.6.7.2 Model fit information: partial models vs. the full model

Variance/ Covariance Structure	−2LL	AIC	AICC	SBC
AR(1)	4304.8	4308.8	4308.8	4315.4
CS	4214.1	4218.1	4218.1	4224.7
HF	4205.5	4221.5	4221.6	4248.0
RIS	4193.4	4201.4	4201.4	4214.6
UN*	4167.0	4223.0	4224.3	4315.8
TOEP	4188.8	4202.8	4202.9	4226.0
RIS-AR(1)	4187.2	4197.2	4197.3	4213.8

Notes:
* Because the UN model has a saturated residual variance/covariance matrix, its −2LL statistics are usually lower than those of other models. The information criteria indices, which are adjusted for the number of model parameters should be used to compare this model with others.

***SAS Program 4.6.7.3;**

```
...
Timec=Time;
Proc Mixed covtest noclprint IC;
  Class ID Timec;
  Model QOL = Trt|Time / S DDFM = KR notest;
  Random Int Time/subject = ID G type = UN;
  Repeated Timec/subject = ID R type = AR(1);
Run;
```

where the *RANDOM* statement specifies the *RIS* model, the *REPEATED* statement specifies residual variance/covariance structure. The selected results of the *RIS-AR*(1) model are shown in SAS Output 4.6.7.3.

SAS Output 4.6.7.3 Selected results of the RIS-AR(1) model

```
                    Covariance Parameter Estimates
                                          Standard        Z
Cov Parm      Subject    Estimate          Error      Value       Pr Z

 UN(1,1)         ID        1.2366         0.1874       6.60     <.0001
 UN(2,1)         ID      -0.06644        0.02816      -2.36     0.0183
 UN(2,2)         ID       0.01107       0.006648       1.67     0.0479
   AR(1)         ID        0.1174        0.04886       2.40     0.0163
Residual                   0.9847        0.06010      16.38     <.0001


                         Information Criteria
Neg2LogLike    Parms        AIC       AICC      HQIC       BIC      CAIC
     4187.2        5     4197.2     4197.3    4203.9    4213.8    4218.8
```

Solution for Fixed-Effects

Effect	Estimate	Standard Error	DF	t Value	Pr>|t|
Intercept	4.2238	0.1356	204	31.16	<.0001
Trt	-0.1825	0.1880	202	-0.97	0.3331
Time	0.05621	0.02337	199	2.41	0.0171
Trt*time	0.04269	0.03252	199	1.31	0.1908

The *Covariance Parameter Estimates* section of SAS Output 4.6.7.3 shows that random-intercept, random slope and their covariance are statistically significant. The *AR* (1) parameter is also statistically significant. This indicates that it is beneficial to model between-subject variation, using random effects, and to model within-subject variation, using the *AR* (1) structure, simultaneously. Although the full model (the *RIS-AR* (1) model) fits data better, the estimates of the fixed-effects from the full model are not that much different from those of the random intercept-slope (*RIS*) model, indicating that the *RIS* model in this example is fairly robust for estimating the fixed-effects.

4.6.8 Time-varying covariates

Growth models can readily accommodate time-varying covariates. Theoretically speaking, the score of variable *SATS* (Substance Abuse Treatment Scale) may be positively associated with *QOL*. In our data, the score of *SATS* varies over time, thus, it can be treated as a time-varying covariate. In the following SAS program, *SATS* is added to the *RIS-AR* (1) model (for simplicity, background covariates are not included in the model):

***SAS Program 4.6.8.1**;

```
...
Timec = Time;
Proc Mixed covtest noclprint IC;
  Class ID Timec;
  Model QOL = Trt|Time SATS/ S DDFM = KR notest;
  Random Int Time/subject = ID G type = UN;
  Repeated Timec/subject = ID R type = AR(1);
Run;
```

SAS Output 4.6.8.1 Selected results of the RIS-AR(1)model with time-varying covariate

Covariance Parameter Estimates

Cov Parm	Subject	Estimate	Standard Error	Z Value	Pr Z
UN(1,1)	ID	1.2253	0.1844	6.65	<.0001
UN(2,1)	ID	-0.05997	0.02719	-2.21	0.0274
UN(2,2)	ID	0.009505	0.006407	1.48	0.0690

```
AR(1)       ID              0.1127      0.04879      2.31    0.0209
Residual                    0.9636      0.05861     16.44    <.0001

                         Information Criteria
Neg2LogLike  Parms    AIC      AICC     HQIC      BIC      CAIC

4151.9       5       4161.9   4161.9   4168.6   4178.4   4183.4

                    Solution for Fixed-Effects
                               Standard
Effect          Estimate       Error      DF    t Value    Pr > |t|

Intercept         3.7891      0.1576     346      24.04     <.0001
Trt              -0.1726      0.1868     202      -0.92     0.3565
Time             0.01417     0.02413     232       0.59     0.5577
Trt*Time         0.03878     0.03180     197       1.22     0.2240
SATS              0.1357     0.02564     885       5.29     <.0001
```

The *SATS* score has a significant positive effect on the outcome measure *QOL* (fixed effect = 0.14, $p < 0.0001$). This suggests that doing well in substance abuse treatment would increase satisfaction with life in general. Note that once the time-varying measurement *SATS* was included in the model, the effect of time per se became non-significant. This indicates that improvement in life satisfaction over time may be explained by improvement in substance abuse treatment outcome (*SATS*) rather than simply the passage of time.

When a time-varying covariate is included in a model, an assumption is made that the between and within-subjects effect of this variable are equal (Hedeker & Gibbons, 2006). To test this assumption, it is possible to partition a time-varying covariate into two variables and re-run the model. A time-varying covariate can be separated into an individual's mean score on this covariate (a between-subject variable) and his or her deviation score around this mean (within-subject variable). In our example, the individual's mean *SATS* score is calculated across seven *SATS* measures taken from baseline to 36 months, and the deviation score is obtained by subtracting this mean score from individual *SATS* scores. Incorporating the mean *SATS* permits us to measure how individual average *SATS* scores are associated with individual average general life satisfaction scores. The effect of the deviation score (within-subject effect) indicates how this deviation score is related to change in the general life satisfaction score. The degree of this relationship may vary at different time points.

To test the assumption of homogeneity of between and within-subject effects of the time-varying covariate *SATS*, we re-run the model with the original time-varying covariate *SATS* (the same model as displayed in SAS output 4.6.8.1), and the model with the both mean *SATS* score and the deviation *SATS* score. The SAS program follows:

***SAS Program 4.6.8.2**;

```
...
Data Data1;
 Set Lib.ML_Chap4_data;
Timec=Time;
Proc SQL;
  Create table data2 as select*, mean(SATS) as Mean_SATS,
     SATS-mean(SATS) as Dev_SATS from data1
```

```
  Group by ID;
Quit;

Proc Mixed data = data2 METHOD = ML covtest noclprint;
Class ID Timec;
Model QOL = Trt|Time SATS/ S DDFM = KR;
Random Int Time/subject=ID G type = UN;
Repeated Timec/subject=ID R type = AR(1);
Run;

Proc Mixed data = data2 METHOD = ML covtest noclprint;
Class ID Timec;
Model QOL = Trt|Time Mean_SATS Dev_SATS/S DDFM = KR;
Random Int Time/subject = ID G type = UN;
Repeated Timec/subject = ID R type = AR(1);
Run;
```

The *PROC SQL* procedure creates new variables *Mean_SATS* and *Dev_SATS*, and saves them in the temporary SAS data file *Data2*. *PROC MIXED* estimates the *RIS-AR* (1) model with the original *SATS* measure, and the *RIS-AR*(1) model with the average and deviation *SATS* variables. As these two models have the same random effects and variance/covariance structure, but different fixed-effects, the models need to be estimated under ML for the purpose of model comparison. The ML model fit statistics for the two models are reported in the following SAS output.

SAS Output 4.6.8.2 Testing the assumption of homogeneity of the between- and within-subject effects

```
Fit Statistics for the Model with Covariate SATS

        -2 Log Likelihood            4129.5
        AIC (smaller is better)      4149.5
        AICC (smaller is better)     4149.6
        BIC (smaller is better)      4182.6

  Fit Statistics for Model with the Both Covariates
                 Mean-SATS and Dev-SATS

        -2 Log Likelihood              4127.1
        AIC (smaller is better)        4149.1
        AICC (smaller is better)       4149.3
        BIC (smaller is better)        4185.5
```

The −2LL statistic and all information criteria indices are almost the same for the two models. Therefore, the between-subject and within-subject effects of *SATS* score can be assumed equal. The model with both mean and deviation of the *SATS* score is not necessary.

4.7 Curvilinear growth models

So far, our discussions have focused on linear growth models. In many circumstances, however, changes over time may not be linear. For instance, the rate of change of the outcome may not be constant over time. In this case, curvilinear or non-linear growth models should be considered.

4.7.1 Polynomial growth model

When change occurs in a nonlinear fashion, polynomial growth models are often considered. The order of the polynomial function, such as linear, quadratic, cubic, or higher order, depends on patterns of change in outcome over time. Theoretically speaking, the order of polynomial function can go as high as ($T-1$) where T is the total number of observation time points. However, higher order models are more difficult to fit and more difficult to interpret. In this chapter we compare linear and quadratic (2nd order of the polynomial time function) growth models. Quadratic growth is characterized by one departure (either curve up or curve down) from linearity.

To fit the quadratic growth model we add a squared term of the time variable (i.e., $Time^2$ or $Time * Time$) to the linear growth model. The effect of $Time^2$ could be specified as either fixed or random depending upon model fit. As indicated previously change in our outcome measure over time, *QOL*, does not appear to be linear, yet upon analysis (not shown), the quadratic component is not significant.

For the purpose of illustration, we altered the values of the baseline outcome measure, by recoding the values of variable *QOL* greater than 5 to 5 at the baseline, to make the growth curves appear more curvilinear (See Figure 4.7.1.1). We name the new artificial outcome variable C_QOL. To test the non-linearity of outcome growth, we employ the following model:

$$\mathrm{C\_QOL}_{ij} = \beta_{0j} + \beta_{1j}\mathrm{Time}_{ij} + \beta_{2j}\mathrm{Time}^2{}_{ij} + e_{ij} \tag{4.7.1.1}$$

$$\beta_{0j} = \gamma_{00} + \gamma_{01}\mathrm{Trt}_j + u_{0j} \tag{4.7.1.2}$$

$$\beta_{1j} = \gamma_{10} + \gamma_{11}\mathrm{Trt}_j + u_{1j} \tag{4.7.1.3}$$

$$\beta_{2j} = \gamma_{20} + \gamma_{21}\mathrm{Trt}_j + u_{2j} \tag{4.7.1.4}$$

The combined model becomes:

$$\begin{aligned}\mathrm{C\_QOL}_{ij} &= \gamma_{00} + \gamma_{10}\mathrm{Time}_{ij} + \gamma_{20}\mathrm{Time}^2_{ij} + \gamma_{01}\mathrm{Trt}_j \\ &+ \gamma_{11}\mathrm{Trt}_j * \mathrm{Time}_{ij} + \gamma_{21}\mathrm{Trt}_j * \mathrm{Time}^2_{ij} \\ &+ (u_{0j} + u_{1j}\mathrm{Time}_{ij} + u_{2j}\mathrm{Time}^2_{ij} + e_{ij})\end{aligned} \tag{4.7.1.5}$$

The corresponding SAS program is presented below:

***SAS Program 4.7.1.1**;

```
...
C_QOL = QOL;
if Time = 0 and QOL>5 then C_QOL = 5;
Time2 = Time*Time;

Proc Mixed covtest noclprint IC;
Class ID;
Model C_QOL = Trt|Time Trt|Time2/S DDFM = KR notest;
Random Int Time Time2/subject = ID G type = UN;
```

```
Run;
```

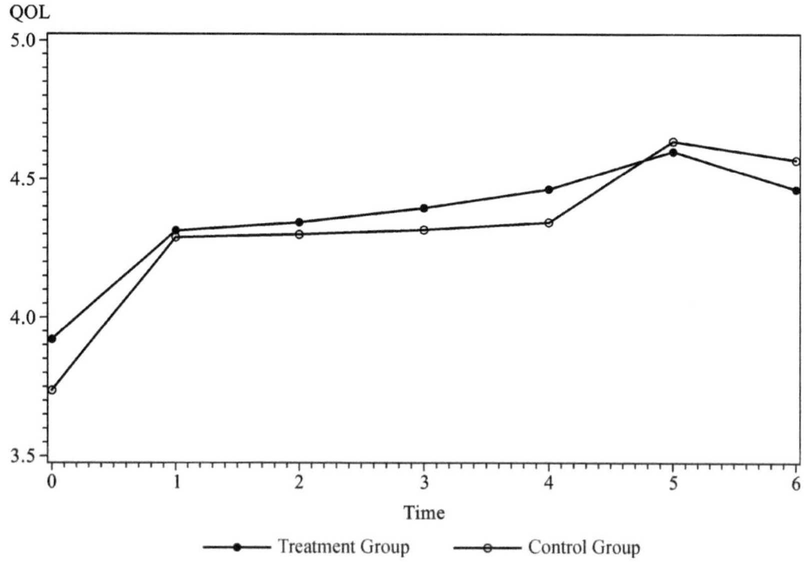

Figure. 4.7.1.1 Average general life satisfaction score over time based on altered data

where the baseline outcome values greater than 5 were recoded to 5 to create an artificial outcome variable *C_QOL*; and the new variable *Time*2 is the quadratic term of *Time*.

The hypothetical model results show that both linear and quadratic effects of *Time* are statistically significant ($\hat{\gamma}_{10} = 0.23$, $p = 0.0013$; $\hat{\gamma}_{20} = -0.02$, $p = 0.0284$), indicating a non-linear growth trajectory of the outcome. That is, the average *C_QOL* score improved over time, but not at a constant rate. The rate of improvement gradually leveled off. The variances of the slope coefficients for *Time* ($\hat{\sigma}^2_{u1} = 0.04$, $p = 0.2337$) and *Time*2 ($\hat{\sigma}^2_{u2} = 0.00$, $p = 0.3652$) are not statistically significant, indicating that both the linear and quadratic time effects may not be random. However, when we specified *Time* and *Time*2 in the *RANDOM* statement separately, eacd of the random effects was statistically significant (results not reported here). This might be an indication of potential collinearity between these two time measures.

SAS Output 4.7.1.1 Selected results of quadratic growth model

Covariance Parameter Estimates

Cov Parm	Subject	Estimate	Standard Error	Z Value	Pr Z
UN(1,1)	ID	0.8681	0.1601	5.42	<.0001
UN(2,1)	ID	0.06887	0.06844	1.01	0.3143
UN(2,2)	ID	0.03716	0.05114	0.73	0.2337
UN(3,1)	ID	-0.01923	0.01012	-1.90	0.0575
UN(3,2)	ID	-0.00341	0.007848	-0.43	0.6644
UN(3,3)	ID	0.000439	0.001275	0.34	0.3652
Residual		0.8911	0.04622	19.28	<.0001

```
                         Information Criteria
Neg2LogLike   Parms     AIC      AICC      HQIC      BIC      CAIC

   4144.7       7     4158.7   4158.8    4168.1   4181.9   4188.9

                    Solution for Fixed-Effects
                                Standard
  Effect          Estimate        Error     DF   t Value   Pr> |t|

  Intercept         3.9878       0.1277    205     31.23    <.0001
  Trt              -0.1296       0.1766    202     -0.73    0.4636
  Time              0.2284      0.07006    204      3.26    0.0013
  Trt*Time         0.01830      0.09672    202      0.19    0.8501
  Time2           -0.02423      0.01098    201     -2.21    0.0284
  Trt*Time2       0.002468      0.01522    201      0.16    0.8714
```

4.7.2 Dealing with collinearity in higher order polynomial growth model

A common problem with polynomial measures of time is that its linear, quadratic and higher order terms are likely to be correlated. Hedeker & Gibbons (2006) illustrate in a quadratic trend model with three time points that the linear time measure (i.e., 0, 1, 2) and the quadratic time measure (i.e., 0, 1, 4) are nearly perfectly correlated.

One way to deal with the collinearity is to center the time variable or choose a mid-point of the observation period as the reference point for time. By centering *Time* the linear and quadratic time measures become as: −1, 0, 1 and 1, 0, 1, respectively, and the correlation between the two time measures can be removed.

In our example, the original time variable was coded as 0, 1, 2, 3, 4, 5, 6. By centering time (i.e., the 18th month follow-up in this example), the new time variable *Time_mid* takes the values of −3, −2, −1, 0, 1, 2, 3; and its quadratic term *Time_mid2* has values of 9, 4, 1, 0, 1, 4, 9. The model with the centered time variables was run with the following SAS program:

```
*SAS Program 4.7.2.1;
...
C_QOL=QOL;
if Time = 0 and QOL > 5 then C_QOL= 5;
Time_mid=Time-3;
Time_mid2=Time_mid*Time_mid;

Proc Mixed METHOD=REML covtest noclprint IC;
Class ID;
Model C_QOL= Trt|Time_mid Trt|Time_mid2/S DDFM=KR notest;
Random Int Time_mid Time_mid2/subject=ID G type=UN;
Run;
```

SAS Output 4.7.2.1 presents selected results from this hypothetical model. The variance of the slope coefficient for the centered linear time measure, *Time_mid,* is significant ($\hat{\sigma}^2_{u1} = 0.01$, $p =$ 0.0068). However, the variance of the slope coefficient for quadratic time variable, *Time_mid2* remained non-significant ($\hat{\sigma}^2_{u2} = 0.00$, $p = 0.3653$). This indicates that *Time_mid2* may not have a random effect. For fixed-effects, the quadratic time main effect remained the same as in the previous model; but the linear time main effect is different now because of controlling for collinearity between the linear and quadratic time measures. In short, centering time at the midpoint of the observation period or shifting the model intercept into the middle, resolves the collinearity between time measures.

SAS Output 4.7.2.1 Selected results of quadratic growth model with centered time score

Covariance Parameter Estimates

Cov Parm	Subject	Estimate	Standard Error	Z Value	Pr Z
UN(1,1)	ID	1.1213	0.1443	7.77	<.0001
UN(2,1)	ID	-0.00325	0.01804	-0.18	0.8571
UN(2,2)	ID	0.01211	0.004907	2.47	0.0068
UN(3,1)	ID	-0.02549	0.01061	-2.40	0.0163
UN(3,2)	ID	-0.00077	0.001623	-0.47	0.6355
UN(3,3)	ID	0.000439	0.001275	0.34	0.3653
Residual		0.8911	0.04622	19.28	<.0001

Information Criteria

Neg2LogLike	Parms	AIC	AICC	HQIC	BIC	CAIC
4144.7	7	4158.7	4158.8	4168.1	4181.9	4188.9

Solution for Fixed-Effects

Effect	Estimate	Standard Error	DF	t Value	Pr > |t|
Intercept	4.4551	0.1215	201	36.68	<.0001
Trt	-0.05253	0.1686	201	-0.31	0.7557
Time_mid	0.08304	0.02177	198	3.82	0.0002
Trt*Time_mid	0.03311	0.03030	198	1.09	0.2759
Time_mid2	-0.02423	0.01098	201	-2.21	0.0284
Trt*Time_mid2	0.002468	0.01522	201	0.16	0.8714

Another approach for avoiding collinearity in polynomial time functions is to use orthogonal polynomials. For equal time intervals, the values of standard integer orthogonal polynomials can be found in some statistics texts. In our example, for data with 7 time points (0, 1, 2, 3, 4, 5, 6), the coefficients of *integer* orthogonal polynomials are:

```
Ioptime0  Ioptime1  Ioptime2

    1        -3         5
    1        -2         0
    1        -1        -3
    1         0        -4
    1         1        -3
    1         2         0
    1         3         5
```

where *Ioptime*0, *Ioptime*1 and *Ioptime*2 are zero degree, first degree, and second degree orthogonal polynomial coefficients, respectively. Their corresponding regression coefficients would be the model intercept, linear effect, and quadratic effect of *Time*. Replacing the time variables *Time_mid* and *Time_mid*2 in SAS Program 4.7.2.1 with *Ioptime*1 and *Ioptime*2—the integer orthogonal polynomials of *Time* and *Time*2 (see SAS Program 4.7.2.2), the time effects remain the same as the model with a centered time variable (see SAS Output 4.7.2.1).

***SAS Program 4.7.2.2**;

```
...
Timec = Time;
C_QOL = QOL;
if Time = 0 and QOL > 5 then C_QOL = 5;
Time2 = Time*Time;

* Linear term;
if Time = 0 then Ioptime1 = −3;
if Time = 1 then Ioptime1 = −2;
if Time = 2 then Ioptime1 = −1;
if Time = 3 then Ioptime1 = 0;
if Time = 4 then Ioptime1 = 1;
if Time = 5 then Ioptime1 = 2;
if Time = 6 then Ioptime1 = 3;

* Quadratic term;
if Time = 0 then Ioptime2 = 5;
if Time = 1 then Ioptime2 = 0;
if Time = 2 then Ioptime2 = −3;
if Time = 3 then Ioptime2 = −4;
if Time = 4 then Ioptime2 = −3;
if Time = 5 then Ioptime2 = 0;
if Time = 6 then Ioptime2 = 5;

Proc Mixed METHOD = REML covtest noclprint IC;
  Class ID Timec;
  Model C_QOL = Trt|Ioptime1 trt|Ioptime2/S DDFM = KR;
  Random Int Ioptime1 Ioptime2/subject=ID G type = UN;
Run;
```

SAS Output 4.7.2.2 Selected results of quadratic growth model with integer orthogonal polynomial coefficients

Covariance Parameter Estimates

Cov Parm	Subject	Estimate	Standard Error	Z Value	Pr Z
UN(1,1)	ID	0.9244	0.1059	8.73	<.0001
UN(2,1)	ID	-0.00633	0.01551	-0.41	0.6833
UN(2,2)	ID	0.01211	0.004907	2.47	0.0068
UN(3,1)	ID	-0.02373	0.007984	-2.97	0.0030
UN(3,2)	ID	-0.00077	0.001623	-0.47	0.6355
UN(3,3)	ID	0.000439	0.001275	0.34	0.3653
Residual		0.8911	0.04622	19.28	<.0001

Information Criteria

Neg2LogLike	Parms	AIC	AICC	HQIC	BIC	CAIC
4144.7	7	4158.7	4158.8	4168.1	4181.9	4188.9

Solution for Fixed-Effects

Effect	Estimate	Standard Error	DF	t Value	Pr > |t|
Intercept	4.3581	0.1042	202	41.82	<.0001
Trt	-0.04266	0.1448	202	-0.29	0.7686
Ioptime1	0.08304	0.02177	198	3.82	0.0002
Trt*Ioptime1	0.03311	0.03030	198	1.09	0.2759
Ioptime2	-0.02423	0.01098	201	-2.21	0.0284
Trt*Ioptime2	0.002468	0.01522	201	0.16	0.8714

The integer orthogonal polynomial coefficients used in SAS Program 4.7.2.2 are valid when time intervals are equal. A more general approach of calculating orthogonal polynomial coefficients for equal time or unequal time intervals is discussed and illustrated by Hedeker & Gibbons (2006). We can use the *ORPOL* function in SAS *IML* to compute these general orthogonal polynomial coefficients. The following SAS program can be used to generate orthogonal second order polynomials for the 7 time points in our example:

***SAS Program 4.7.2.3**;

```
Proc IML;
Time = {0,1,2,3,4,5,6};
Op_Time = ORPOL(Time,2);
mattrib Op_time format = 6.4;
Print Op_time;
Quit;
```

where a vector *Time*, the original time variable, is created first, and the *ORPOL* function in SAS interactive matrix language *(IML)* procedure is used to generate orthogonal polynomial coefficients. The option "2" specified in the *ORPOL* function requests SAS to produce orthogonal polynomial coefficients with the quadratic term as the highest order.

SAS Output 4.7.2.3 Orthogonal second order polynomial coefficients generated using SAS *ORPOL* function

```
          OP_TIME
0.3780    -.5669    0.5455
0.3780    -.3780    -.0000
0.3780    -.1890    -.3273
0.3780    0.0000    -.4364
0.3780    0.1890    -.3273
0.3780    0.3780    -.0000
0.3780    0.5669    0.5455
```

The corresponding SAS program for this quadratic growth model follows:

***SAS Program 4.7.2.4**;

```
...
C_QOL=QOL; if Time = 0 and QOL > 5 then C_QOL = 5;
optime0 = 0.3780;
if Time =.then Optime0 =.;

* Linear term;
if Time =0 then Optime1 = −.5669;
if Time =1 then Optime1 = −.3780;
if Time = 2 then Optime1 = −.1890;
if Time = 3 then Optime1 = 0;
if Time = 4 then Optime1 = 0.1890;
if Time = 5 then Optime1 = 0.3780;
if Time = 6 then Optime1 = 0.5669;

* Quadratic term;
if Time = 0 then Optime2 = 0.5455;
if Time = 1 then Optime2 = 0;
if Time = 2 then Optime2 = −.3273;
if Time = 3 then Optime2 = −.4364;
if Time = 4 then Optime2 = −.3273;
if Time = 5 then Optime2 = 0;
if Time = 6 then Optime2 = 0.5455;

Proc Mixed METHOD=REML covtest noclprint IC;
  Class ID ;
  Model C_QOL = Optime0 Trt|Optime1 Trt|Optime2/S DDFM = KR NOINT notest;
  Random Int Optime1 Optime2/subject=ID G type=UN;
Run;
```

where option *NOINT* in the *MODEL* statement is used to suppress model default intercept and the coefficient of the zero-order time measure, *Optime*0, will be the intercept.[5] The model output follows.

SAS Output 4.7.2.4 Selected results of quadratic growth model with general orthogonal polynomial coefficients

Covariance Parameter Estimates

Cov Parm	Subject	Estimate	Standard Error	Z Value	Pr Z
UN(1,1)	ID	0.9243	0.1059	8.73	<.0001
UN(2,1)	ID	-0.03350	0.08204	-0.41	0.6830
UN(2,2)	ID	0.3389	0.1374	2.47	0.0068
UN(3,1)	ID	-0.2175	0.07317	-2.97	0.0029
UN(3,2)	ID	-0.03730	0.07873	-0.47	0.6357
UN(3,3)	ID	0.03681	0.1071	0.34	0.3656
Residual		0.8911	0.04622	19.28	<.0001

Information Criteria

Neg2LogLike	Parms	AIC	AICC	HQIC	BIC	CAIC
4127.2	7	4141.2	4141.3	4150.6	4164.4	4171.4

Solution for Fixed-Effects

Effect	Estimate	Standard Error	DF	t Value	Pr > |t|
optime0	11.5295	0.2757	202	41.83	<.0001
Trt	-0.04266	0.1448	202	-0.29	0.7686
Optime1	0.4394	0.1152	198	3.82	0.0002
Trt*Optime1	0.1752	0.1603	198	1.09	0.2759
Optime2	-0.2221	0.1006	201	-2.21	0.0284
Trt*Optime2	0.02262	0.1395	201	0.16	0.8714

Note that the intercept estimate for this model, $\hat{\gamma}_{00} = 11.5295$ is much larger than the previous estimate (e.g., $\hat{\gamma}_{00} = 4.3581$) in SAS Output 4.7.2.2. This is because the non-integer orthogonal polynomial coefficients are rescaled measures of *Time*. In this example, the number of time points is 7, thus the zero-order polynomial coefficient is 0.3780, which is the square root of 1/7. The values in column 1 of SAS Output 4.7.2.3 are normalized, and their sum of squares equals 1.0 (see Hedeker & Gibbons, 2006). Multiplying the intercept (i.e., $\hat{\gamma}_{00} = 11.5295$) in SAS Output 4.7.2.4 by 0.3780, we have $\hat{\gamma}_{00} = 4.3582$, which is the same as the coefficient of *Time* in SAS Output 4.7.2.2. Similarly, the linear and quadratic main effects of time ($\hat{\gamma}_{10} = 0.4394$ and $\hat{\gamma}_{20} = -0.2221$) in SAS Output 4.7.2.4 are different from the previous estimates in

[5] To specify a random intercept in the model, we may also replace *Int* in the *RANDOM* statement with variable *optime*0. Either way, results will be identical.

the model with integer orthogonal polynomial coefficients ($\hat{\gamma}_{10} = 0.08304$ and $\hat{\gamma}_{20} = -0.02423$). We can multiply the coefficients by a constant, the ratio of corresponding non-integer orthogonal polynomial coefficient to the integer orthogonal polynomial coefficient, to get slope coefficient estimates (e.g., $\hat{\gamma}_{10} = 0.4394 * (0.5669/3) = 0.0830$, and $\hat{\gamma}_{20} = -0.2221 * (0.5455/5) = -0.0242$) that are almost identical to those in the previous model. The interaction effects can be translated the same way.

The time effects, including the linear and quadratic main effects, as well as their interactions with treatment estimated from the orthogonal polynomials model, are identical to those estimated from the model with centering time at the mid-time point of the observation periods. However, the intercept and treatment main effects are different between the two models. This is because the intercept in the orthogonal polynomial model represents the average outcome across time for the control group, and the treatment main effect is the difference between the treatment and control groups in average outcome across time (Hedeker & Gibbons, 2006). In the time-centered quadratic models, the intercepts and main treatment effects are the average outcome for the control group and the difference between treatment and control groups at the midpoint of the observation periods, respectively.

In conclusion, when outcome changes over time show a non-linear pattern, curvilinear growth models are recommended. In this example, a new outcome variable, *C_QOL* was created to mimic non-linear growth for the purpose of demonstrating such models. Several strategies for dealing with collinearity between polynomial time measures have been discussed. Our examples were confined to quadratic models, though in some instances higher polynomial terms may be justified.

4.7.3 Piecewise (linear spline) growth model

When change appears to have a curvilinear pattern, another approach is to model it using piecewise linear growth models. This approach breaks up the curvilinear growth into separate linear segments or pieces. Each segment will have a separate slope, but is joined or tied to adjacent segment at fixed time points. The locations where the lines meet or are tied together are known as "knots". This approach is particularly appealing in the following situations:

(1) when outcome growth process cannot be approximated by polynomial functions. (Fitzmaurice et al., 2004)

(2) when analysts are interested in comparing growth rates for two or more periods based on a substantial interest, such as growth rates before and after an intervention. (Fitzmaurice et al., 2004)

A piecewise growth curve is sometimes referred to as a linear spline. In fact, splines do not have to be linear. Spline models have an extensive body of research in statistics including automated choices for the knot locations and complex smoothing functions. In this book we restrict our discussion to linear spline models only.

To fit a piecewise model, it will be helpful to inspect the curvilinear growth pattern first to gain an idea for how many segments we should consider for modeling. In our case, Figure 4.7.1.1 suggests there are three segments in the growth process of the outcome variable (*QOL* scores): a sharp increase from the baseline to 6 month, a steady increase from 6 to 30 months, and a slight decrease from 30 to 36 months. The difference between the 2nd segment (6 to 30 month) and the 3rd segment does not seem to be substantial. For simplicity, we will ignore it and only fit the two-segment linear growth model (baseline to 6 month vs. 6 month to 36 months).

Once the number of segments is determined we need to recode the time variable to correspond to the growth trajectories for each segment. We fit the model with separate growth trajectories for each segment by including *Time*1 and *Time*2 (see Table 4.7.3.1) into the model. *Variable Time*1 measures linear change from the baseline to the 6-month follow-up. Variable *Time2* measures linear change from 6 month to 36 months. That is, the slope coefficients of *Time*1 and *Time*2 would measure the linear growth rates for segment one (baseline to 6-month follow-up) and segment two (6-month to 36-month follow-ups), respectively. Alternatively, we may include variables *Time* and *Time*2 in the model. In this case, the slope coefficient of variable *Time* would represent the base growth in the overall period, while the slope coefficient of the variable *Time*2 would represent an incremental (or decremental) growth rate in the 2nd period. In other words, the slope of *Time*2 would represent an additional change in the period from 6-month to 36-month on top of the change represented by variable *Time*.

Table 4.7.3.1 Time score coding for a piecewise model

Time Point	Time Scores		
	Time	Time1	Time2
1	0	0	0
2	1	1	0
3	2	1	1
4	3	1	2
5	4	1	3
6	5	1	4
7	6	1	5

The following SAS program fits the model with separate growth rates for the period between the baseline and the 6-month follow-up, and the period between the 6-month and the 36-month follow-ups, respectively. Again, the variable *Time*1 represents change from the baseline to 6 month, and *Time2* represents change from 6 month to 36 month. The dependent variable is *C_QOL* in which the values of the original outcome measure are altered to create a non-linear growth process.

***SAS Program 4.7.3.1**;

```
...
C_QOL = QOL;
if Time = 0 and QOL > 5 then C_QOL = 5;
If Time =0 then do;
  Time1 = 0;
  Time2 = 0;
End;
Else if Time in (1,2,3,4,5,6) then do;
  Time1 = 1;
  Time2 = Time-1;
End;

Proc Mixed covtest noclprint noitprint IC;
Class ID;
Model C_QOL = Trt|Time1 Trt|Time2
```

```
    / S DDFM = KR OUTPM = Pred_out1 (keep = Time Trt C_QOL Pred);
Random int Time1 Time2/subject = ID type = UN;
Run;
```

where the option *OUTPM* used in the *MODEL* statement saves the variables *Time*, *Trt*, and *C_QOL*, as well as the model predicted outcome *Pred*, in a temporary SAS data file, *Pred_out1* that will be used later for plotting. Here is the selected output of the piecewise model.

SAS Output 4.7.3.1 Selected results of piecewise model

Covariance Parameter Estimates

Cov Parm	Subject	Estimate	Standard Error	Z Value	Pr Z
UN(1,1)	ID	0.6819	0.1628	4.19	<.0001
UN(2,1)	ID	0.2711	0.1239	2.19	0.0287
UN(2,2)	ID	0.09264	0.1646	0.56	0.2868
UN(3,1)	ID	-0.06762	0.02470	-2.74	0.0062
UN(3,2)	ID	-0.02501	0.02736	-0.91	0.3605
UN(3,3)	ID	0.02118	0.008129	2.61	0.0046
Residual		0.8719	0.04543	19.19	<.0001

Fit Statistics

-2 Res Log Likelihood	4121.0
AIC (smaller is better)	4135.0
AICC (smaller is better)	4135.1
BIC (smaller is better)	4158.2

Null Model Likelihood Ratio Test

DF	Chi-Square	Pr > ChiSq
6	529.11	<.0001

Information Criteria

Neg2LogLike	Parms	AIC	AICC	HQIC	BIC	CAIC
4121.0	7	4135.0	4135.1	4144.4	4158.2	4165.2

Solution for Fixed-Effects

Effect	Estimate	Standard Error	DF	t Value	Pr > |t|
Intercept	3.9170	0.1281	202	30.58	<.0001
Trt	-0.1789	0.1767	200	-1.01	0.3125
Time1	0.4049	0. 1249	197	3.24	0.0014
Trt*Time1	0.1001	0.1718	195	0.58	0.5608
Time2	0.04425	0.02781	189	1.59	0.1132
Trt*Time2	0.02387	0.03884	190	0.61	0.5396

In the piecewise model, the entire growth period of the outcome under study was broken into two pieces: 1) linear growth from the baseline to the 6-month follow-up; and 2) linear growth from the 6-month to the 36-month follow-up. The main effect of *Time*1($\hat{\gamma}_{10}$ = 0.41, p = 0.0014) in *Solution for Fixed-Effects* of the SAS output is the slope or linear growth rate during the first growth period, while the main effect of *Time*2 ($\hat{\gamma}_{20}$ = 0.04, p = 0.1132) was the slope or linear growth rate during the second growth period. The results indicate that the study participants in the control group experienced a significant increase in outcome only during the first 6-month period; and improvement was not statistically significant after then. The interaction effects, i.e., *Time*1**Trt* ($\hat{\gamma}_{11}$ = 0.10, p= 0.5608) and *Time2***Trt* ($\hat{\gamma}_{21}$ = 0.02, p = 0.5396), represent the additional amount of change for the participants in the treatment group (*Trt* = 1) in the two growth periods, respectively. However, these additional changes were not statistically significant.

Alternatively, we may include the original time variable *Time* and the new time variable *Time*2 into the model. Using this approach, *Time*2 would measure additional change from the 6th month to the 36th month on top of the linear change represented by *Time*. The corresponding SAS program follows:

***SAS Program 4.7.3.2;**

```
...
C_QOL=QOL;
if Time=0 and QOL > 5 then C_QOL = 5;
If Time=0 then do;
Time1=0;
Time2=0;
End;
Else if Time in (1,2,3,4,5,6) then do;
Time1 = 1;
Time2 = Time-1;
End;
Proc Mixed covtest noclprint noitprint IC;
Class ID;
Model C_QOL = Trt|Time Trt|Time2
 /S DDFM = KR OUTPM = Pred_out2(keep = ID Time Time1 Time2 Trt C_QOL
Pred);
Random Int Time Time2/subject = ID type = UN;
Run;
```

where the effect of variable *Time* reflects base rate of change in the outcome during the entire observation period, and the effect of variable *Time*2 reflects the incremental or decremented outcome change in the 2nd observation period from the 6th month to the 36th month.

We can save the predicted outcome data from SAS Program 4.7.3.2 for plotting. The following SAS program was used to generate the plot shown in Figure 4.7.3.1[6].

***SAS Program 4.7.3.3;**

```
...
C_QOL=QOL;
if Time = 0 and QOL > 5 then C_QOL= 5;
If Time = 0 then do;
```

[6] Using data file *Pred_out*1 created by SAS Program 4.7.3.1 and data file *Pred_out*2 by SAS Program 4.7.3.2 will generate an identical plot.

```
    Time1 = 0;
    Time2 = 0;
  End;
```

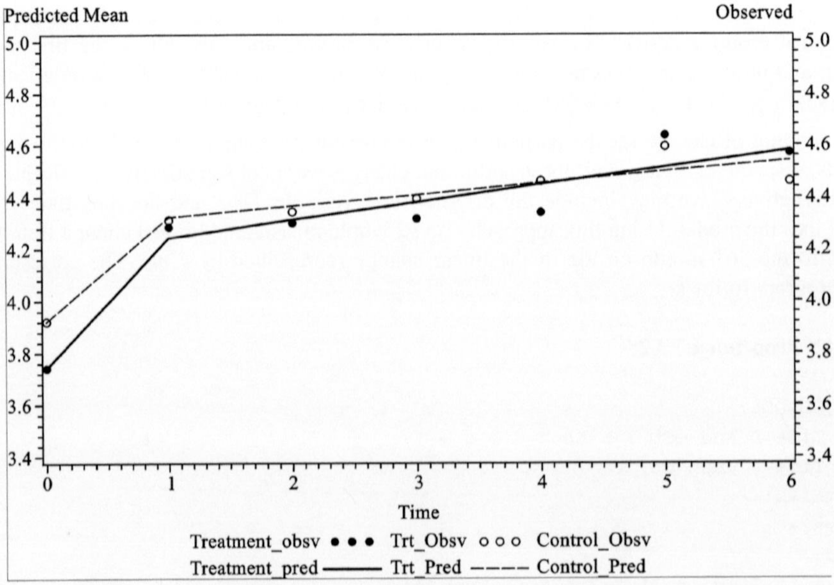

Figure 4.7.3.1 Fitted mean life satisfaction scores over time with piecewise model

```
  Else if Time in (1,2,3,4,5,6) then do;
    Time1 = 1;
    Time2 = Time-1;
  End;

  Proc Mixed covtest noclprint noitprint IC;
  Class ID;
  Model C_QOL = Trt|Time Trt|Time2
    /S DDFM = KR OUTPM = Pred_out2(keep=ID Time Time1 Time2 Trt C_QOL Pred);
  Random Int Time Time2/subject = ID type = UN;
  Run;

  Proc Means Data=Pred_out2 noprint nway;
  Var C_QOL  Pred;
  Class Trt Time;
  Output OUT=predmean mean=C_QOL_M  Predm;
  Run;

  Proc Print Data=predmean (drop= _type_ _freq_);
  Format _numeric_ f4.2; *clearing format of numerical variables and refor-
mat them;
  Run;

  Data Pdata;
```

```
Set Predmean;
group_ob = Trt; group_pr = Trt;
Proc Format;
Value group_ob 0 = 'Control_Obsv' 1 = 'Trt_Obsv';
Value group_pr 0 = 'Control_Pred' 1 = 'Trt_Pred';
Proc GPLOT data = Pdata;
Format group_ob group_ob.;
Format group_pr group_pr.;
Label group_pr = 'Treatment_pred'; label group_ob = 'Treatment_obsv';
Label C_QOL_M ='Observed';
Symbol1 c = blue i=j l = 2;
Symbol2 c = red i= j l = 1;
Symbol3 c = blue i = none v = circle;
Symbol4 c = red i = none v = dot;
Plot predm*time = group_pr / haxis = 0 to 6 by 1 vaxis = 3.4 to 5 by .2;
Plot2 C_QOL_M*Time = group_ob /vaxis = 3.4 to 5 by .2;
Title'';
Title2 'Piecewise linear growth model';
Run;
Quit;
```

In this chapter we presented applications of multilevel modeling to longitudinal data. The typical characteristics of longitudinal data and the advantages of applying multilevel models to such data were discussed. After introducing the growth model, we have demonstrated how to run a series of models in SAS for longitudinal data. While focusing on linear growth models, we also demonstrated various curvilinear growth models, including polynomial and piecewise growth models.

The models we have presented in Chapters 3 and 4 are multilevel models for continuous outcome measures. In next chapter, we will turn to multilevel models for discrete outcomes.

Chapter 5
Multilevel models for discrete outcome measures

5.1 Introduction to generalized linear mixed models

In previous chapters, we discussed multilevel models for continuous outcome variables, which are also referred to as multilevel linear models (MLM) or linear mixed models. In this chapter we will extend MLM to generalized linear mixed models that involve discrete outcomes. We begin with reviewing generalized linear models, and then briefly describe their extension—generalized linear mixed models. Following the presentation of the general framework, we will discuss multilevel modeling for different types of discrete outcomes: binary, ordinal, nominal and count outcomes, respectively. The models are applicable for both cross-sectional multilevel data and longitudinal data; however, we will demonstrate the applications of these models mainly using cross-sectional multilevel data. As an example, however, we will also demonstrate the application of logistic regression models to the longitudinal binary outcome.

5.1.1 Generalized linear models

Traditional linear models, including linear regression and analysis of variance (ANOVA), are suitable for normally distributed, continuous outcomes only.

$$Y = XB + \varepsilon \qquad (5.1.1.1)$$

where $\varepsilon \sim N(0, R)$, and R or Var(ε) is assumed to be $\sigma^2 I$. That is, error terms are assumed to be normally and independently distributed with zero mean and constant variance. The expected value of Y or the mean structure part of the model, denoted by μ, is

$$\mu = XB \qquad (5.1.1.2)$$

The generalized linear model (GLM) (McCullagh & Nelder, 1989; Nelder & Wedderburn, 1972) is a generalization of ordinary least squares (OLS) regression. It allows the outcome probability distribution to be any member of an exponential family of distributions. By selecting an appropriate *link function* and outcome probability distribution, many commonly used statistical models can be subsumed under the name of generalized linear models. The GLM family of models is a collection of regression models including standard linear models with normal errors, *logistic* and *probit* models for binary and multinomial data, and log-linear models for counts, plus many other statistical models such as gamma, exponential, etc., through its general model formulations.

A generalized linear model consists of three components: 1) *random component*, 2) *systematic component*, and 3) *link function* (Fitzmaurice et al., 2004).

The *random component* specifies distributional assumptions for GLM. The classical linear regression model only has one distributional assumption, that is the normal distribution. GLM extends the single normal distribution to a range of distributions called *exponential family distributions*. Many distributions including normal, Bernoulli/binomial, and Poisson distributions belong to this family. This family of distributions shares some common statistical properties. For example, the variance of outcomes is functionally related to the mean of the responses through a variance function:

$$\text{Var}(y) = \phi v(\mu) \tag{5.1.1.3}$$

where ϕ is a scale or dispersion parameter, which is either a known constant or unknown parameter that must be estimated; and $v(\mu)$ is a variance function, which describes how variance of the outcome is functionally related to the mean of the outcome. For the normal distribution, ϕ is an unknown parameter, and the variance function, $v(\mu)$ =1, thus the outcome variance $\text{Var}(Y) = \phi$. For the Bernoulli and Poisson distribution, ϕ = 1 (known constant), and the variance functions are $v(\mu) = \mu(1-\mu)$ and $v(\mu) = \mu$, respectively; thus the variance of Bernoulli outcomes is $\text{Var}(Y) = \mu(1-\mu)$, and the variance of Poisson outcome is $\text{Var}(Y) = \mu$.

The *systematic component* is defined just as it is for traditional linear regressions:

$$\eta = X\beta \tag{5.1.1.4}$$

This quantity is a linear combination of the unknown regression coefficient, β, and a set of covariates, *Xs*. This component is also termed the *linear predictors* denoted by η.

The third component of GLM is the *link function*. The link function, denoted as $g(\mu)$, connects the *linear predictors* to expected values of the outcome measure. Specifically, the *link function* links the linear predictors to the mean of the outcome distribution. For normal distribution, it has an identity link (i.e., $g(\mu)$ = 1). For Bernoulli /binomial and Poisson distribution, the associated links are logit link and log link, respectively. The following link functions show how the linear predictors are linked to the mean/transformed means of outcomes for different distributions[1]:

Normal: $$\mu = \eta \tag{5.1.1.5}$$

Bernoulli/Binomial: $$\log\left(\frac{\mu}{1-\mu}\right) = \eta \tag{5.1.1.6}$$

Poisson: $$\log(\mu) = \eta \tag{5.1.1.7}$$

Although the family of exponential distributions includes a variety of distributions, we will restrict our discussion to the three most frequently used distributions in this chapter. Link functions, mean functions and variance functions for these three types of outcomes were displayed in Table 5.1.1.1. All link functions listed in the table are called canonical link functions.

Link functions for the binomial distribution can also be expressed as a *probit* link, that is, $\Phi^{-1}(\mu) = \eta$, where Φ is the cumulative standard normal distribution function (CDF). The probit link function is non-canonical.

Ordinal outcomes and nominal outcomes have multinomial distributions. The link functions for the former is the cumulative *logit* function, and for the latter is the generalized *logit* function. We did not list these two types of functions as separate categories in Table 5.1.1.1 because they are variations of the binomial distribution link function, and can be easily derived from the binomial distribution.

---

[1] As the natural log function in SAS is denoted as "log," we use "log" to represent natural logarithm through out of the book.

Table 5.1.1.1 Components of generalized linear models

Distribution	Link Function g(μ)	Mean Function	Variance Function
Normal	Identity: $\mu = \eta$	$\mu = XB$	$v(\mu) = 1$
Binomial	Logit: $\log\left(\frac{\mu}{1-\mu}\right) = \eta$	$\mu = \frac{1}{1+e^{-X\beta}}$ or $\mu = \frac{e^{X\beta}}{1+e^{X\beta}}$	$v(\mu) = \mu(1-\mu)$
Poisson	Log: $\log(\mu) = \eta$	$\mu = XB$	$v(\mu) = \mu$

5.1.2 Generalized linear mixed models

The generalized linear model is a collection of fixed-effect models that assumes all observations on outcome measures are independent of each other. As discussed in Chapter 2, this assumption is inappropriate for multilevel or hierarchically structured data. As such, the generalized linear model is further extended to two modeling frameworks: 1) marginal models called *generalized estimating equations* (GEE), and 2) *generalized linear mixed effects models* (GLMM). GLMM can also be viewed as extensions of the linear multilevel model or linear mixed model. In this chapter we focus on GLMM. Readers, who are interested in GEE models, are referred to Diggle et al. (1998).

Unlike GLM, GLMM can take into account the *intraclass correlation* coefficient (ICC) in hierarchically structured data by allowing random coefficients in the model. The link functions for GLMM are the same as for GLM; however, the outcome distributions are now conditional distributions or distributions given random effects, *U*. GLMM can be written as

$$g\{E(Y \mid U)\} = \eta = X\beta + ZU \tag{5.1.2.1}$$

where $g(\cdot)$ is a link function, β and U are vectors of fixed and random effects parameters, and X and Z are design matrix for fixed and random effects, respectively. As in multilevel linear models, the random effects, U, are usually assumed to have a normal distribution with zero mean and covariance matrix G.

The outcome variance, conditional on the random effect, is

$$\mathrm{Var}(Y \mid U) = \upsilon\{(E(Y \mid U)\}\phi \tag{5.1.2.2}$$

where $\upsilon(\cdot)$ is a known variance function or a function of the conditional mean, $E(Y|U)$; ϕ is a dispersion parameter (i.e., residual variance).

Alternatively, the conditional variance of the outcome can be formulated as:

$$\mathrm{Var}\ (Y / U) \ = \ A^{1/2}RA^{1/2} \tag{5.1.2.3}$$

The matrix A is a diagonal matrix containing the variance function. The matrix R is a residual variance matrix (SAS Institute Inc., 2008).

In GLMM the transformed outcome means are linked to linear predictors via various link functions, thus models with different outcome distributions and more complex variance structures can be fitted. The linear multilevel models discussed prior to this chapter are a special case of GLMM with normally distributed continuous outcome measure and an *identity link function* (i.e., $g(\mu) = 1$). In this section, rather than covering statistical theories and computational details of GLMM, we will limit our discussion to conceptual descriptions and applications with some frequently used discrete outcome measures using SAS *PROC GLIMMIX* and *PROC NLMIXED.*

5.2 SAS Procedures for multilevel modeling with discrete outcomes

As in the case of MLM, estimation methods for GLMM are based on some form of likelihood principle. However, the maximum likelihood (ML) estimation of GLMM is much more complex, it maximizes the marginal or integrated likelihood as follows:

$$L(\beta,\theta,Y) = \int f(Y \mid u) p(u) \mathrm{d}u \tag{5.2.1}$$

where β denotes fixed-effects; θ denotes unknown parameters of variances/covariances; $f(y|u)$ is the distribution of the outcome measure conditional on the random effect U; and $p\ (u)$ is the distribution of the random effects. The estimates are obtained by integrating out or averaging over the distribution of random effects. Parameter estimates are those that maximize this marginal likelihood function.

In the linear model case, this integral can be solved in closed form, and the resulting likelihood or restricted likelihood can be maximized directly. For nonlinear multilevel models, however, the integral is usually unknown and must be approximated. Many methods have been proposed for such maximization approximation. Two basic methods are: 1) *linearization*, which approximates the integrated likelihood function using techniques such as *Taylor series expansion*, 2) *integral approximation* with numerical methods. These approaches are implemented in two SAS procedures, *PROC GLIMMIX* and *PROC NLMIXED* and two macros, *%GLIMMIX* and *%NLMIXED*, respectively.

Prior to the current version of SAS (SAS 9.2) (SAS Institute Inc., 2008), *PROC GLIMMIX* is solely based on linearization methods. In version 9.2 of *PROC GLIMMIX,* linearization is the default estimation method, and two numerical integration methods—Laplace approximation method and adaptive Gauss-Hermite quadrature have been added as options. The *linearization* method is also called a *pseudo-likelihood method*, in which pseudo-data are generated from the original data, and likelihood function is approximated using Taylor series expansions (Schabenberger, 2005). The essential idea of the *linearization* method is to approximate GLMM using normal linear mixed model estimates repeatedly. Among the various linearization methods available in the procedure, the default method is the restricted or residual pseudo-likelihood (REPL) (Wolfinger & O'Connell, 1993). The maximization of the pseudo-likelihood can be carried out by various optimization techniques in *PROC GLIMMIX*. The default optimization technique is the Newton-Raphson algorithm.

The major advantages of linearization-based methods include: First, they can fit models for which the joint distribution is difficult or impossible to ascertain. Second, compared with numerical integration methods, they allow a larger number of random effects to be estimated in the model. Third, the variance/covariance structure of the level-1 residual matrix (i.e., R matrix) can be readily accommodated. Fourth, the model is iteratively estimated based on the linear mixed model, thus both ML and REML are available for model estimation (Schabenberger, 2005). In addition, linearization based models are much faster to run.

The disadvantages of linearization-based methods include: First, they are based on iterative model estimation using *pseudo-data* constructed from the original data; as such, they do not have a real likelihood, and therefore −2LL or deviance statistic cannot be used for model comparisons. Second, *PROC GLIMMIX* does not support a broad array of variance/covariance structures of the R matrix that you can draw on with the *PROC MIXED* procedure (Schabenberger, 2005).

Numerical integration methods approach maximum likelihood estimation by approximating the integrated marginal likelihood function (i.e., the likelihood integrated over the random effects). SAS version 9.2 *PROC GLIMMIX* incorporates two numerical integration methods, Laplace approximation method and adaptive Gauss-Hermite quadrature. *PROC NLMIXED* is exclusively based on numerical methods. Among the various methods of numerical integration available in the procedure (e.g., the first-order method, Hardy quadrature, Gaussian quadrature and adaptive importance sampling methods), the default in *PROC NLMIXED* procedure is adaptive *Gaussian quadrature* (Pinheiro & Bates, 1995). The resulting marginal likelihood function can be maximized using different optimization techniques available in *PROC NLMIXED*. The default is the dual quasi-Newton algorithm.

The key advantage of the numerical integration methods is that these techniques provide a likelihood function from the original data for optimization. As such, the deviance statistic or −2LL provided by this approach can be used for comparisons among nested models using the likelihood ratio (LR) test. However, there are some limitations as well: First, it is difficult in practice to manage more random effects in a multilevel nonlinear model. Second, with numerical methods, the variance/covariance structure of level-1 residuals (i.e., *R* matrix) cannot be accommodated in either *PROC GLIMMIX* or *PROC NLMIXED*. Finally, the models based on numerical integration cannot be implemented in REML, but only in ML (SAS Institute Inc., 2003).

In early versions of SAS, two SAS macros were often used to fit multilevel generalized linear models and nonlinear models: *%GLIMMIX* and *%NLMIXED*. These two SAS macros were developed prior to *PROC GLIMMIX* and *PROC NLMIXED*. For details about the differences between the macros and procedures readers are referred to Littell, Miliken et al. (2006). In this chapter, we will only illustrate the applications of *PROC GLIMMIX* and *PROC NLMIXED*. We also need to point out that the class of models with numerical methods using Laplace approximation method and adaptive Gauss-Hermite quadrature implemented in *PROC GLIMMIX* is much smaller than the pseudo-likelihood methods (SAS Institute Inc., 2008). In addition, pseudo-likelihood methods are computationally much more efficient (faster). Thus, in our demonstration using *PROC GLIMMIX*, we will rely on the default pseudo-likelihood method for most applications, and will only briefly illustrate the application of numerical integration methods.

5.3 Multilevel models for binary outcomes

Binary measures, such as "yes" vs. "no", "success" vs. "failure", "event occurrence" vs. "no event occurrence", are probably the most common outcome measures in many social, health and medical science studies. Logistic regression or logit models are widely used for such binary outcomes. In this section, we begin with a brief review of fixed-effects logistic regression models, and then proceed to formulate the mixed-effect or multilevel logistic regression model, followed by applications using real world data.

5.3.1 Logistic regression models

In logistic regression for binary outcome measures in non-hierarchically structured data, the probability of "event" (i.e., $p = \Pr(y = 1)$) is usually converted to odds, $p/(1-p)$. The log-odds (i.e., $\log [p / (1-p)]$), or *logit*, is then treated as a linear function of a set of explanatory variables, resulting in the following logistic regression or *logit* model:

$$\operatorname{logit}(p) = \log\left(\frac{p}{1-p}\right) = \beta_0 + \sum_{k=1}^{K} \beta_k x_k \tag{5.3.1.1}$$

where β_k is the regression slope coefficient of the explanatory variable x_k. With this logit transformation of the outcome, the nonlinear relationship between the outcome and covariates is converted to a linear relationship. A regression coefficient in eq. 5.3.1.1 can be interpreted as the amount of change in the *logit* (or log-odds) per unit change in the corresponding explanatory variable, controlling for other covariates.

Note that the *logit* model in eq. 5.3.1.1 does not have an error term as in a linear model. The variance of the outcome measure is determined by its own mean, i.e., p or the probability of "yes":

$$\mathrm{Var}(y) = p(1-p) \tag{5.3.1.2}$$

The *logit* model can be expressed in terms of the probability of an event occurrence (e.g., "yes" response in the binary outcome measure):

$$p = \Pr(y_i = 1 \mid X) = \frac{\exp(z)}{1+\exp(z)} \tag{5.3.1.3}$$

or

$$p = \frac{1}{1+\exp(-z)} \tag{5.3.1.4}$$

where $z = \beta_0 + \sum_{k=1}^{K} \beta_k x_k$. Unlike eq. 5.3.1.1, which depicts the linear relationship between the *logit* of the outcome and covariates, eqs. 5.3.1.3 and 5.3.1.4, called the logistic function, have a S-shaped distribution, which signifies a non-linear relationship between outcome probability and covariates.

Viewed as a generalized linear model (GLM), the log-transformed outcome measure (*logit*) in eq. 5.3.1.1 is called the *logit link*. Equations 5.3.1.3 and 5.3.1.4 can be viewed as a mean *function*,[2] and Equation 5.3.1.2 can be seen as a variance function with a scale parameter $\phi = 1.0$:

$$\mathrm{Var}(y_i) = \phi\, \upsilon(\mu_i) = \phi p_i(1-p_i) \tag{5.3.1.5}$$

However, in many cases, due to unobserved heterogeneity, the number of "yes" events may have a variability that exceeds that predicted by the binary distribution. If this occurrs, it results in a so-called *over-dispersion*. In that case, ϕ would be greater than 1.0. To account for this extra variability, the scale parameter ϕ is often set free and needs to be estimated. Otherwise, estimated standard errors would be biased downward, and statistical tests would be inflated.

5.3.2 Probit models

If the outcome probability is transformed using the *probit link* function, instead of the *logit link* function, we would have the *probit* regression model. The *probit* model is an alternative to logistic regression for dichotomous, ordinal and nominal outcomes. For dichotomous outcomes, the probability of "yes" is expressed as:

$$P = \Pr(y_i = 1 \mid X) = F\left(\beta_0 + \sum_{k=1}^{K} \beta_k x_{ki}\right) \tag{5.3.2.1}$$

similarly, the *probit* model can be written as

$$F^{-1}(P) = \beta_0 + \sum_{k=1}^{K} \beta_k x_{ki} \tag{5.3.2.2}$$

where F is the cumulative distribution function (CDF) of the standard normal distribution, and $F^{-1}(P)$ is the inverse of the CDF, which is the *link function* for the *probit* model. The CDF of the

---

[2] The mean of a binary measure represents the probability p.

standard normal distribution and the logistic distribution are similar to each other. However, the logistic distribution has a wider spread or a longer tail because it has a fixed variance equal to $\pi^2/3 \approx 3.29$, while the standard normal distribution has a variance equal to 1. In this chapter, we focus on the *logit* model, which is more often used for analysis of binary outcome measures.

5.3.3 Unobserved latent variables and observed binary outcome measures

A binary variable can be conceptualized in terms of an underlying latent variable. That is, it is assumed that there is an unobserved continuous variable y_i^* that underlies the observed dichotomous variable y_i. The observed dichotomous variable, y_i, can be viewed as a categorization of this underlying latent variable, y_i^*, either below or above a certain threshold, which is a constant (e.g., $c = 0$):

$$y_i = 1, \text{ if } y_i^* > c \quad (5.3.3.1)$$

$$y_i = 0, \text{ if } y_i^* \leq c \quad (5.3.3.2)$$

That is,when the unobserved variable y_i^* is greater than c, we observe an event (e.g., "yes"), and $y_i = 1$; otherwise $y_i = 0$.

The model with a latent continuous outcome variable can be written in the same form as the regular linear regression model. For a simple model with only one explanatory variable x,

$$y_i^* = \alpha + \beta x_i + e_i \quad (5.3.3.3)$$

Setting the threshold c=0, we have:

$$\begin{aligned} P(y_i = 1 \mid x_i) &= P[(\alpha + \beta x_i + e_i) > 0] \\ &= P[e_i > (-\alpha - \beta x_i)] \\ &= P[e_i \leq (\alpha + \beta x_i)] \\ &= F(\alpha + \beta x_i) \end{aligned} \quad (5.3.3.4)$$

where F is a cumulative distribution function (CDF) of e_i. Assuming a logistic distribution of e_i, the CDF of e_i follows the logistic function, and it would result in a logistic regression model. Assuming a normal distribution of e_i would result in a *probit* model. Since y_i^* is unobserved, its scale cannot be determined for either logistic regression models or *probit* models. The error term e_i is normalized by fixing it at $\pi^2/3 \approx 3.29$ (or standard deviation $\pi/\sqrt{3} \approx 1.8$) for logistic regression models and at 1 for *probit* models.

Since the latent variable y_i^* has a larger variance for the logistic distribution than the standard normal distribution, the regression coefficient of the logistic regression is larger than in the *probit* model by a constant. Hedeker & Gibbons (2006) reviewed research by different authors and concluded that the value of these coefficients differ by an approximate scale difference of 1.6—1.8.

This brief review on the relationship between the unobserved latent variable and the observed binary outcome variable is helpful for understanding the association and differences between the *logit* and *probit* models. This will prove helpful for formulating *ordinal* logistic regression models discussed in later sections of this chapter.

5.3.4 Multilevel logistic regression models

Multilevel logistic regression models or multilevel logit models are an extension of fixed effect logistic regression, incorporating random effects into the model to deal with the intraclass correlation coefficient (ICC) that arises in multilevel data. Similar to multilevel linear models discussed in previous chapters, the multilevel logit model has mixed effects, i.e., both fixed-effects and random effects. The model is specified in matrix format in SAS:

$$\log\left(\frac{p}{1-p}\right) = X\beta + ZU \tag{5.3.4.1}$$

where X is the design matrix, corresponding to fixed-effect parameter vector β; and Z is the design matrix corresponding to the random effect parameter vector U. This equation is similar to eq. 2.2.10, except there is no error term in the equation here.

Multilevel logistic regression models can be represented in the 2-stage format used in linear multilevel models as discussed in previous chapters. Suppose we have one level-1 explanatory variable and one level-2 explanatory variable, then the level-1 equation of the model is described as:

$$\log\left(\frac{p_{ij}}{1-p_{ij}}\right) = \beta_{0j} + \beta_{1j}x_{1ij} \tag{5.3.4.2}$$

The level-2 equations are:

$$\beta_{0j} = \gamma_{00} + \gamma_{01}w_{1j} + u_{0j} \tag{5.3.4.3}$$

$$\beta_{1j} = \gamma_{10} + \gamma_{11}w_{1j} + u_{1j} \tag{5.3.4.4}$$

where both the level-1 random intercept β_{0j} and random slope β_{1j} are treated as linear functions of level-2 explanatory variable w_{1j}. Thus, the combined model follows:

$$\log\left(\frac{p_{ij}}{1-p_{ij}}\right) = \gamma_{00} + \gamma_{01}w_{1j} + \gamma_{10}x_{1ij} + \gamma_{11}x_{1ij}w_{1j} + (u_{0j} + u_{1j}x_{1ij}) \tag{5.3.4.5}$$

where $(\gamma_{00} + \gamma_{01}w_{1j} + \gamma_{10}x_{1ij} + \gamma_{11}x_{1ij}w_{1j})$ and $(u_{0j} + u_{1j}x_{1ij})$ are the fixed and random components of the model, respectively.

5.3.5 Application of multilevel logistic regression models

In this section, we demonstrate how to run a multilevel logistic regression model or a random-effects logistic regression model using both *PROC GLIMMIX* and *PROC NLMIXED*.

Data and variable description:
The data set used in this chapter was extracted from the same national database described and used in Chapter 3. The SAS data file with file name *ML_data_Chap*5 contains the following variables:

Level-1 Variables:
Outcome variable:

- *Heavy_inj*—dichotomous measure: 1—heavy drug injection (i.e., injected drugs 100 or more times in the past 30 days); 0 — non-heavy drug injection (i.e., injected less than 100 times in the past 30 days). This dichotomous outcome measure is recoded from the variable, *Inject*, used in the models of Chapter 3.
- *Inject_c* — categorical measures: 1—injected drug 100+ times in past 30 days; 2—inject drug 31 to 100 times; 3 — inject drugs 1 to 30 times.

- *Jailyear* — count data: total number of years ever spent in jail.
  Covariates:
- *Gender* — dummy variable (1-Male; 0-Female).
- *Ethnic* — dummy variable (1-Black; 0-White).
- *Gdmc_age* — grand-mean centered measure of age.
- *Highsch* — dummy variable (1-High school or above; 0-Less than high school).

Level-2 Contextual variable:

- *HIV_Region* — dummy variable (1-Higher HIV prevalence region; 0-Lower HIV prevalence region).

Unconditional random intercept model and intraclass correlation coefficient (ICC):
We begin with the empty model that has only a random intercept, and no covariates. As described in Chapter 3, the empty model allows us to examine intraclass correlation coefficient (ICC), and it often serves as a base for building more elaborated models. The levels 1 and 2 equations of the model follow:

$$\log\left(\frac{p_{ij}}{1-p_{ij}}\right) = \gamma_{0j} \tag{5.3.5.1}$$

$$\gamma_{0j} = \gamma_{00} + u_{0j} \tag{5.3.5.2}$$

The combined model is:

$$\log\left(\frac{p_{ij}}{1-p_{ij}}\right) = \gamma_{00} + u_{0j} \tag{5.3.5.3}$$

where p_{ij} is the probability of heavy injection; γ_{00} is the overall average log-odds or *logit* of heavy injection, analogous to the grand mean in linear multilevel models; and u_{0j} is the random variation in the level-1 intercepts across groups (i.e., project sites in this study), which represents the deviations of the jth group's mean *logit* from the grand mean or overall mean *logit.*

We will demonstrate how to use SAS *PROC GLIMMIX* and *PROC NLMIXED* to fit the model. We start with *PROC GLIMMIX.*

***SAS Program 5.3.5.1**;

```
LIBNAME lib 'd:\MLM\data';

Data ML_logit;
 Set lib.ML_data_Chap5;
PROC GLIMMIX METHOD=RSPL;
Class site;
Model heavy_inj (event = '1')= /S DIST = BIN LINK = LOGIT DDFM = BW;
Random Int /subject = site;
NLOPTIONS TECH = NRRIDG;
COVTEST  'H: var[random int] = 0'  0;
Run;
```

The syntax in *PROC GLIMMIX* is very similar to the syntax in *PROC MIXED.* By default, *PROC GLIMMIX* uses *restricted/residual psuedo likelihood* (*RSPL*) for parameter estimation. Other estimation methods, including numerical integration may be chosen using the "*METHOD* = " option.

In the *MODEL* statement, following the specification of the dichotomous outcome variable (i.e., *Heavy_inj* in this study), the option *event* = "1" specified in parentheses requests SAS to model probability of being 1 (i.e., the probability of being a heavy injector in this study). No explanatory variables are specified on the right side of the *MODEL* statement for this empty model. The *DIST* option in the *MODEL* statement specifies the type of outcome distribution. In this example, the *DIST* = *BIN*, standing for *Binary* distribution, is specified. The default option for *DIST* is *Binomial*, which produces identical results. However, the (event =) option only works for the binary distribution that allows us to choose to model the probability of either "1" or "0". With the *Binomial* option, *PROC GLIMMIX* always models the probability of the last category of the dichotomous outcome measure (i.e., 1 in this example).

The *LINK* option in this *MODEL* statement specifies the *logit* link function. Alternatively, a *probit* link function can be specified in the *LINK* option for *probit* models. The *DDFM* option specifies the method for computing denominator d.f. for fixed-effects; the *BW* (*BETWITHIN*) option, which decomposes the residual d.f. into between-group d.f., and within-group d.f., is used here (see Section 3.6 in Chapter 3 for discussion on option *DDFM*). Specification in the *RANDOM* statement is the same as in Chapter 3. By specifying *SUBJECT* = *SITE*, we indicate that the research project sites are the group or level-2 units, and that the individual project participants are level-1 units nested within the project sites.

The *TECH* option in the *NLOPTIONS* statement enables us to choose the optimization technique in parameter estimation. The default method is *Dual Quasi-Newton* optimization, which works well for most distributions, but may run into convergence problems for some distributions, especially the binomial distribution. In that case, the *ridge-stabilized Newton Raphson algorithm* (*NRRIDG*) is recommended. In this example, the *NRRIDG* method is chosen. For more information about the optimization methods used in SAS, interested readers are referred to SAS Institute Inc. (2008).

The *COVTEST* statement in *PROC GLIMMIX* is a new addition in SAS 9.2 that provides significance testing for variances and covariances of the level-1 random effects.[3] This statement works for both pseudo-likelihood methods and likelihood methods (Schabenberger, 2007). Unlike the *COVTEST* option in *PROC MIXED*, the *COVTEST* statement in *PROC GLIMMIX* is separate and can be specified in several alternative ways. The statement "*COVTEST* '*H: var*[*random int*]=0' 0" in the above SAS program is one way to specify the test that the intercept variance is significantly different from zero.

SAS Output 5.3.5.1 Results of the empty multilevel logistic regression model using SAS *PROC GLIMMIX*

```
                 The GLIMMIX Procedure

                  Model Information

Data Set                         WORK.ML_LOGIT
Response Variable                Heavy_inj
Response Distribution            Binary
Link Function                    Logit
Variance Function                Default
```

[3] When the pseudo-likelihood method is used in *PROC GLIMMIX*, the COVTEST statement provides significance testing for random effects using the pseudo-data from the last iteration or the final pseudo-data to compute the likelihood functions for both the full model and the reduced model. As such, the likelihood functions for both models are based on the same data and therefore a likelihood ratio test is possible.

Variance Matrix Blocked By	site
Estimation Technique	Residual PL
Degrees of Freedom Method	Between-Within

Number of Observations Read	9824
Number of Observations Used	9824

Response Profile

Ordered Value	heavy_inj	Total Frequency
1	0	7966
2	1	1858

The GLIMMIX procedure is modeling the probability that heavy_inj='1'.

Dimensions

G-side Cov. Parameters	1
Columns in X	1
Columns in Z per Subject	1
Subjects (Blocks in V)	20
Max Obs per Subject	1311

Optimization Information

Optimization Technique	Newton-Raphson with Ridging
Parameters in Optimization	1
Lower Boundaries	1
Upper Boundaries	0
Fixed-effects	Profiled
Starting From	Data

Iteration Histor

Iteration	Restarts	Subiterations	Objective Function	Change	Max Gradient
0	0	6	46199.980631	0.85638941	0.000018
1	0	4	47175.254697	0.24369751	1.243E-6
2	0	2	47861.188665	0.06145186	0.00149
3	0	1	48018.095904	0.00538430	0.001091
4	0	1	48029.156603	0.00017766	1.174E-6
5	0	1	48029.268369	0.00000130	6.26E-11
6	0	0	48029.268782	0.00000000	1.157E-7

```
Convergence criterion (PCONV = 1.11022E-8) satisfied.

                      Fit Statistics

          -2 Res Log Pseudo-Likelihood     48029.27
          Generalized Chi-Square            9646.63
          Gener. Chi-Square / DF               0.98

                 Covariance Parameter Estimates

                                                 Standard
          Cov Parm      Subject     Estimate        Error

          Intercept     site          0.7942       0.2808

                     Solutions for Fixed-effects

                                 Standard
       Effect        Estimate       Error        DF     t Value      Pr > |t|

       Intercept      -1.6936      0.2036        19       -8.32        <.0001

                         Tests of Covariance Parameters
                    Based on the Residual Pseudo-Likelihood
  Label                   DF   -2 Res Log P-Like      ChiSq     Pr > ChiSq    Note
  H: var[random int]=0    1                48412     382.29         <.0001      MI
MI: P-value based on a mixture of chi-squares.
```

The *Model Information* section in SAS Output 5.3.5.1 provides information about model fitting and estimation. The default *Variance Function* for binary distribution is $\mu(1-\mu)$. The *Response Profile* summarizes the total number of observation in each response category (0-non-heavy injection; 1-heavy injection). It also indicates that the probability of event "1" was modeled.

The *Dimensions* section lists the size of the relevant matrices. As there is only one random effect (random intercept), the number of variance/covariance parameters in G, as well as the number of columns in the Z matrix, are all 1. And, there is only 1 column corresponding to levels of the fixed-effect, i.e., intercept only. The number of groups (i.e., project sites) is 20, and the largest group size (i.e., the number of cases in a project site) is 1 311 cases.

The *Optimization Information* section shows that the optimization technique, *NRRIDG*, was used, and the *Iteration History* section displays information about progress of optimization process. For this example, *PROC GLIMMIX* required 6 iterations after the initial optimization to converge.

Note the word "pseudo" in the "–2 *Res Log Pseudo-Likelihood*" in the *Fit Statistics* panel. This indicates that this model fit index is computed from *pseudo* data, and it is a *pseudo-likelihood*, rather than the true likelihood. The ratio of the generalized Chi-square statistic to its degrees of freedom is 0.98, which is close to 1, indicating that the variability in these data may be adequately modeled, and that there may be no residual over-dispersion (Littell et al., 2006).

The *Covariance Parameter Estimates* section displays the estimated variance of the random intercept on the *logit* scale: $\hat{\sigma}^2_{u0} = 0.7942$. Note that the standard error estimates shown in this section are biased and not appropriate for significance testing for random effects (Schabenberg, 2005). The SAS 9.2 *COVTEST* statement provides correct significance testing for random effects in the section of *Tests of Covariance Parameters*. In this example, the variance of $\hat{\sigma}^2_{u0}$ is statistically significant (χ^2 = 382.29, and p-value <0.0001). The note "MI" in the output indicates that the p-value of this likelihood ratio test is based on a mixture of Chi-square distributions, that is, the average of the p-values from two distributions. When variance parameters under the null hypothesis fall on the boundary of the parameter space (e.g., a zero for variance), the distribution of the likelihood ratio statistics can be a mixture of Chi-square distributions (Schabenberger, 2007), as is the case in this example.

Between-group variation in the multilevel *logit* model can also be assessed using the estimated intraclass correlation coefficient (ICC), which is an indicator of between-group heterogeneity or within-group homogeneity discussed in Chapter 2. ICC represents the proportion of group-level variance in the total variance. For the logistic regression model, the residual variance is $\pi^2/3$, thus,

$$\text{ICC} = \frac{\hat{\sigma}^2_{u0}}{\hat{\sigma}^2_{u0} + \pi^2/3} = \frac{0.7942}{0.7942 + 3.289} = 0.19 \tag{5.3.5.4}$$

ICC = 0.19 shows a moderately large between-group heterogeneity or within-group homogeneity, and the variance of $\hat{\sigma}^2_{u0}$ is statistically significant. Thus the multilevel modeling approach should be applied to this data.

Finally, the estimates of the fixed-effects are listed in the *Solutions for the Fixed-effects* section of the SAS output 5.3.5.1. For this model, there is only one fixed-effect, i.e., $\hat{\gamma}_{00} = -1.6936$ ($p < 0.0001$), indicating that the overall mean of log-odds of heavy injection was about – 1.6936. The corresponding probability of heavy injection on average in the population can be estimated as:

$$\hat{p} = \frac{\exp(-1.6936)}{1 + \exp(-1.6936)} \approx 0.16$$

As an alternative to the pseudo-likelihood estimation method, we can fit the model with maximum likelihood estimation using either the Laplace or adaptive Gauss-Hermite quadrature method in *PROC GLIMMIX* (SAS 9.2). If in the Gauss-Hermite quadrature method the number of quadrature points is not specified, then this number is determined "adaptively" by evaluating the log likelihood at the starting value until it reaches a tolerance value. Estimation using Gauss-Hermite quadrature with one quadrature point has the same result as estimation using the Laplace method. However, the Laplace method is computationally simpler. In addition, although only a small class of models can be fitted with either the Laplace or quadrature method (e.g., the *R*-matrix is not available for either method). The Laplace method can be applied to a larger class of models than the quadrature method. For example, some more sophisticated models (e.g., crossed random effect models when subjects are nested within more than one higher units) can be handled with the Laplace approximation, but not with the quadrature method (Schabenberger, 2007).

To fit the same empty model with the Laplace approximation method, we only need to replace the default method with METHOD = LAPLACE:

***SAS Program 5.3.5.2;**

```
PROC GLIMMIX METHOD = LAPLACE;
Class site;
Model Heavy_inj (event = '1') = /S DIST= BIN LINK = LOGIT DDFM = BW;
Random Int /subject = site;
NLOPTIONS TECH = NRRIDG;
```

```
COVTEST  'H: var[random int] = 0'  0;
Run;
```

SAS Output 5.3.5.2 Selected results of the empty multilevel logistic regression model using SAS *PROC GLIMMIX* procedure with Laplace method

```
                    The GLIMMIX Procedure

                      Model Information

         Data Set                          WORK.M1
         Response Variable                 Heavy_inj
         Response Distribution             Binary
         Link Function                     Logit
         Variance Function                 Default
         Variance Matrix Blocked By        site
         Estimation Technique              Maximum Likelihood
         Likelihood Approximation          Laplace
         Degrees of Freedom Method         Between-Within

           Number of Observations Read              9824
           Number of Observations Used              9824

                       Response Profile

         Ordered                             Total
           Value      heavy_inj          Frequency

                          1    0              7966
                          2    1              1858

  The GLIMMIX procedure is modeling the probability that heavy_inj='1'.

                          Dimensions

           G-side Cov. Parameters                  1
           Columns in X                            1
           Columns in Z per Subject                1
           Subjects (Blocks in V)                 20
           Max Obs per Subject                  1311

                   Optimization Information

  Optimization Technique           Newton-Raphson with Ridging
  Parameters in Optimization       2
  Lower Boundaries                 1
  Upper Boundaries                 0
  Fixed-effects                    Not Profiled
  Starting From                    GLM estimates
```

```
                         Fit Statistics

              -2 Log Likelihood                8996.27
               AIC  (smaller is better)        9000.27
               AICC (smaller is better)        9000.27
               BIC  (smaller is better)        9002.26
               CAIC (smaller is better)        9004.26
               HQIC (smaller is better)        9000.66

          Fit Statistics for Conditional Distribution

               -2 log L(heavy_inj | r. effects)        8905.44
               Pearson Chi-Square                      9626.35
               Pearson Chi-Square / DF                    0.98

                   Covariance Parameter Estimates
                                                        Standard
                Cov Parm      Subject     Estimate         Error

                Intercept     site          0.7709        0.2684

                      Solutions for Fixed-effects

                              Standard
      Effect       Estimate      Error        DF     t Value     Pr > |t|

      Intercept     -1.7044     0.2011        19       -8.48       <.0001

                 Tests of Covariance Parameters
                    Based on the Likelihood

   Label                  DF    -2 Log Like      ChiSq    Pr > ChiSq     Note
H: var[random int] = 0    1         9528.44     532.17        <.0001       MI

MI: P-value based on a mixture of chi-squares.
```

The estimated fixed and random parts of the intercept (–1.7044 and 0.7709) are fairly close to the pseudo-likelihood estimates (–1.6936 and 0.7942). An important difference is that the estimation is an approximated maximum likelihood using Laplace. Accordingly, the model fit statistic now is – 2 Log Likelihood rather than – 2 residual Log pseudo-likelihood, thus, we can use LR tests for nested model comparisons.

To run the same model with adaptive Gauss-Hermite quadrature, we would specify METHOD = QUAD. For this example, one quadrature point is adaptively selected, and the result of the estimation is exactly the same as that by the Laplace method displayed in SAS Output 5.3.5.2. The model results are not reported here.

The same empty model or unconditional random intercept model can be fitted using *PROC NLMIXED*, which is more flexible for dealing with different distributions in some special models (see Chapter 6). Unlike the default method (pseudo-likelihood) in *PROC GLIMMIX*, *PROC*

NLMIXED implements true likelihood estimation via numerical quadrature for model estimation. As such, the model fit statistics produced by *PROC NLMIXED* enable us to conduct likelihood ratio tests for comparing nested models. The following SAS program runs the same empty multilevel *logit* model using *PROC NLMIXED*:

***SAS Program 5.3.5.3**;

```
...
PROC NLMIXED;
PARMS B0 = 0 V_u0 = 1;
Logodds = B0 + u0j;
Odds = exp(logodds);
P = odds/(1+odds);
Model Heavy_inj~binary(P);
RANDOM u0j~normal(0, V_u0) subject = site;
ESTIMATE 'ICC' V_u0/(V_u0 + 3.29);
Run;
```

The SAS syntax used in *PROC NLMIXED* is somewhat different from the syntax in *PROC GLIMMIX*, and requires more statistical analysis and SAS programming experience. The *PARMS* (standing for parameters) statement is used to specify model parameters and their starting values. The parameters that are not specified in the statement will be automatically assigned a starting value of 1. With poor starting values, a model may fail to converge on an optimal solution or may fail to converge at all. Specifying appropriate starting values in *PROC NLMIXED* is a challenge, and is one of the major drawbacks of the current version of this procedure (Molenberghs & Verbeke, 2005). Our experience suggests that for less complex models, estimation is robust to the specified starting values.

The next line in the above SAS program specifies the empty model in which the dependent variable is log odds, and the model has a random intercept; the parameter B_0 is the overall-mean intercept across groups (i.e., project sites in this study); and u_{0j} represents the variation of level-1 random intercept. Next, the odds and probability of event occurrence are defined as odds = exp(logodds), and p = odds/(1 + odds), respectively (see eq. 5.3.1.3).

The *MODEL* statement specifies the outcome variable and its conditional distribution (i.e., binary distribution in this example); and the *RANDOM* statement defines the random effects and their variances/covariances (i.e., G matrix). For the current model, a random intercept is specified with mean 0 and variance of V_u0.[4] The *SUBJECT* option specifies the variable that represents groups, or higher level units (i.e., project site in this study). The last optional statement, *ESTIMATE* is used to estimate intraclass correlation coefficient (ICC).

The selected results of this model using *PROC NLMIXED* are shown in SAS Output 5.3.5.3:

SAS Output 5.3.5.3 Selected results of the empty multilevel logit model using SAS PROC NLMIXED

```
                    Specifications

     Data Set                                        WORK.ML_LOGIT
```

[4] The variances of random effects may also be specified in terms of the standard deviation, e.g., *SD_u0j* SD_u0j*, where *SD_u0j* is the standard deviation of the random intercept. Coding in such a way would help stabilize the numerical process when the variance component is close to zero (Agresti et al., 2000).

Dependent Variable	Heavy_inj
Distribution for Dependent Variable	Binary
Random Effects	u0
Distribution for Random Effects	Normal
Subject Variable	site
Optimization Technique	Dual Quasi-Newton
Integration Method	Adaptive Gaussian Quadrature

Dimensions

Observations Used	9824
Observations Not Used	0
Total Observations	9824
Subjects	20
Max Obs Per Subject	1311
Parameters	2
Quadrature Points	1

Parameters

B0	V_u0	NegLogLike
0	1	4526.68353

Iteration History

Iter	Calls	NegLogLike	Diff	MaxGrad	Slope
1	2	4509.87748	16.80604	17.29444	-257.126
2	3	4504.08748	5.790006	19.43058	-26.3301
3	4	4498.28514	5.802341	2.050419	-34.2964
4	6	4498.1575	0.127635	0.63569	-0.14677
5	8	4498.14347	0.014028	0.396622	-0.01596
6	10	4498.13557	0.007898	0.05284	-0.00853
7	12	4498.13545	0.000124	0.003439	-0.00026
8	14	4498.13545	4.491E-7	0.000077	-9.15E-7

NOTE: GCONV convergence criterion satisfied.

Fit Statistics

-2 Log Likelihood	8996.3
AIC (smaller is better)	9000.3
AICC (smaller is better)	9000.3
BIC (smaller is better)	9002.3

Parameter Estimates

Parameter	Estimate	Standard Error	DF	t Value	Pr > |t|	Alpha	Lower	Upper	Gradient
B0	-1.7044	0.2011	19	-8.48	<.0001	0.05	-2.1253	-1.2835	-.00007

```
V_u0       0.7709 0.2684 19    2.87   0.0098   0.05    0.2091   1.3327   0.000077
```

```
                         Additional Estimates

                   Standard
Label  Estimate     Error   DF    t Value    Pr > |t|    Alpha    Lower       Upper
ICC    0.1898    0.05355   19    3.55        0.0022      0.05   0.07776      0.3019
```

The *Specifications* panel in the above SAS output lists some basic information about the model we have specified. Included are the input data set, dependent variables, random effects, related distributions, and optimization technique. The default optimization technique, *Dual Quasi-Newton* was used.

The *Dimension* panel lists various counts related to the model, including the number of observations, subjects, and parameters. Also listed is the number of quadrature points that *PROC NLMIXED* has selected based on the evaluation of the log likelihood at the starting values of the parameter. For this analysis, two parameters, one fixed, and one random effect were estimated. Thus, one quadrature point is necessary for this model to fit. By default, the number of quadrature points is selected adaptively in the process of model estimation. However, users may disable the adaptive search by specifying a specific number of quadrature points in the *QPOINT* option.

The *Parameters* panel lists the parameters to be estimated, their starting values, and the negative log likelihood evaluated at the starting values. The starting value for the fixed intercept is specified as 0, and for the random intercept is specified as 1.

The *Iteration* History panel in the SAS output reports the steps for minimizing the negative log likelihood. For each iteration of *quasi-Newton optimization*, values are listed for the number of function calls, the value of the negative log likelihood, the difference from the previous iteration, the absolute value of the largest gradient, and the slope of the search direction. The process stops when the convergence criterion is satisfied.

The *Fit Statistics* panel shows the final maximized value of the log likelihood, as well as the information criteria (e.g., AIC, AICC, and BIC), which can be used for model comparisons as they are defined in terms of the maximized likelihood with numerical integration based on the actually observed data.

The *Parameter Estimates* panel lists the maximum likelihood estimates of the two parameters and their approximate standard errors. Approximate *t*-values and Wald-type confidence limits are also provided, with degrees of freedom equal to the number of groups (i.e., project sites in this study) minus the number of random effects. The last column in this panel of SAS output is the gradient vector at the optimized solution. Each of elements is sufficiently small indicating a stationary point. The two parameter estimates ($\hat{\gamma}_{00} = -1.70$, and $\hat{\sigma}^2_{u0} = 0.77$) are exactly the same as those estimated with the Laplace and quadrature methods, and fairly close to those estimated with pseudo-likelihood method in *PROC GLIMMIX* ($\hat{\gamma}_{00} = -1.69$, and $\hat{\sigma}^2_{u0} = 0.79$).

Instead of defining odds, log odds, and probability, as well as specifying conditional distribution (e.g., binary distribution) as in SAS Program 5.3.5.3, alternative SAS syntax can be used in *PROC NLMIXED* to fit the same model. The SAS program follows:

***SAS Program 5.3.5.4**;

```
...
PROC NLMIXED;
PARMS B0= 0 V_u0= 1;
Z = B0 + u0j;
If (Heavy_inj= 1) then P = 1/(1 + exp(- Z));
```

```
else P = 1 - (1/(1 + exp(- Z)));
LL = Log(P);
MODEL Heavy_inj ~ GENERAL(LL);
RANDOM u0j~normal(0, V_u0) subject= site;
ESTIMATE 'ICC' V_u0/(V_u0+3.289868134);
Run;
```

The probability of event occurrence P is defined based on eqs. 5.3.1.3 and 5.3.1.4; the LL = Log(P) statement defines the log likelihood function for the model, and the conditional probability distribution is specified in terms of a general log-likelihood function, using option *GENERAL*() in the *MODEL* statement. The option *GENERAL*() denotes the *general distribution.* For SAS-built in distributions (e.g., binomial distribution), we can either use the SAS-built in distribution (e.g., *Heavy_inj~binary*(*P*) in SAS Program 5.3.5.3) or by specifying corresponding log likelihood function using option *GENERAL*() in the *MODEL* statement. However, distributions that are not built-in within SAS (e.g., the zero-inflated Poisson (ZIP) to be discussed in Chapter 6) must be specified using option *GENERAL*().

Random intercept and slope logit models:

As discussed in Chapter 3, the empty model with random intercept-only is a necessary first step for model development. Subsequent steps usually include determining the appropriate *fixed-effects component* (i.e., conditional "mean structure") and the *random-effects component* (i.e., residual variance/covariance structure) of the model.

The fixed component of the model is determined by adding appropriate covariates at both macro and micro levels to explain the between- and within-group variations, respectively. The *random component* of the model is determined by assessing which level-1 explanatory variables have random slopes. Although it is a common practice to examine the mean structure of the model first, model fitting is a dynamic process between the fixed-effect and random-effect parts because multilevel models are conditional models. That is, estimation of the fixed-effect part is dependent on specification of the random-effect part of the model.

Rather than going through the lengthy model selection process demonstrated in Chapter 3, we will fit the "*final*" model developed in chapter 3. This model has a level-1 random intercept, and only one level-1 covariate (i.e., *Ethnic*) has a random slope. All other level-1 covariates, such as *Gender*, *Gdmc_age*, and *Highsch*, have fixed-effects. There is only one level-2 covariate. i.e., *HIV_Region.* The outcome measure is also a measure of drug injection, except that the continuous outcome measure (i.e., *Inject*) is re-coded to a dichotomous variable: *Heavy_inj* (1-heavy injection; 0- non-heavy injection). Thus, the multilevel *logit* model follows:

$$\ln\left(\frac{p_{ij}}{1-p_{ij}}\right) = \beta_{0j} + \beta_{1j}\,\text{Ethnic}_{ij} + \beta_2\,\text{Gender}_{ij} + \beta_3\,\text{Gdmc\_age}_{ij} + \beta_4\,\text{Highsch}_{ij} \qquad (5.3.5.5)$$

$$\beta_{0j} = \gamma_{00} + \gamma_{01}\,\text{HIV\_Region}_j + u_{0j} \qquad (5.3.5.6)$$

$$\beta_{1j} = \gamma_{10} + \gamma_{11}\,\text{HIV\_Region}_j + u_{1j} \qquad (5.3.5.7)$$

The combined model is:

$$\ln\left(\frac{p_{ij}}{1-p_{ij}}\right) = \gamma_{00} + \gamma_{01}\,\text{HIV\_Region}_j + \gamma_{10}\,\text{Ethnic}_{ij} + \gamma_{11}\,\text{HIV\_Region}_j * \text{Ethnic}_{ij}$$
$$+ \beta_2\,\text{Gender}_{ij} + \beta_3\,\text{Gdmc\_age}_{ij} + \beta_4\,\text{Highsch}_{ij} + (u_{0j} + u_{1j} * \text{Ethnic}_{ij}) \qquad (5.3.5.8)$$

The level-2 residuals are assumed to follow normal distributions with zero mean and covariance matrix G (see eq. 2.3.6). The SAS program for the model follows:

***SAS Program 5.3.5.5;**

```
...
Reg_Eth = HIV_Region*Ethnic;
PROC NLMIXED;
PARMS B0 = -1.6 B1 = -.15 B2 = 0.17 B3 = 0 B4 = 0 B5 = 0.8 B6 = 0 V_u0
=0.5 V_u1 = 0.1 Cov_u01 = 0;
Z = B0+B1*Ethnic+B2*Gender+B3*Gdmc_age+B4*Highsch+B5*HIV_Region
 +B6*Reg_Eth+(u0j+u1j*Ethnic);
If (Heavy_inj = 1) then P = 1/(1+exp(-Z));
Else P = 1-(1/(1+exp(- Z)));
LL = LOG(P);
MODEL Heavy_inj ~ GENERAL(LL);
RANDOM u0j u1j~normal([0,0],[V_u0,Cov_u01,V_u1]) SUBJECT=site;
Run;
```

SAS codes in this program are similar to those in SAS Program 5.3.5.4, except that more covariates are added, and one random slope (i.e., random slope of level-1 variable *Ethnic*) is defined by including level-2 residual term u_{1j} in the model. Note that a new variable (i.e., *Reg_Eth*) was created before *PROC NLMIXED*, which denotes the cross-level interaction between the contextual variables *HIV_region* and the level-1 variable *ethnic*.

In the *PARMS* statement, the starting values we specified for the model parameters were based on our exploratory modeling. In the *RANDOM* statement, both random effects are specified as normally distributed with means of zero and the variance and covariance matrix *G*, using the option *~normal*([ ],[ ]), where the means are specified in the first [ ] and variances and covariances are specified in the second [ ].

The output of this model are not reported here. The estimated covariance between the two level-2 residuals (i.e., *u0j* and *u1j*) is not statistically significant (Cov_*u*01 = −0.04, *P* = 0.6968), indicating that the random intercept and slope are not significantly related, that is, the relationship between ethnicity and the outcome is not dependent on the group specific outcome means.

PROC NLMIXED takes a long time to run when the data set is large, and when many model parameters, especially many random effects, are estimated. When an unstructured variance/covariance matrix (or full *G* matrix) is specified (like the *TYPE=UN* option in *PROC MIXED* discussed in Chapter 3), the model often takes a long time to run and may encounter estimation problems (e.g., model estimation may not achieve convergence). There are some tips for specifying models in *PROC NLMIXED* to ensure stabilization of estimation of the unstructured *G* matrix. One recommendation is to use Cholesky[5] parameterization to specify the unstructured variance/covariance matrix.

Here we briefly describe the Cholesky decomposition of the *G* matrix. In our current model there are two random effects (i.e., *u0j* and *u1j*). Let *T* be the lower triangular matrix such that *TT′*=G (Littell et al., 2006):

$$G = TT' = \begin{pmatrix} t_{11} & 0 \\ t_{12} & t_{22} \end{pmatrix} \begin{pmatrix} t_{11} & t_{12} \\ 0 & t_{22} \end{pmatrix} = \begin{pmatrix} t_{11}^2 & t_{11}t_{12} \\ t_{11}t_{12} & t_{12}^2 + t_{22}^2 \end{pmatrix} = \begin{pmatrix} \sigma_{u0}^2 & \sigma_{u01}^2 \\ \sigma_{u01}^2 & \sigma_{u1}^2 \end{pmatrix} \tag{5.3.5.9}$$

In the following SAS program, the Cholesky parameterization method is applied to stabilize the estimation of the variances and covariance of the two random effects.

---

[5] Given a symmetric positive-definite matrix *V*, the Cholesky decomposition is a lower triangular matrix *L* such that $V = LL'$.

***SAS Program 5.3.5.6**;

```
...
Reg_Eth=HIV_Region*ethnic;
PROC NLMIXED;
PARMS B0 = -1.6 B1 = -.15 B2 = 0.17 B3 = 0 B4 = 0 B5= 0.8 B6 = 0 t11 = 0.7
t22 = 0.3 t12 = 0;
Z=B0+B1*Ethnic+B2*Gender+B3*Gdmc_age+B4*Highsch+B5*HIV_Region
  +B6*Reg_Eth+(u0j+u1j*Ethnic);
If (Heavy_inj=1) then P=1/(1+exp(-Z));
Else P=1-(1/(1+exp(-Z)));
LL=LOG(P);
V_u0=t11*t11;
Cov_u01=t11*t12;
V_u1=t12*t12+t22*t22;
MODEL heavy_inj ~ GENERAL(LL);
RANDOM u0j u1j~normal([0,0],[V_u0,Cov_u01,V_u1]) SUBJECT=site;
estimate 'Var(u0)' V_u0;
estimate 'Cov(u0, u1)' Cov_u01;
estimate 'Var(u1)' V_u1;
Run;
```

The fixed-effects estimated from SAS Program 5.3.5.5 and 5.3.5.6 were identical, whereas the variances and covariances of random effects were slightly different (the output of SAS program 5.3.5.5 and 5.3.5.6 are not reported here). The results of the two SAS programs show that the covariance between the two random effects was not statistically significant. Therefore, the covariance between u_{0j} and u_{1j} is set to 0 in the following SAS program.

***SAS Program 5.3.5.7**;

```
...
Reg_Eth=HIV_Region*ethnic;
PROC NLMIXED;
PARMS B0=-1.6 B1HIV_Region=0.8 B2Ethnic=-0.5 B3Reg_Eth=0.4 B4Gender=-0.2
 B5Gdmc_age=0.02 B6Highsch=0.03 V_u0=0.5 V_u1=0.19;
Z=B0+B1HIV_Region*HIV_Region+B2Ethnic*Ethnic+B3Reg_Eth*Reg_Eth+B4Gender*
 Gnder+B5Gdmc_age*Gdmc_age+B6Highsch*Highsch+(u0j+u1j*Ethnic);
If (heavy_inj=1) then P=1/(1+exp(-Z));
Else P=1- (1/(1+exp(-Z)));
LL=LOG(P);
MODEL Heavy_inj ~ GENERAL(LL);
RANDOM u0j u1j~normal([0,0],[V_u0,0,V_u1]) SUBJECT=site;
Run;
```

SAS Output 5.3.5.7 Selected results of multilevel logistic regression model using *PROC NLMIXED*

```
                Fit Statistics

-2 Log Likelihood                 8950.1
AIC (smaller is better)           8968.1
```

```
AICC (smaller is better)          8968.1
BIC (smaller is better)           8977.1
```

Parameter Estimates

parameter	Estimate	Standard Error	DF	t Value	Pr>|t|	Alpha	Lower	Upper	Gradient
B0	-1.6120	0.1956	18	-8.24	<.0001	0.05	-2.0228	-1.2011	0.00585
B1HIV_Region	0.8380	0.3946	18	2.12	0.0478	0.05	0.009039	1.6670	0.000641
B2Ethnic	-0.4570	0.1489	18	-3.07	0.0066	0.05	-0.7698	-0.1441	-0.01586
B3Reg_Eth	0.4163	0.3040	18	1.37	0.1877	0.05	-0.2223	1.0549	-0.01591
B4Gender	-0.1703	0.05964	18	-2.86	0.0105	0.05	-0.2956	-.04501	-0.03058
B5Gdmc_age	0.002717	0.003745	18	0.73	0.4775	0.05	-0.00515	0.01059	-0.10351
B6Highsch	0.02588	0.05524	18	0.47	0.6450	0.05	-0.09017	0.1419	0.051014
V_u0	0.4404	0.1730	18	2.55	0.0203	0.05	0.07695	0.8038	0.016379
V_u1	0.1624	0.08667	18	1.87	0.0772	0.05	-0.01965	0.3445	0.016573

Parameter V_u0 is the variance of the random intercept, and V_u1 is the variance of the slope of level-1 explanatory variable *Ethnic*. The variation of intercept across project sites is statistically significant ($p = 0.0203$), while the variation of the random slope of *Ethnic* is only marginally significant ($p = 0.0772$). Overall, this model fits the data much better than the empty model or the random intercept only model: the difference in –2LL (or deviance statistic) between the two models is 8996.3–8950.1 = 46.2 with d.f. = 7, which is highly significant ($p < 0.0001$).

The level-1 explanatory variables (i.e., *Ethnic* and *Gender*) have statistically significant effects on the dichotomous variable heavy injection. The odds ratio (OR) of Black injection drug users (IDUs) (i.e., *Ethnic* = 1) vs. White IDUs (i.e., *Ethnic* = 0) was approximately exp (–0.4570)≈ 0.63 (95% C.I.: 0.43, 0.93). The odds ratio of male (*Gender* = 1) IDUs vs. female (*Gender* = 0) IDUs was approximately exp(– 0.1703) ≈ 0.84 (95% C.I.: 0.47, 0.85).

IDUs in the higher HIV prevalence region (i.e., *HIV_Region* = 1) were more likely to be heavy injectors, compared with those in the lower HIV prevalence region (i.e., *HIV_Region*= 0): the OR = exp(0.8380) ≈ 2.31 (95% C.I.: 1.07, 5.00). There was no significant interaction between the contextual variable *HIV_Region* and the individual level variable *Ethnic*; in other words, the effect of *Ethnic* does not significantly vary across HIV prevalence regions (*B3Reg_Eth* = 0.4163, $p = 0.1877$).

The same model can also be estimated using *PROC GLIMMIX*. The SAS program follows:

*** SAS Program 5.3.5.8;**

```
...
Proc GLIMMIX noitprint noclprint;
Class site;
Model Heavy_inj (event='1')= HIV_Region|Ethnic Gender Gdmc_age Highsch
 /S DIST=BINARY LINK=LOGIT DDFM=BW;
RANDOM int Ethnic /subject=site TYPE=CHOL;
NLOPTIONS TECH=NRRIDG;
covtest 'H: var[random int]=0' 0;
covtest 'H: var[random cov]=0' . 0;
covtest 'H: var[random slope]=0' . . 0;
Run;
```

SAS Output 5.3.5.8 Selected results of multilevel logistic regression model using *PROC GLIMMIX*

Fit Statistics

-2 Res Log Pseudo-Likelihood	48134.84
Generalized Chi-Square	9586.15
Gener. Chi-Square / DF	0.98

Covariance Parameter Estimates

Cov Parm	Subject	Estimate	Standard Error
CHOL(1,1)	site	0.7179	0.1534
CHOL(2,1)	site	-0.07450	0.1691
CHOL(2,2)	site	0.4369	0.1159

Solutions for Fixed-effects

Effect	Estimate	Standard Error	DF	t Value	Pr > |t|
Intercept	-1.6036	0.2064	18	-7.77	<.0001
HIV_Region	0.8290	0.4231	18	1.96	0.0657
Ethnic	-0.4460	0.1517	9799	-2.94	0.0033
HIV_Region*ethnic	0.4077	0.3242	9799	1.26	0.2086
Gender	-0.1700	0.05957	9799	-2.85	0.0043
Gdmc_age	0.002724	0.003741	9799	0.73	0.4666
Highsch	0.02533	0.05516	9799	0.46	0.6461

Tests of Covariance Parameters
Based on the Residual Pseudo-Likelihood

Label	DF	-2 Res Log P-Like	ChiSq	Pr > ChiSq	Note
H: var[random int]=0	1	48226	91.45	<.0001	MI
H: var[random cov]=0	1	48135	0.20	0.6573	DF
H: var[random slope]=0	1	48156	21.12	<.0001	MI

DF: P-value based on a chi-square with DF degrees of freedom.
MI: P-value based on a mixture of chi-squares.

The SAS code is the same as in SAS Program 5.3.5.1, with the added level-1 covariates in this *MODEL* statement, and the random slope for variable *Ethnic* in the *RANDOM* statement. The option *Type=Chol* in the *RANDOM* statement requests Cholesky decomposition for the unstructured variance-covariance matrix (i.e., *G* matrix) in model estimation. This method is employed to insure that the *G* matrix has all nonnegative eigenvalues, thus the parameter estimates are correct. The three separate "COVTEST" statements were included to test significance of the three random effects, respectively, where "0" means we are testing the null hypothesis that the variations for each component equal to zero. In the *G*-matrix, the first component is variance of the intercept, the second component is covariance between the intercept and slope, and the third one is variance of the slope. We added one dot "." and two dots ". ." for the second and the third statements, respectively, so that each statement tests a variance or covariance component individually.

The SAS Output shows that the main effect of the contextual variable *HIV_Region* is only marginally significant ($p = 0.0657$) in this model. Note that the *p*-value for the covariance test was based on the standard Chi-square distribution, while *p*-values for intercept and slope variances were based on a mixture of Chi-square distributions because these two parameters might hit a boundary (e.g., 0 variance) of the parameter space. As the covariance between the model random intercept and the random slope of variable *Ethnic* was not statistically significant (*p*-value = 0.6573), we refit the model by removing the option *Type* = *Chol* from the above SAS program. The main effect of contextual variable *HIV_Region* remained marginally significant ($p = 0.0583$).

5.3.6 Application of multilevel logit models to longitudinal data

The multilevel logistic regression model can also be applied to longitudinal data with individuals as level-2 units and repeated measures for each individual as level-1 units. In fact, all of the models discussed in this chapter can be applied to either clustered cross-sectional or longitudinal data. Due to the space limitations, we only demonstrate how to apply the multilevel logistic model to longitudinal data.

The data used for the application (SAS data file *ML_data2_Chap5*) were retrieved from the same database (i.e., the same New Hampshire Dual Disorder Study database) used in Chapter 4. For this demonstration, our outcome is a dichotomous measure: ever been vs. never been in psychiatric hospital in a 6-month interval during a 3-year observation period. The following variables are used for the study:

Outcome Measure:

- *Hospdum* — binary variable (1-in psychiatric hospital in the past 6 months; 0-not in psychiatric hospital in the past 6 months)

Level-1 Explanatory Variables:

- *Time* — time variable coded as 0, 1, 2, 3, 4, 5, 6, representing the baseline, 6-month, 12-month, … , 36-month interview time points, respectively.

Level-2 Explanatory Variables:

- *Trt* —dummy variable (1-Treatment group; 0-Control group).
- *Gender* — dummy variable (1-Male; 0-Female).
- *Age* — continuous variable, participant's age at the time of the baseline interview.
- *Edu*—education, dummy variable (1-High school completion or above; 0-Less than high school completion).

In this example, we only fit a random intercept-slope model with a level-2 covariate *Trt*, while controlling for individual background variables (e.g., *Gender*, *Age*, and *Edu*). The random intercept-slope model is described as:

$$\ln\left(\frac{p_{ij}}{1-p_{ij}}\right) = \beta_{0j} + \beta_{1j}\text{Time}_{ij} \tag{5.3.6.1}$$

$$\beta_{0j} = \gamma_{00} + \gamma_{01}\text{Trt}_j + \gamma_{02}\text{Gender}_j + \gamma_{03}\text{Age}_j + \gamma_{04}\text{Edu}_j + u_{0j} \tag{5.3.6.2}$$

$$\beta_{1j} = \gamma_{10} + \gamma_{11}\text{Trt}_j + u_{1j} \tag{5.3.6.3}$$

The combined model is

$$\ln\left(\frac{p_{ij}}{1-p_{ij}}\right) = \gamma_{00} + \gamma_{01}\,\text{Trt}_j + \gamma_{02}\,\text{Gender}_j + \gamma_{03}\,\text{Age}_j + \gamma_{04}\,\text{Edu}_j$$

$$+ \gamma_{10}\,\text{Time}_{ij} + \gamma_{11}\,\text{Time}_{ij} * \text{Trt}_j + (u_{0j} + u_{1j} * \text{Time}_{ij}) \tag{5.3.6.4}$$

According to the findings of our previous analyses, the individual background variables (i.e., *Gender*, *Age*, and *Edu*) only affect the initial level of log-odds of hospitalization, while treatment (i.e., *Trt*) affects both the initial level and the rate of change in the log-odds over time. The fixed and random effects in the model are interpreted in the same way as in the models demonstrated in Chapter 4. The SAS program, using *PROC GLIMMIX* follows:

***SAS Program 5.3.6.1**;

```
...
LIBNAME lib 'D:\MLM\data';

Data data1;
 Set lib.ML_data2_Chap5;
Gmc_age=age- 33.70;
PROC GLIMMIX;
Class ID;
Model Hospdum (event='1')=Trt|Time Gmc_age Gender Edu
 /S DIST=BINARY LINK=LOGIT DDFM=BW;
RANDOM int Time/subject=ID TYPE=Chol;
NLOPTIONS TECH=NRRIDG;
covtest 'H: var[random int]=0' 0;
covtest 'H: var[random cov]=0' . 0;
covtest 'H: var[random slope]=0' . . 0;
Run;
```

where data are retrieved from SAS data file *ML_data2_Chap5*; variable *Age* is transformed to grand-mean centered age, G*mc_Age*; the option *TYPE* = *Chol* in the *RANDOM* statement is used to specify the *G*-matrix (i.e., the variance/covariance matrix of the random effects) using Cholesky decomposition. Selected results of this model are:

SAS Output 5.3.6.1 Selected results of random intercept & slope logit model for longitudinal data using *PROC GLIMMIX*

```
                    Fit Statistics
-2 Res Log Pseudo-Likelihood        6025.92
Generalized Chi-Square              1066.44
Gener. Chi-Square / DF                 0.80

              Covariance Parameter Estimates
```

Cov Parm	Subject	Estimate	Standard Error
CHOL(1,1)	ID	0.8093	0.1801
CHOL(2,1)	ID	-0.00555	0.08474
CHOL(2,2)	ID	0.2251	0.05141

Solutions for Fixed-effects

Effect	Estimate	Standard Error	DF	t Value	Pr > |t|
Intercept	-0.08180	0.2577	198	-0.32	0.7512
Trt	0.4697	0.2410	198	1.95	0.0527
Time	-0.2067	0.05163	1138	-4.00	<.0001
Trt*Time	-0.01895	0.07150	1138	-0.27	0.7910
Gmc_age	-0.01269	0.01165	198	-1.09	0.2775
Gender	-0.1501	0.2132	198	-0.70	0.4820
EDU	-0.1145	0.1913	198	-0.60	0.5503

Tests of Covariance Parameters
Based on the Residual Pseudo-Likelihood

Label	DF	-2 Res Log P-Like	ChiSq	Pr > ChiSq	Note
H: var[random int]=0	1	6042.67	16.75	<.0001	MI
H: var[random cov]=0	1	6025.92	0.00	0.9482	DF
H: var[random slope]=0	1	6028.31	2.40	0.0608	MI

DF: P-value based on a chi-square with DF degrees of freedom.
MI: P-value based on a mixture of chi-squares.

The model results provided by *PROC GLIMMIX* can be used as starting values for the corresponding parameters in *PROC NLMIXED* for the same model. In SAS Output 5.3.6.1 the covariance between random intercept and random slope is not significant (p-value = 0.9482), thus we set the covariance to zero in the following *PROC NLMIXED* program. The resulting parameter estimates are similar to earlier models.

*** SAS Program 5.3.6.2**;

```
...
Gdmc_age=age-33.70;
PROC NLMIXED;
PARMS b0=-.08 b_trt=.47 b_t=-.21 b_trt_t=-.02 b_Gdmc_age=-0.15 b_gender=
-0.15 b_edu=-0.12
   V_u0=.64 V_u1=.05;
logodds=b0+b_Trt*Trt+b_t*Time+b_Trt_t*Trt*Time+b_Gdmc_age*Gdmc_age+b_Gender*
Gender+b_Edu*Edu+(u0+u1*Time);
odds=exp(logodds);
p=odds/(1+odds);
```

```
MODEL Hospdum~binary(p);
RANDOM u0 u1 ~ normal([0,0],[ V_u0,0,V_u1]) subject=ID;
Run;
```

SAS Output 5.3.6.2 Selected results of multilevel logit model for longitudinal data using *PROC NLMIXED*

Fit Statistics

-2 Log Likelihood	1587.0
AIC (smaller is better)	1605.0
AICC (smaller is better)	1605.1
BIC (smaller is better)	1634.8

Parameter Estimates

Parameter	Estimate	Standard Error	DF	t Value	Pr> |t|	Alpha	Lower	Upper	Gradient
b0	0.04406	0.2974	201	0.15	0.8824	0.05	-0.5424	0.6305	0.002413
b_Trt	0.5160	0.2639	201	1.96	0.0519	0.05	-.00440	1.0364	0.003639
b_t	-0.2964	0.06726	201	-4.41	<.0001	0.05	-0.4291	-0.1638	0.015921
b_Trt_t	-0.00451	0.08783	201	-0.05	0.9591	0.05	-0.1777	0.1687	-0.00502
b_Gdmc_age	-0.01157	0.01370	201	-0.84	0.3991	0.05	-0.03858	0.01543	-0.03685
b_Gender	-0.2308	0.2505	201	-0.92	0.3579	0.05	-0.7247	0.2631	0.004451
b_Edu	-0.1782	0.2247	201	-0.79	0.4286	0.05	-0.6213	0.2649	-0.005
V_u0	0.8610	0.2954	201	2.91	0.0040	0.05	0.2785	1.4435	0.011006
V_u1	0.1199	0.03794	201	3.16	0.0018	0.05	0.04506	0.1947	0.038132

5.4 Multilevel models for ordinal outcomes

Logistic regression for dichotomous variables can be extended to modeling ordinal outcome measures, which are also quite common in behavioral and health studies. Examples of ordinal measures include *Likert Scales* employed in public opinion surveys (e.g., 1-strongly disagree; 2-disagree; 3-neutral; 4-agree, 5-strongly agree), measures of severity of illness (e.g., 1-mild; 2-moderate; 3-severe), and scales of substance use (1-abstinent; 2-use without impairment; 3-abuse; 4-dependence, 5-dependence with institutionalization).

This section discusses and illustrates multilevel models for ordinal outcomes. SAS *PROC GLIMMIX* and SAS *PROC NLMIXED* will be used for model demonstration.

5.4.1 Cumulative logit models

Logistic regression for ordinal outcome measures is often formulated as *cumulative logit models* or *proportional odds models* (Agresti, 1989). The *cumulative logit model* is an extension of the binary logistic regression model, which is defined as the following:

$$y^* = \alpha + \sum_{k=1}^{K} \beta_k x_x + e \tag{5.4.1.1}$$

where y^* is the unobserved measure of the underlying ordinal outcome measure while e is the error term.

Suppose the observed outcome measure has M categories ($m = 1, 2, \ldots, M$) with values of ($y = 1$), ($y = 2$), ..., ($y = $ M), and ($y = 1$) < ($y = 2$) < ⋯< ($y = M$), then, there would be $M-1$ unknown thresholds or cut-off points that separate the adjacent categories. That is:

$$\begin{aligned} &y = 1, \text{ if } y^* < C_1 \\ &y = 2, \text{ if } C_1 \le y^* < C_2 \\ &\vdots \\ &y = M-1, \text{ if } C_{M-2} \le y^* < C_{M-1} \\ &y = M \text{ if}, C_{M-1} \le y^* \end{aligned} \tag{5.4.1.2}$$

where C_m has $M-1$ values and $C_1 < C_2 < \cdots < C_{M-1}$. The *cumulative logit model* can be expressed in the following form:

$$\log\left(\frac{p(y \le m)}{1 - p(y \le m)}\right) = \beta_{0m} + \sum_{k=1}^{K} \beta_k x_k \tag{5.4.1.3}$$

The model is formed to take into account the rank order of the outcome measure (Agresti, 1996).

For M categories, a total of ($M - 1$) cumulative logits will be formulated, where β_{0m} is the intercept of the mth logit[6], and β_k is the slope coefficient of covariates x_k. The important feature of this model is that the $M-1$ intercepts are different, but the slope parameters are the same in each logit (Agresti, 1989). Because β has a constant value for each of the ($M - 1$) cumulative logits, this model is also called *proportional odds model.*

Suppose we have a 3-level ordinal outcome measure, then there would be two cumulative logits:

$$\log\left(\frac{p(y \le 1)}{1 - p(y \le 1)}\right) = \log\left(\frac{p_1}{p_2 + p_3}\right) = \beta_{01} + \sum_{k=1}^{K} \beta_k x_k \tag{5.4.1.4}$$

$$\log\left(\frac{p(y \le 2)}{1 - p(y \le 2)}\right) = \log\left(\frac{p_1 + p_2}{p_3}\right) = \beta_{02} + \sum_{k=1}^{K} \beta_k x_k \tag{5.4.1.5}$$

Letting linear predictors $\eta = \sum_{k=1}^{K} \beta_k x_k$, we can define the probabilities of category membership as:

$$\begin{aligned} \Pr(y = 1) &= \Pr(y \le 1) = \frac{1}{1 + \exp[-(\beta_{01} + \eta)]} \\ \Pr(y = 2) &= \Pr(y \le 2) - \Pr(y \le 1) \\ &= \frac{1}{1 + \exp[-(\beta_{02} + \eta)]} - \frac{1}{1 + \exp[-(\beta_{01} + \eta)]} \\ \Pr(y = 3) &= 1 - \Pr(y \le 2) \end{aligned} \tag{5.4.1.6}$$

The proportional odds assumption in the cumulative logit model may not be always satisfied. When running cumulative *logit* models with data that are not hierarchically structured, the proportional odds assumption can be readily tested using *PROC LOGISTIC*. If the proportional odds assumption does not hold, a partial proportional odds model may be considered in which the proportional odds assumptions are relaxed for a subset of covariates (Peterson & Harrell,

---

[6] Note that the threshold C_m is integrated into the intercept β_{0m} in SAS.

1990). In this section, we demonstrate the multilevel cumulative *logit* model by assuming the odds are proportional. Currently, SAS does not provide statistical testing for the proportionality assumption for multilevel cumulative logit modeling.

5.4.2 Multilevel cumulative logit models

Formulation for multilevel cumulative *logit* models is similar to multilevel binary logit models, except a total of $(M-1)$ *logits*, instead of a single *logit*, are simultaneously estimated. The multilevel cumulative *logit* model can be expressed as:

$$\log\left(\frac{p(y \le m)}{1-p(y \le m)}\right) = \beta_{0m_j} + X\beta + ZU \tag{5.4.2.1}$$

This is similar to the multilevel *logit* model for binary outcome data shown in eq. 5.3.4.1, except that it has multiple intercepts, corresponding to each of the $(M-1)$ *logits*. As before, X is the design matrix, containing covariates that have fixed-effects, while Z is the design matrix, containing covariates that have random effects.

As the $M-1$ intercepts differ by incremental constant velues, the model only estimates one variance for the intercepts when the intercepts are treated as random coefficients. In addition, there is only one set of slope coefficients in the model; the vector β consists of the fixed slopes while vector U consists of the random slopes.

To demonstrate how to run multilevel cumulative *logit* models in SAS, we use the same database employed for the models in Chapter 3. The outcome measure is recoded to a 3 category ordinal measure of drug injection frequency in the past 30 days prior to the baseline interview:

- *Inject_c* (1—injected drug more than 100 times; 2—injected drug between 31 to 100 times; and 3—injected drug between 1 to 30 times).

We may label these 3 categories as heavy injection, moderate injection and light injection. The covariates used in the model are the same as those we used in the multilevel logit model for binary data. For a 3-level ordinal outcome measure, there are two *cumulative logits* to be estimated, as shown in eqs. 5.4.1.4 and 5.4.1.5.

The same set of covariates used in the binary multilevel logit model (i.e., *HIV_Region*, *Ethnic*, *Gender*, *Gdmc_age*, and *Highsch*) are used in the multilevel cumulative logistic regression model. The model specification is the same as eqs. 5.3.5.5 — 5.3.5.8, except $(3-1)=2$ logits rather than one logit, are modeled simultaneously.

The model will be fitted using *PROC GLIMMIX* and *PROC NLMIXED*. We start with *PROC GLIMMIX*.

***SAS Program 5.4.2.1**;

```
LIBNAME lib 'D:\MLM\data';

Data ML_logit;
 Set lib.ML_data_Chap5;

Proc GLIMMIX noclprint;
class site;
Model Inject_c=HIV_Region|Ethnic Gender Gdmc_age Highsch
```

```
/S DIST=MULTI LINK=CLOGIT DDFM=BW;
Random int Ethnic/Subject=site TYPE=Chol;
nloptions tech=nrridg;
covtest 'H: var[random int]=0' 0;
covtest 'H: var[random cov]=0' . 0;
covtest 'H: var[random slope]=0' . . 0;
Run;
```

The SAS codes are basically the same as those for modeling dichotomous outcome, except the assumed distribution and link function are different. As the ordinal outcome measure (*Inject_c*) has more than two categories, multinomial distribution and cumulative logit link function are specified with options *DIST=MULTI* and *LINK=CLOGIT* in the *MODEL* statement. The *G*-matrix (i.e., the variance/covariance matrix of the random effects) is specified with option *TYPE=Chol* that uses the Cholesky decomposition method to ensure stable estimation.

SAS Output 5.4.2.1 Selected results of multilevel cumulative logit using *PROC GLIMMIX*

```
                         Model Information

          Data Set                        WORK.ML_LOGIT
          Response Variable               Inject_c
          Response Distribution           Multinomial (ordered)
          Link Function                   Cumulative Logit
          Variance Function               Default
          Variance Matrix Blocked By      site
          Estimation Technique            Residual PL
          Degrees of Freedom Method       Between-Within

          Number of Observations Read          9824
          Number of Observations Used          9824

                        Response Profile

           Ordered                             Total
             Value     inject_c            Frequency

                 1      1                       1858
                 2      2                       3837
                 3      3                       4129

The GLIMMIX procedure is modeling the probabilities of levels of in-
                     ject_c having lower
          Ordered Values in the Response Profile table.

                           Dimensions

            G-side Cov. Parameters              3
            Columns in X                        8
```

```
Columns in Z per Subject          2
Subjects (Blocks in V)           20
Max Obs per Subject            1311
```

Optimization Information

Optimization Technique	Newton-Raphson with Ridging
Parameters in Optimization	3
Lower Boundaries	2
Upper Boundaries	0
Fixed-effects	Profiled
Starting From	Data

Fit Statistics

-2 Res Log Pseudo-Likelihood	72084.49

Covariance Parameter Estimates

Cov Parm	Subject	Estimate	Standard Error
CHOL(1,1)	site	0.6971	0.1300
CHOL(2,1)	site	-0.02839	0.1209
CHOL(2,2)	site	0.3279	0.08894

Solutions for Fixed-effects

Effect	Drug injection: 1-100+ times; 2-30 to 100 times; 3<30 times	Estimate	Standard Error	DF	t Value	Pr > |t|
Intercept	1	-1.6492	0.1916	18	-8.61	<.0001
Intercept	2	0.3292	0.1908	18	1.73	0.1016
HIV_Region		0.9008	0.3955	9798	2.28	0.0228
Ethnic		-0.6445	0.1080	9798	-5.97	<.0001
HIV_Region*ethnic		0.3865	0.2496	9798	1.55	0.1215
Gender		-0.09457	0.04508	9798	-2.10	0.0359
Gdmc_age		0.01450	0.002733	9798	5.30	<.0001
Highsch		0.04271	0.04049	9798	1.05	0.2915

Tests of Covariance Parameters
Based on the Residual Pseudo-Likelihood

Label	DF	-2 Res Log P-Like	ChiSq	Pr > ChiSq	Note

```
H: var[random int]=0    1            72306    221.02        <.0001    MI
H: var[random cov]=0    1            72085      0.06        0.8127    DF
H: var[random slope]=0 1             72105     20.31        <.0001    MI

DF: P-value based on a chi-square with DF degrees of freedom.
MI: P-value based on a mixture of chi-squares.
```

The *Model Information* section of the SAS output provides information about the fitted model and estimation procedure. The multinomial (ordinal) distribution and cumulative logit link function were indicated. The *Response Profile* section summarizes the total number of observation in each response category. It also indicates that the probabilities of lower levels of the ordinal outcome *Inject_c*, which Indicate heavier drug injection in this study, were modeled.

The *Covariance Parameter Estimates* section displays the estimated variance and covariance matrix on the logit scale. We specify option *Type = Chol* in the *RANDOM* statement to estimate the full *G*-matrix. Since the covariance between the random intercept and random slope is not statistically significant (*p*-value = 0.8127), this covariance could be removed from the model.

The *Solutions for Fixed-effects* section shows the estimated fixed parameters in logit scale. The estimated parameter for the two intercepts, −1.6492 and 0.3292, represent the logit of being a heavy injection user, and the logit of being either a heavy injection user or moderate injection user, respectively, for an individual who is in the reference group with a zero value on all co-variates[7]. Note that all slope coefficients are the same for the *logits* in a cumulative logit model. The interpretations of these regression coefficients are the same as those in the binary logistic regression model.

Recall that the ordinal outcome measure in this illustrative model is a 3-level ordinal measure of drug injection in the past 30 days: 1—heavy drug injection; 2— moderate drug injection; and 3—light drug injection; thus two logits are modeled： 1) the odds of heavy injection vs. moderate/light injection (i.e., 1 vs. 2-3); and 2) the odds of heavy/moderate injection vs. light injection (1-2 vs. 3). As such, a positive slope coefficient indicates a higher likelihood to be in the heavier injection categories, and a negative slope indicates a lower likelihood of being in the heavier injection categories. For example, the slope of *HIV_Region* is 0.90 ($p = 0.0228$), indicating that IDUs in the higher HIV prevalence regions were more likely to be involved in heavier drug injection. We leave interpretation of the other parameter estimates to the readers.

For the purpose of illustration we will also fit the same model with *PROC NLMIXED*, which is more flexible in regard to model specifications. One way to specify the random intercept model is described as the following:

***SAS Program 5.4.2.2**;

```
...
Reg_Eth=HIV_Region*Ethnic;
PROC NLMIXED;
PARMS    B1HIV_Region=0.9    B2Ethnic=-0.6    B3Reg_Eth=0.4    B4Gender=-0.1
B5Gdmc_age=0.02 B6Highsch=0.04 V_u0=0.70 V_u1=0.32;
Bounds b>0;
eta=B1HIV_Regio  * HIV_Region+B2Ethnic * Ethnic+B3Reg_Eth * Reg_Eth+B4Gender *
Gender
```

---

[7] Note that the zero value of a grand-centered covariate (e.g., *Gdmc_age*) represents the mean value of the original measure (e.g., *Age*).

```
    +B5gdmc_age*Gdmc_age+B6Highsch*Highsch+u0 + u1*Ethnic;
IF (Inject_c=1) THEN P=1/(1+EXP(- (a+eta)));
ELSE IF (Inject_c=2) THEN P=(1/(1+EXP(- (a+b+eta))))-(1/(1+EXP(- (a+eta))));
ELSE IF (Inject_c=3) THEN P=1- (1/(1+EXP(- (a+b+eta))));
if (P>1e-8) then LL=log(P); else LL=-1e20;
MODEL Inject_c ~ GENERAL(LL);
RANDOM u0 u1 ~ NORMAL([0,0], [V_u0,0,V_u1]) SUBJECT=SITE;
ESTIMATE 'Intercept in the 2d logit'  a+b;
Run;
```

The linear predictor, *eta*, is defined as a function of a set of covariates; the starting values for slope coefficients of the covariates were defined in the *PARMS* statement, and these starting values are from the results of *PROC GLIMMIX* shown in SAS Program Output 5.4.2.1.

The probability denoted as *P* is specified based on eq. 5.4.1.6. The value of *P* is checked during model estimation. If the value of *P* is numerically too close to zero, the program automatically create a extremely large negative value for the corresponding log likelihood (i.e., LL = (−1e20)) (Sheu, 2002).

Note that no intercept is specified in the linear predictor *eta*. The intercept for the first *logit* "*a*" is specified in the statement:

IF (Inject_c = 1) THEN P = 1/(1+EXP(− (a+eta)));

And "*b*", which is the incremental value in the intercept, is specified in the statements

ELSE IF (Inject_c = 2) THEN P = (1/(1+EXP(− (a+b+eta)))) − (1/(1+EXP(− (a+eta))));
ELSE IF (Inject_c = 3) THEN P = 1− (1/(1+EXP(− (a+b+eta))));

where "*a+b*" is the intercept for the second logit in the cumulative model. To ensure a positive increment in intercept, the value of "*b*" is restricted to be greater than 0 by using statement *BOUNDS b*>0. The *ESTIMATES* statement requests that the intercept in the second logit (i.e., "*a+b*") be computed by adding the increment "*b*" to the intercept "*a*" in the first logit.

Finally, in the *RANDOM* statement, a bivariate normal distribution with zero means and variances of *V_u*0, and *V_u*1 were specified for the two random effects (i.e., *u*0 and *u*1). According to the results of SAS Program 5.4.2.1, the covariance between *u*0 and *u*1 was not statistically significant ($p = 0.8127$), thus the covariance is specified as 0 in SAS Program 5.4.2.2.

The parameter estimates from *PROC NLMIXED* are shown in the following SAS output:

SAS Output 5.4.2.2 Selected results of multilevel cumulative logit using *PROC NLMIXED*

```
                     Fit Statistics

          -2 Log Likelihood                 19145
          AIC (smaller is better)           19165
          AICC (smaller is better)          19165
          BIC (smaller is better)           19175


                   Parameter Estimates
```

Parameter	Estimate	Standard Error	DF	t Value	Pr > |t|	Alpha	Lower	Upper	Gradient
B1HIV_Region	0.9099	0.3726	18	2.44	0.0251	0.05	0.1272	1.6926	0.000672
B2Ethnic	-0.6436	0.1037	18	-6.21	<.0001	0.05	-0.8614	-0.4258	0.002899
B3Reg_Eth	0.3802	0.2352	18	1.62	0.1234	0.05	-0.1140	0.8745	0.000703
B4Gender	-0.09456	0.04519	18	-2.09	0.0508	0.05	-0.1895	0.000383	0.001115
B5Gdmc_age	0.01453	0.002730	18	5.32	<.0001	0.05	0.008792	0.02026	0.001883
B6Highsch	0.04293	0.04048	18	1.06	0.3029	0.05	-0.04211	0.1280	0.000364
V_u0	0.4256	0.1470	18	2.90	0.0096	0.05	0.1167	0.7345	0.000181
V_u1	0.08821	0.04756	18	1.85	0.0801	0.05	-0.01172	0.1881	-0.00123
b	1.9794	0.02860	18	69.22	<.0001	0.05	1.9193	2.0395	0.00111
a	-1.6508	0.1809	18	-9.13	<.0001	0.05	-2.0308	-1.2708	0.000456

Additional Estimates

Label	Estimate	Standard Error	DF	t Value	Pr > |t|	Alpha	Lower	Upper
Intercept in 2d logit	0.3286	0.1801	18	1.82	0.0847	0.05	-0.04976	0.7070

where the intercept estimate for the first logit (a) is -1.6508, and the second logit ($a+b$) is $(-1.6508+1.9794) = 0.3286$.

5.5 Multilevel models for nominal outcomes

In Section 5.4 we discussed the multilevel logit model for ordinal outcome measures. In this section we will introduce the multilevel multinomial logit model for nominal outcome measures. Before demonstrating how to run such a model using both *PROC GLIMMIX* and *PROC NLMIXED*, we briefly review the basics of the multinomial logit model.

5.5.1 Multinomial logit models

Many frequently used response variables in social and health sciences are discrete measures with three or more unordered categories. These are called nominal measures. Examples include political candidates, transportation choices and academic majors. For nominal outcome measures the *multinomial logit model* or generalized logit model should be considered, which is also an extension of the binary logit model.

For a nominal outcome measures with $m = 1, 2, \ldots, M$ categories, the multinomial logit model can be specified in the following logit form:

$$\log\left(\frac{P(y=m)}{P(y=M)}\right) = \alpha_m + \sum_{k=1}^{K} \beta_{mk} x_k \tag{5.5.1.1}$$

Where the model is formulated to make pairwise comparisons of non-redundant response categories, which is analogous to estimating several binary logistic regressions simultaneously (Agresti, 1989, 1996). With M response categories, there would be $(M-1)$ logits to be estimated for the multinomial logit model. In contrast with the cumulative logistic regression model, which has $(M-1)$ intercepts but only one set of slopes, the multinomial logit model has $(M-1)$ intercepts, as well as $(M-1)$ sets of slope coefficients as shown below:

$$\log\left(\frac{\Pr(y=1)}{\Pr(y=M)}\right)=\alpha_1+\sum_{k=1}^{K}\beta_{1k}x_k$$
$$\vdots$$
$$\log\left(\frac{\Pr(y=m)}{\Pr(y=M)}\right)=\alpha_m+\sum_{k=1}^{K}\beta_{mk}x_k \tag{5.5.1.2}$$
$$\vdots$$
$$\log\left(\frac{\Pr(y=M-1)}{\Pr(y=M)}\right)=\alpha_{(M-1)}+\sum_{k=1}^{K}\beta_{(M-1)k}x_k$$

The subscript 1, m, and $(M-1)$ in eq. 5.5.1.2 denotes the 1st, the mth, and the $(M-1)^{\text{th}}$ logit, respectively. There are two ways to specify the logit models for unordered categories. One is to choose an arbitrary comparison category, and model the probability or odds of response to any other category against this reference category. The reference category is arbitrarily chosen. This approach is referred to as the baseline category logit model (Agresti, 1996). Another approach allows any possible set of $(M–1)$ pair-wise comparisons among the M categories rather than choosing one reference category (Hedeker & Gibbons, 2006).

In this chapter, we only discuss a frequently used baseline-category logit model with $(M–1)$ comparisons.

Because $\Pr(y=1)+\Pr(y=2)+\cdots+\Pr(y=M)=1$, we have:

$$\begin{aligned}&\Pr(y=1)+\Pr(y=2)+\cdots+\Pr(y=M)\\&=\Pr(y=M)\left(1+\sum_{m=1}^{M-1}\exp(\alpha_m+\sum_{k=1}^{K}\beta_{mk}x_k)\right)\\&=1\end{aligned} \tag{5.5.1.3}$$

Thus, the probabilities of membership in specific categories are:

$$\Pr(y=1)=\frac{\exp\left(\alpha_1+\sum_{k=1}^{K}\beta_{1k}x_k\right)}{1+\sum_{m=1}^{M-1}\exp\left(\alpha_m+\sum_{k=1}^{K}\beta_{mk}x_k\right)}$$
$$\vdots$$
$$\Pr(y=m)=\frac{\exp\left(\alpha_m+\sum_{k=1}^{K}\beta_{mk}x_k\right)}{1+\sum_{m=1}^{M-1}\exp\left(\alpha_m+\sum_{k=1}^{K}\beta_{mk}x_k\right)} \tag{5.5.1.4}$$
$$\vdots$$
$$\Pr(y=M)=\frac{1}{1+\sum_{m=1}^{M-1}\exp\left(\alpha_m+\sum_{k=1}^{K}\beta_{mk}x_k\right)}$$

5.5.2 Multilevel multinomial logit models

For a nominal outcome variable with M categories, multilevel multinomial logit models can be formulated in terms of generalized linear mixed-effects:

$$\log\left(\frac{\Pr(y=m)}{\Pr(y=M)}\right)=X\beta_m+ZU_m \tag{5.5.2.1}$$

where $m = 1, 2, \dots, (M-1)$, β_m is a vector of unknown fixed-effects to be estimated with known design matrix X, and U_m is a vector of unknown random effects to be estimated with known design matrix Z. As in other multilevel models, the random effects are usually assumed to have a normal distribution with a 0 mean and a variance/covariance matrix G, i.e., $u \sim N(0, G)$.

Using a nominal outcome measure with 3 categories as an example, there are two logits to be estimated for a 3-level nominal outcome measure:

$$\log\left(\frac{\Pr(y=1)}{\Pr(y=3)}\right) = \eta_1 = X\beta_1 + ZU_1 \qquad (5.5.2.2)$$

$$\log\left(\frac{\Pr(y=2)}{\Pr(y=3)}\right) = \eta_2 = X\beta_2 + ZU_2 \qquad (5.5.2.3)$$

where the last category (i.e., category 3) is treated as the reference category. There are two sets of fixed-effects (i.e., β_1 and β_2) and two sets of random effects (i.e., U_1 and U_2) to be estimated *simultaneously* for the model.

The probability of membership in a specific category, conditional on the random effect u_m, can be estimated as:

$$\Pr(y_{ij} = 1 \mid u_1) = \frac{\exp(\eta_1)}{1 + \sum_{m=1}^{M-1} \exp(\eta_m)} \qquad (5.5.2.4)$$

$$\Pr(y_{ij} = 2 \mid u_2) = \frac{\exp(\eta_2)}{1 + \sum_{m=1}^{M-1} \exp(\eta_m)} \qquad (5.5.2.5)$$

$$\Pr(y_{ij} = 3 \mid u_3) = \frac{1}{1 + \sum_{m=1}^{M-1} \exp(\eta_m)} \qquad (5.5.2.6)$$

Note that y_{ij} denotes the outcome observation for the ith level-1 units in the jth level-2 unit.

5.5.3 Application of multilevel multinomial logit models

To demonstrate the application of the multilevel multinomial logit model for nominal outcome measures, we will analyze the same data used for ordinal outcomes in Section 5.4. The outcome variable is still the three-category measure of drug injection, *Inject_c*. However, for the purpose of demonstration, we assume that the proportionality for the cumulative logits does not hold. The same covariates that were used in the cumulative *logit* model are used here. Model specification is the same as eqs. 5.3.5.5 — 5.3.5.8, except that $(3-1)=2$ logits rather than 1 will be modeled simultaneously. The SAS program using *PROC GLIMMIX* follows:

```
*SAS Program 5.5.3.1;
...
PROC GLIMMIX noitprint noclprint;
Class site Inject_c;
MODEL Inject_c=HIV_Region|Ethnic Gender Gdmc_age Highsch
/S DIST=MULTI LINK=GLOGIT DDFM=BW;
RANDOM int Ethnic/subject=site group=Inject_c;
NLOPTIONS TECH= NRRIDG;
```

```
covtest 'H: var[random int_1]=0' 0;
covtest 'H: var[random slope_1]=0' . 0;
covtest 'H: var[random int_2]=0' . . 0;
covtest 'H: var[random slope_2]_2=0' . . . 0;
Run;
```

In this program, instead of the *LOGIT* link function, the *GLOGIT* link function, standing for *generalized logit*, is used in the *MODEL* statement for nominal outcome measures. In addition, the nominal outcome variable must be specified in the *CLASS* statement. The procedure also requires the "*GROUP* = outcome" syntax in the *RANDOM* statement to identify groups by which to vary the covariance parameters. Each new level of the grouping effect produces a new set of covariance parameters for each (M–1) pair of linked logit models. *COVTEST* statements are used to test elements of the *G* matrix. According to the results of our exploratory modeling, no covariances between random intercepts and slopes are specified for each logit in the model.

SAS Output 5.5.3.1 Selected results of multilevel multinomial logit model using *PROC GLIMMIX*

```
                          Model Information

Data Set                          WORK.DATA1
Response Variable                 Inject_c
Response Distribution             Multinomial (nominal)
Link Function                     Generalized Logit
Variance Function                 Default
Variance Matrix Blocked By        site
Estimation Technique              Residual PL
Degrees of Freedom Method         Between-Within

Number of Observations Read            9824
Number of Observations Used            9824

                    Response Profile

Ordered                              Total
Value     inject_c               Frequency

    1     1                           1858
    2     2                           3837
    3     3                           4129

In modeling category probabilities, inject_c='3' serves as the reference
category.

                          Dimensions

    G-side Cov. Parameters              6
    Columns in X                       14
    Columns in Z per Subject            6
```

```
Subjects (Blocks in V)          20
Max Obs per Subject           1311

                    Optimization Information

Optimization Technique         Newton-Raphson with Ridging
Parameters in Optimization     6
Equality Constraints           2
Lower Boundaries               6
Upper Boundaries               2
Fixed-effects                  Profiled
Starting From                  Data

   Convergence criterion (PCONV=1.11022E-8) satisfied.

  Estimated G matrix is not positive definite.

                     Fit Statistics

     -2 Res Log Pseudo-Likelihood    75299.34

               Covariance Parameter Estimates

                                                             Standard
     Cov Parm     Subject     Group          Estimate          Error

     Intercept    site        Inject_c 1       0.8052         0.3067
     Ethnic       site        Inject_c 1       0.1882         0.1035
     Intercept    site        Inject_c 2       0.3958         0.1480
     Ethnic       site        Inject_c 2      0.07683        0.05667
     Intercept    site        Inject_c 3            0              .
     Ethnic       site        Inject_c 3            0              .

                 Solutions for Fixed-effects

                Drug
                Injection:
                1-100+
                times;
                2-30 to
                100 times;              Standard

Effect          3 < 30 times  Estimate       Error      DF   t Value    Pr > |t|
Intercept       1              -0.9002      0.2517      18     -3.58      0.0022
Intercept       2             -0.07734      0.1785      18     -0.43      0.6700
HIV_Region      1               1.4047      0.5187    9792      2.71      0.0068
HIV_Region      2               0.9622      0.3808    9792      2.53      0.0115
Ethnic          1              -0.7523      0.1536    9792     -4.90      <.0001
Ethnic          2              -0.6928      0.1035    9792     -6.69      <.0001
```

HIV_Region*Ethnic	1	0.3218	0.3426	9792	0.94	0.3475
HIV_Region*Ethnic	2	0.09899	0.2607	9792	0.38	0.7042
Gender	1	-0.1638	0.06725	9792	-2.44	0.0149
Gender	2	0.01015	0.05563	9792	0.18	0.8553
Gdmc_age	1	0.01545	0.004169	9792	3.71	0.0002
Gdmc_age	2	0.02351	0.003311	9792	7.10	<.0001
Highsch	1	0.05257	0.06150	9792	0.85	0.3927
Highsch	2	0.05132	0.04912	9792	1.04	0.2961

Tests of Covariance Parameters
Based on the Residual Pseudo-Likelihood

Label	DF	-2 Res Log P-Like	ChiSq	Pr > ChiSq	Note
H: var[random int_1]=0	1	75434	135.07	<.0001	--
H: var[random slope_1]=0	1	75318	19.04	<.0001	--
H: var[random int_2]=0	1	75417	117.39	<.0001	--
H: var[random slope]_2=0	1	75305	5.47	0.0194	--

--: Standard test with unadjusted p-values.

In the above SAS output, the *Model Information* panel shows that the distribution of the response variable is multinomial (nominal), and a generalized logit link function is applied to the model. The *Response Profile* panel shows that Category 3 (i.e., injected drug less than 30 times in the past 30 days in this example) of the outcome variable (i.e., *Inject_c*) is used as the reference category. By default, SAS treats the last category of a categorical variable as the reference category.

The *Dimensions* panel in the SAS output indicates there are 6 random effects, which are the variance/covariance parameters in the *G*-matrix; however, the *Covariance Parameter Estimates* panel shows only 4 such random effects since there are only two logits in the model, and random effect parameters are not applicable to the reference group (i.e., *Inject_c* = 3). The 4 random effects are the random intercept and random slope coefficient of covariate *Ethnic* in each of the two *logits*.

The random intercepts and slopes in both logits are significant. Note that the tests of significance are based on the standard Chi-square distribution rather than an adjusted test based on a mixture of Chi-square distributions because the values under the null hypothesis did not fall on the boundary of the parameter space for the random effects variance in this example.

Also note the message "*Estimated G matrix is not positive definite*" in the SAS output. This message should be ignored in this case because the random effects in the reference category are set to zero for this model.

In the *Solutions for Fixed-effects* panel of the output, the total number of fixed parameter estimates is 14 (7 for each of the 2 *logits*). We may view the 14 parameters as they are for two related binary logistic regressions. For example, there are two intercepts, corresponding to the first and the second *logits*, respectively. The negative value for the first intercept (−0.9) indicates that a randomly selected participant was less likely to be in category 1 (i.e., heavy drug injection) than in category 3 (light drug injection), and the related odds ratio is exp(−0.90) = 0.41 ($p = 0.022$). The 2nd intercept is also negative (– 0.08), and the related odds ratio is exp(−0.08) = 0.92. However, there is no significant difference in the odds of being in Category 2 rather than in Category 3 ($p = 0.6700$). These results are consistent with the observed frequency distributions of the outcome variable, *Inject_c* in the *Response Profile* panel of the output. In terms of this distribution, the chances of being in either category 2 or 3 are higher than the chances of being in category 1. About 19% of participants were in Category 1, while the corresponding figures in Categories 2 and 3 are about 39% and 42%, respectively. However, the difference in the chance of being in either Category 2 or Category 3 is very small.

The contextual variable *HIV_region* has a significant effect on both of the two *logits* (i.e., Category 1 vs. Category 3, and Category 2 vs. Category 3) ($p = 0.0068$ and $p = 0.0115$ respectively. In other words, IDUs in the higher HIV prevalence regions were more likely to be in the heavier injection categories (i.e., categories 1 and 2). However, this contextual variable does not significantly moderate the effect of the individual level variable *Ethnic* on the two *logits.* The interaction between *HIV_region* and *Ethnic* is not statistically significant in either of the two *logits* ($p = 0.3475$ and $p = 0.7042$, respectively).

The other fixed-effects are interpreted in the same way as for the multilevel binary logit model. Attention should be paid to the effects of covariates on specific *logits*.

In the following SAS program, we fit the same model with *PROC NLMIXED.*

***SAS Program 5.5.3.2;**

```
...
Reg_Eth=HIV_Region*Ethnic;
PROC NLMIXED;
PARMS
GA0 = -0.90   GA1HIV_Region =   1.40  GA2Ethnic = -0.75  GA3Reg_Eth =   0.32
GA4Gender= -0.16 GA5Gdmc_age =  0.02 GA6Highsch =  0.05 V_u0a =  0.81 V_u1a =  0.18
GB0 = - 0.08 GB1HIV_Region =    0.96 GB2Ethnic = -0.69 GB3Reg_Eth =   0.10
GB4Gender   = 0.01 GB5Gdmc_age =  0.02 GB6Highsch =   0.05V_u0b = 0.40 V_u1b =  0.08;

eta1 = GA0+GA1HIV_Region*HIV_Region+GA2Ethnic*Ethnic+GA3Reg_Eth*Reg_Eth
  +GA4Gender*Gender+GA5Gdmc_age*Gdmc_age+GA6Highsch*Highsch+(u0a+u1a*Ethnic);
eta2 = GB0+GB1HIV_Region*HIV_Region+GB2Ethnic*Ethnic+GB3Reg_Eth*Reg_Eth
  +GB4Gender*Gender+GB5Gdmc_age*Gdmc_age+GB6Highsch*Highsch+(u0b+u1b*Ethnic);

IF (Inject_c=3) THEN P= 1/(1+EXP(eta1)+EXP(eta2));
ELSE IF (Inject_c=1) THEN P= EXP(eta1)/(1+EXP(eta1)+EXP(eta2));
ELSE IF (Inject_c=2) THEN P= EXP(eta2)/(1+EXP(eta1)+EXP(eta2));
LL = LOG(P);
MODEL Inject_c ~ GENERAL(LL);
RANDOM  u0a  u1a  u0b  u1b~NORMAL([0,0,0,0],[V_u0a,0,V_u1a,0,0,V_u0b,0,0,0,
V_u1b])
SUBJECT=site;
Run;
```

The parameters specified in the *PARMS* statement with prefixes of *GA* and *GB* are starting values for regression coefficients in the first and second *logits*, and these starting values are from the results of *PROC GLIMMIX* (see SAS Program 5.5.3.1 Output). The linear predictors in the two *logits* are defined with *eta1* and *eta2*. In the *IF THEN* statements, the probabilities of membership are specified based on eqs. 5.5.2.4—5.5.2.6.

The variance/covariance matrix of the 4 random effects (i.e., *u0a* and *u1a* for *logit* 1; *u0b* and *u1b* for *logit* 2) are specified in the *RANDOM* statement. Specification of the zero means of the random effects is straightforward, while specification of their variances/covariances is not. In the above SAS program, the variance/covariance of the random effects (i.e., *u0a*, *u1a*, *u0b*, and $u1b$) is defined as a lower triangle of the G matrix with variances on the diagonal and zero covariances off the diagonal.

The model results show that almost all fixed-effects estimated using *PROC NLMIXED* were close to those using *PROC GLIMMIX*. However, results of the random effects are some what different from those of SAS Output 5.5.3.1. The variance of the random intercept is statistically significant in both *logit* 1 and *logit* 2 ($V\_u0a = 0.80, p = 0.0210$; $V\_u0b = 0.37, p = 0.0143$); but the variance of the random slope of *Ethnic* is not significant in either of the two *logits* ($V\_u1a = 0.16, p = 0.0827$; $V\_u1b = 0.06, p = 0.2333$).

SAS Output 5.5.3.2 Selected results of multilevel multinomial logit model using *PROC NLMIXED*

```
                 The NLMIXED Procedure

                    Specifications

Data Set                                  WORK.DATA1
Dependent Variable                        inject_c
Distribution for Dependent Variable       General
Random Effects                            u0a u1a u0b u1b
Distribution for Random Effects           Normal
Subject Variable                          site
Optimization Technique                    Dual Quasi-Newton
Integration Method                        Adaptive Gaussian
                                          Quadrature

                      Dimensions

 Observations Used                    9824
 Observations Not Used                   0
 Total Observations                   9824
 Subjects                               20
 Max Obs Per Subject                  1311
 Parameters                             18
 Quadrature Points                       1

                   Iteration History

Iter      Calls   NegLogLike        Diff      MaxGrad       Slope
   1          4   9505.33893    1.776552     302.9489    -5975.95
   2          5   9504.95632    0.382614      25.2883     -0.7393
   3          7   9504.94447    0.011845     10.87327    -0.01258
 ...
  14         32   9504.77388    0.000062     1.124033    -0.00007

      NOTE: GCONV convergence criterion satisfied.

                      Fit Statistics
```

-2 Log Likelihood	19010
AIC (smaller is better)	19046
AICC (smaller is better)	19046
BIC (smaller is better)	19063

Parameter Estimates

Parameter	Estimate	Standard Error	DF	t Value	Pr>|t|	Alpha	Lower	Upper	Gradient
GA0	-0.9051	0.2513	16	-3.60	0.0024	0.05	-1.4378	-0.3723	0.137791
GA1HIV_Region	1.4004	0.5156	16	2.72	0.0153	0.05	0.3074	2.4934	-0.07515
GA2Ethnic	-0.7495	0.1500	16	-5.00	0.0001	0.05	-1.0674	-0.4316	-0.34124
GA3Reg_Eth	0.3208	0.3280	16	0.98	0.3426	0.05	-0.3745	1.0162	0.008273
GA4Gender	-0.1671	0.06733	16	-2.48	0.0246	0.05	-0.3098	-0.02433	-0.41608
GA5Gdmc_age	0.01559	0.004174	16	3.73	0.0018	0.05	0.006740	0.02444	-0.01175
GA6Highsch	0.05312	0.06157	16	0.86	0.4010	0.05	-0.07740	0.1836	-0.20593
V_u0a	0.7951	0.3105	16	2.56	0.0210	0.05	0.1368	1.4534	0.721235
V_u1a	0.1519	0.08207	16	1.85	0.0827	0.05	-0.02208	0.3259	-1.12403
GB0	-0.07768	0.1731	16	-0.45	0.6596	0.05	-0.4446	0.2892	-0.14343
GB1HIV_Region	0.9618	0.3736	16	2.57	0.0204	0.05	0.1699	1.7537	-0.13251
GB2Ethnic	-0.6939	0.09944	16	-6.98	<.0001	0.05	-0.9047	-0.4831	-0.27434
GB3Reg_Eth	0.09297	0.2656	16	0.35	0.7309	0.05	-0.4701	0.6560	0.194274
GB4Gender	0.009054	0.05569	16	0.16	0.8729	0.05	-0.1090	0.1271	-0.04907
GB5Gdmc_age	0.02363	0.003314	16	7.13	<.0001	0.05	0.01660	0.03066	-0.02082
GB6Highsch	0.05292	0.04917	16	1.08	0.2978	0.05	-0.05132	0.1572	0.752094
V_u0b	0.3670	0.1335	16	2.75	0.0143	0.05	0.08394	0.6501	0.541636
V_u1b	0.05953	0.04806	16	1.24	0.2333	0.05	-0.04234	0.1614	0.192796

The significance levels for the random intercept and slope are slightly different from those of *PROC GLIMMIX* due to the use of different estimation methods (maximum likelihood estimation vs. pseudo-likelihood estimation).

5.6 Multilevel models for count outcomes

Count outcomes are also very common in social and health science studies. The number of arrests, car accidents, days of smoking, times intoxicated, times hospitalized, incidence of diseases, incidence of crime, and number of days absent from school are just some examples of count outcomes.

Event counts are sometimes analyzed using the binary *logit* model or the cumulative *logit* model by recoding the original measure into dichotomous categories (e.g., 1-Smoked; 0-Did not smoke) or ordinal measures (1-Light smoking; 2-Moderate smoking; 3-Heavy smoking). By doing so, however, some important information may be lost. A more appropriate strategy for analyzing count data is the Poisson regression model. When count data are hierarchically structured, multilevel Poisson regression models should be considered.

5.6.1 Poisson regression models

Poisson regression model assumes that the distribution of the outcome measures follows a Poisson distribution. The probability density function for the Poisson distribution is:

$$\Pr(Y = y) = \frac{\exp(-\lambda)\lambda^y}{y!}, \qquad y = 0,1,2,... \tag{5.6.1.1}$$

where y is a nonnegative integer denoting the number of event occurrences, λ is the expected value of y, representing the expected number of event occurrences during a given time interval or space.

Count data with a Poisson distribution have some special features for which traditional OLS regression has limitations. First, in populations where events are very rare, the distribution is often skewed to the right, while OLS assumes a symmetric distribution. Second, event counts are always non-negative values, whereas negative values are possible for OLS regression. And finally, in the Poisson count distribution, variance is assumed to equal the mean, i.e., $\text{Var}(y) = \text{E}(y)$, thus the variance increases as the mean increases, whereas OLS does not have this assumption.

Poisson regression is a type of log-linear model. It uses a log transformation of the outcome variables to adjust for the skewness. As the mean count grows, the Poission distribution approaches a normal distribution (see Figure 5.6.1.1). Poisson model are appropriate for analyzing rare event count data. But, as the mean number of events becomes larger, linear model based on normal distribution will become appropriate, and will replace the Poisson model.

Figure 5.6.1.1 can be generated using the following SAS program:

***SAS Program 5.6.1.1**;

```
Data POISSON;
 Do n=0 to 20;
  y1=pdf('POISSON',n,0.5);
  y2=pdf('POISSON',n,1);
  y3=pdf('POISSON',n,2);
  y4=pdf('POISSON',n,3);
  y5=pdf('POISSON',n,4);
  y6=pdf('POISSON',n,5);
  y7=pdf('POISSON',n,6);
  y8=pdf('POISSON',n,7);
  y9=pdf('POISSON',n,8);
  y10=pdf('POISSON',n,9);
  y=n;
  Output;
  End;
Symbol i=spline;
Proc GPLOT;
Plot (y1-y10)*y / overlay legend;
Run;
```

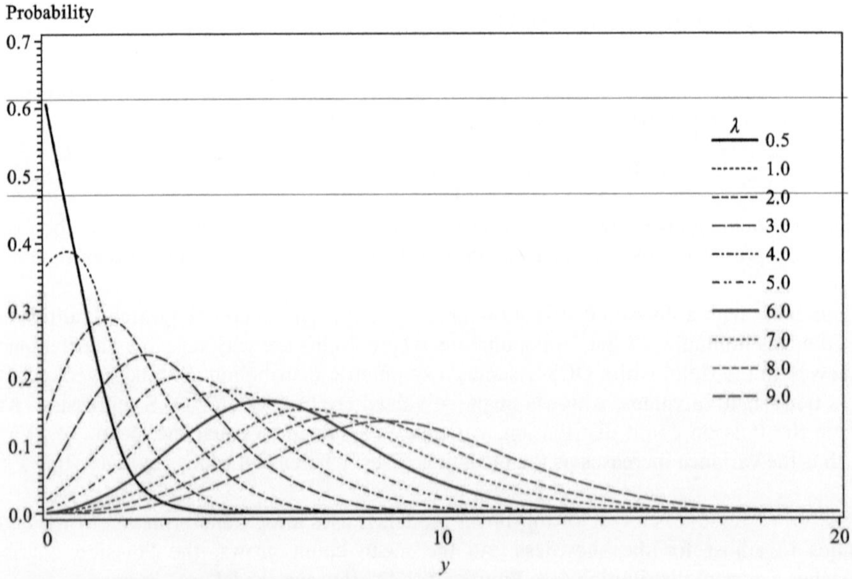

Figure 5.6.1.1 Poisson distribution with different means

In terms of the Generalized Linear Models formulation, the Poisson regression can be written as:

$$\log(\lambda) = X\beta \tag{5.6.1.2}$$

When events occur in time intervals of different lengths or in places of different sizes (e.g., populations in different geographic areas), it is often more appropriate to model the rate of event occurrences rather than absolute counts. Such data are often grouped and organized in contingency tables. In this case, the Poisson regression model shown in eq. 5.6.1.2 would be modified as follows:

$$\log\left(\frac{\lambda}{n_j}\right) = X\beta \tag{5.6.1.3}$$

$$\log(\lambda) = \log(n_j) + X\beta \tag{5.6.1.4}$$

where n_j represents the number of cases in the jth covariate pattern or the size of the jth sub-population. The log of the expected count (i.e., log (λ)) in eq. 5.6.1.2 is replaced with log of the incidence rate (i.e., log (λ / n_j)) in eq. 5.6.1.3. Unlike eq. 5.6.1.2, an additional term log(n_j), which is called the *offset*, is added on the right side of eq. 5.6.1.4. The *offset* is a regressor with a constant coefficient of 1.0. The purpose of including the *offset* in the model is to adjust for different sizes of sub-populations at risk. For example, suppose that we study the number of occurrences of a disease. We need to adjust for the number of subjects at risk in each covariate pattern or sub-population because more exposures will result in more occurrences of the disease.

A similar problem of exposure exists in longitudinal studies. For example, in a study of doctor visits among patients, if some subjects were followed for a year, some for less than a year, and the others for more than a year, then we should adjust or "standardize" different lengths of exposure time by modeling a "rate" (number doctor visits per unit of time). In this chapter, we will only discuss modeling count data rather than incidence rate.

A few points are remindful in regard to inclusion of an *offset* variable in modeling count data: 1) The n_j in log(n_j) is known and it is not treated as a random variable. SAS automatically constrains the parameter of the *offset* variable to 1 in model estimation; 2) the n_j in $\log(n_j)$ should be at least 10 to 100 times larger than the event count variable, otherwise a Poisson distribution assumption would not be appropriate (Liao, 1994); and 3) the regression coefficients would be interpreted as effects on expected rate of event occurrence, rather than expected counts, when an *offset* variable is included in the model.

5.6.2 Poisson regression with over-dispersion and a negative binomial model

As mentioned previously, the variance and mean of the Poisson distributions are assumed to be equal to each other. This rather restrictive assumption is rarely true in real data. When the variance is smaller than the mean it has an under-dispersion. When variance is larger than the mean, it has an over-dispersion. Under-dispersion is possible when observed counts are more closely clustered close to the mean. Over-dispersion is more often the case, occurring when variance is far greater than the mean. Over-dispersion can occur with grouped binary data, but it is far more common for count data.

One way to adjust for over-dispersion is to explicitly include a dispersion parameter or a scale factor in the Poisson variance function:

$$\mathrm{Var}(Y) = \phi\mu \tag{5.6.2.1}$$

If the dispersion parameter equals 1, then the adjusted Poisson model reduces to the standard Poisson model. A similar approach of accounting for either over-dispersion or under-dispersion is termed generalized Poisson model (Consul & Famoye, 1992; Famoye, 1993), where variance for the Poisson distribution is defined as $\mu(1+\alpha\mu)^2$, where α is the dispersion parameter. $\alpha >$ 0 indicates over-dispersion, $\alpha <$ 0 indicates under-dispersion, and $\alpha =$ 0 indicates the unadjusted Poisson distribution.

An alternative and widely used approach for modeling count data, which allows for over-dispersion, is the negative binomial (NB) model. The NB distribution can be viewed as a combination of a Poisson distribution and a Gamma distribution. It assumes that event counts are still generated by a Poisson process and the distribution of the unexplained extra variation underling Poisson process no longer follows the Poisson but the Gamma distribution. The pdf of NB distribution is defined as (Saha & Dong, 1997; Ridout et al., 2001)[8]:

$$\Pr(Y = y) = \frac{\Gamma(y+1/k)}{\Gamma(y+1)\Gamma(1/k)} \cdot \left(\frac{\lambda}{\lambda+1/k}\right)^{y} \cdot \left(\frac{1/k}{\lambda+1/k}\right)^{1/k} \tag{5.6.2.2}$$

where Γ is the Gamma function, λ is the Poisson mean count, and k is a dispersion parameter, which is the variance of the Gamma distribution ($k >$ 0 and does not depend on covariates).

Equation 5.6.2.2 can be rewritten as the following, which will be used in our SAS program to define the pdf of the NB distribution (SAS Institute Inc., 2007):

$$f(Y = y) = \frac{\Gamma(y+1/k)}{\Gamma(y+1)\Gamma(1/k)} \cdot \frac{(k\lambda)^{y}}{(1+k\lambda)^{y+1/k}} \tag{5.6.2.3}$$

The variance of the NB distribution is:

$$\mathrm{Var}(Y) = (\lambda + k\lambda^2) \tag{5.6.2.4}$$

When k approaches 0, the variance of the NB distribution approaches the variance of the Poisson distribution (i.e., λ), indicating no over-dispersion in the count data. As a result, the NB

[8] The index c in Saha & Dong (1997) and Ridout et al. (2001) identifies the particular form of negative binomial distribution. Usually, c=1 is used for an underlying negative binomial distribution.

distribution reduces to the Poisson distribution. Because of the feature of the dispersion parameter k in the NB distribution, the NB model naturally accounts for over-dispersion. In addition, if the model estimated k parameter is significantly greater than 0, then over-dispersion in the count data is significant, in which case the NB is preferred over the Poisson model. The formulation of the NB model is the same as the Poisson model (see eq. 5.6.1.2), except that its variance is defined differently (see eq. 5.6.2.4).

5.6.3 Multilevel Poisson and negative binomial models

To model multilevel count data, the generalized linear model for counts is extended to a generalized linear mixed model by including random components into the model. The conditional mean of Y depends upon fixed and random effects in the following form:

$$\log(\lambda) = X\beta + ZU \qquad (5.6.3.1)$$

where λ, the conditional mean of Y (i.e., $E(Y|U)$) is related to the linear predictor by a log link function. The conditional mean is assumed to equal the conditional variance, i.e., $E(Y \mid U) = Var(Y|U)$, to satisfy the standard Poisson distribution by letting the dispersion parameter $\phi = 1$. As in previous models, β and U denote the fixed-effects and random effects, respectively; and the random effects are assumed to follow a multivariate normal distribution with a mean of 0 and the variance/covariance matrix G (Fitzmaurice et al., 2004). In case of existence of over-dispersion in count data, multilevel negative binomial model should be applied (Cameron & Trivedi, 1986). As the mean function of multilevel negative binomial model is identical to the Poisson, the interpretations of the fixed and level-2 random parameter estimates for multilevel Poisson regression model apply to multilevel negative binomial model.

5.6.4 Application of multilevel Poisson and negative binomial models

The same data set (SAS data file *ML_data_Chap5*) used in previous models in this chapter is used to demonstrate application of multilevel Poisson regression to count data. In this case, the outcome measure is the total number of years ever spent in jail as reported by respondents (variable *Jailyear*). The level-1 covariates are the same as those used for other models, i.e., *Gender* (1-Male; 0-Female), *Ethnic* (1-Black; 0-White), *Gmc_age* (grand mean centered age), and *Highsch* (1-High school or above; 0-Less than high school). The contextual variable, *HIV_region* (1- Higher HIV prevalence region; 0-Lower HIV prevalence region), the macro level predictor we used earlier is no longer meaningful for the current outcome measure. Therefore, we created another contextual variable, *Hinj_Region* (1-Region with higher proportion of heavy injectors; 0-Region with lower proportion of heavy injectors) as a contextual level predictor for the mean number of years in jail.

In this section, we focus on modeling expected counts with the techniques described in this chapter, such as the multilevel Poisson regression model, the multilevel Poisson regression model with adjustment of over-dispersion, and the multilevel negative binomial model.

Model specification for multilevel Poisson regression is nearly the same as the binary logit model shown in eqs. 5.3.5.5—5.3.5.8, except that $\log(\lambda)$, instead of the *logit*, is on the left side of the level-1 equation.

Prior to fitting the Poisson model, it is important to check the distribution of the count data. We plotted the response variable, *Jailyear* using the following SAS program:

***SAS Program 5.6.4.1**;

```
LIBNAME lib 'D:\MLM\data';

Data data1;
 Set lib.ML_data_Chap5;

Proc Means mean std;
Var Jailyear;
Run;

Proc GCHART;
 VBAR3D Jailyear/discrete;
Run;
```

The results of the above SAS program show that the estimated mean and variance of count variable *Jailyear* are about 3.97 and 35.05, respectively indicating that the assumption of equal mean and variance in Poisson distribution does not hold in this data. The distribution of counts (Figure 5.6.4.1) shows that there is a spike at zero (i.e., 44 % of the responses were zero years of jail). Without taking into account the problem of extra zeros, Poisson or adjusted Poisson regression models (e.g., over-dispersed Poisson or negative binomial model) would not fit this data well. However, for the purpose of demonstration we will fit the multilevel adjusted Poisson regression models using this data set. In chapter 6, we will discuss the approaches that are more appealing for modeling count data with extra zeros.

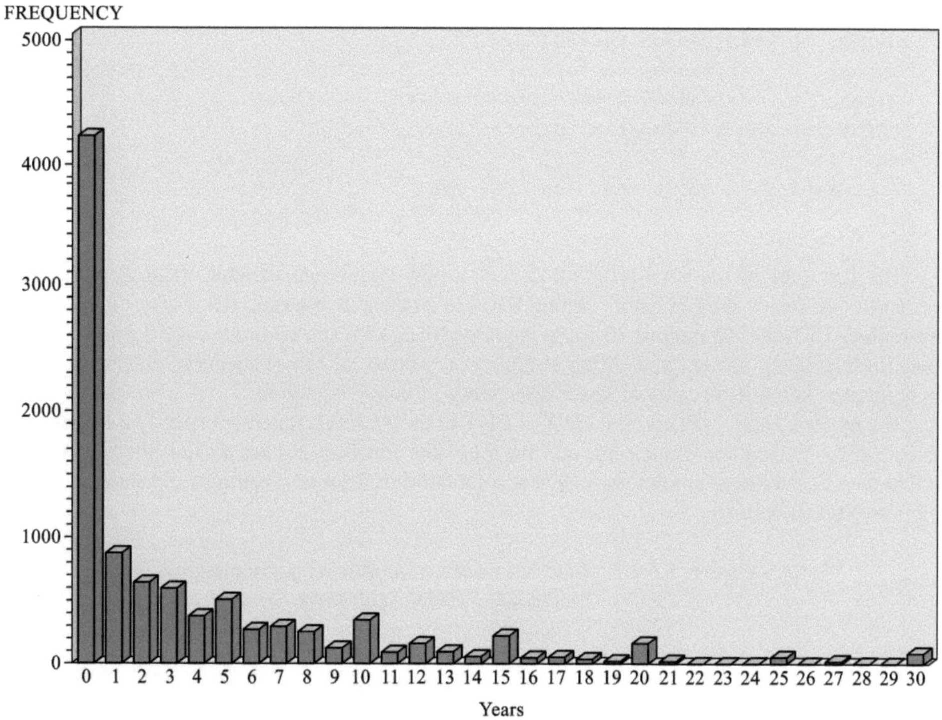

Figure 5.6.4.1 Frequency of number of years in jail in lifetime

From exploratory modeling, we discovered that the level-1 explanatory variable *Gender* has a significant random slope coefficient. Therefore, in the following section, we will fit a multilevel Poisson regression model with two random regression coefficients: random intercept and random slope of *Gender* using *PROC GLIMMIX* and *PROC NLMIXED*. We start with *PROC GLIMMIX*:

***SAS Program 5.6.4.2**;

```
...
PROC SQL;
Create table temp as select site, Heavy_inj,
 100*(mean(Heavy_inj)) as m_Heavy_inj from data1
 Group by site;
 Create table data2 as select distinct site, m_Heavy_inj,
 m_Heavy_inj>15 as Hinj_Region from temp;
Quit;

Data data3;
Merge data1 data2; by site;

 PROC GLIMMIX noitprint noclprint;
 Class site;
 ID site Jailyear;
 MODEL Jailyear=Hinj_Region|Gender Ethnic Gdmc_age Highsch
  /S DIST=POISSON LINK=LOG DDFM=BW;
 RANDOM int Gender/subject=site type=chol;
 NLOPTIONS TECH=NRRIDG;
 covtest 'H: var[random int]=0' 0;
 covtest 'H: var[random cov]=0'. 0;
 covtest 'H: var[random slope of gender]=0' . . 0;
 OUTPUT out=gmout (keep=res) pearson(blup)=res;
 Run;
 PROC Means data=gmout mean std; var res;
 Run;
```

The first part of the SAS program creates a new contextual variable, *Hinj_Region*. If the proportion of heavy drug injectors among IDUs in a research site (i.e., IDUs who injected drugs more than 100 times in the last 30 days) is greater than 15% (an arbitrary cut-off point), the site is defined as being located in a region of higher prevalence of heavy injection; otherwise the site is defined as being in a region of lower prevalence of heavy injection.

The options *DIST = POISSON LINK = LOG* in the *MODEL* statement specify a Poisson distribution for the outcome measure, and the *Log link* function is used to link the mean of the count data to the linear predictors, which is a function of a set of covariates. Selected model results are shown bellow:

SAS Output 5.6.4.2 Selected results of multilevel poisson regression model using *PROC GLIMMIX*

```
                    Model Information

   Data Set                          WORK.DATA3
```

```
Response Variable             Jailyear
Response Distribution         Poisson
Link Function                 Log
Variance Function             Default
Variance Matrix Blocked By    site
Estimation Technique          Residual PL
Degrees of Freedom Method     Between-Within

Number of Observations Read        9824
Number of Observations Used        9685

                  Dimensions

G-side Cov. Parameters             3
Columns in X                       7
Columns in Z per Subject           2
Subjects (Blocks in V)            20
Max Obs per Subject             1305

                  Optimization Information

Optimization Technique        Newton-Raphson with Ridging
Parameters in Optimization    3
Lower Boundaries              2
Upper Boundaries              0
Fixed-effects                 Profiled
Starting From                 Data

       Convergence criterion (PCONV=1.11022E-8) satisfied.

                    Fit Statistics

       -2 Res Log Pseudo-Likelihood     77637.51
       Generalized Chi-Square           71261.04
       Gener. Chi-Square / DF               7.36

              Covariance Parameter Estimates

                                             Standard
       Cov Parm      Subject     Estimate       Error

       CHOL(1,1)     site          0.5124     0.09458
       CHOL(2,1)     site         -0.4467     0.09703
       CHOL(2,2)     site          0.1747     0.03074

                      Solutions for Fixed-effects

                                   Standard
Effect                 Estimate       Error       DF    t Value    Pr > |t|
```

Intercept	0.07834	0.1769	18	0.44	0.6632
Hinj_Region	0.4463	0.2362	18	1.89	0.0750
Gender	1.0456	0.1665	9660	6.28	<.0001
Hinj_Region*Gender	-0.2350	0.2224	9660	-1.06	0.2908
Ethnic	0.1976	0.01434	9660	13.79	<.0001
Gdmc_age	0.03921	0.000664	9660	59.01	<.0001
Highsch	-0.00527	0.01059	9660	-0.50	0.6190

Tests of Covariance Parameters
Based on the Residual Pseudo-Likelihood

Label	DF	-2 Res Log P-Like	ChiSq	Pr > ChiSq	Note
H: var[random int]=0	1	78004	366.50	<.0001	MI
H: var[random cov]=0	1	77671	33.49	<.0001	DF
H: var[random slope of gender]=0	1	77838	200.46	<.0001	MI

DF: P-value based on a chi-square with DF degrees of freedom.
MI: P-value based on a mixture of chi-squares.

Analysis Variable : res Pearson Residual

Mean	Std Dev
-0.0082261	2.7120140

The *Model Information* panel in the above SAS output shows that Poisson distribution for response and log link function is used for this model. The *Dimension* panel indicates the model has 7 fixed parameters, and 3 elements for *G*-matrix. The *Covariance Parameter Estimates* panel lists the variances of the random intercept (*CHOL*(1,1)) and the random slope of *Gender* (*CHOL*(2,2)), as well as the covariance of the two random coefficients (*CHOL*(2,1)). All random components are statistically significant (p-values < 0.0001).

Pearson Chi-Square divided by the degrees of freedom is used to detect over-dispersion or under-dispersion in the Poisson regression model. Values greater than 1.0 indicate over-dispersion (i.e., more variation was observed in the response than what is expected by the Poisson assumption); values smaller than 1.0 indicate under-dispersion (i.e., less variation was observed in the response than what is expected by the Poisson assumption). The *Generalized Chi-square /DF* statistic in the *Fit Statistics* panel is one measure of dispersion in the Poisson regression. In this example, *Chi-square /DF* = 7.36, indicating an over-dispersion in the data. With over-dispersion the standard errors of the Poisson regression coefficient estimates would be biased although the coefficient estimates may not be (Cox, 1983).

Some insight into whether assessment of over-dispersion can be gained by examining the Pearson-type residuals in *PROC GLIMMIX* (Littell et al., 2006). As suggested by these authors, we added the statement, "*OUTPUT out* = *gmout* (*keep* = *res*) *pearson*(*blup*) = *res*" in the SAS program to output the residuals using the BLUPs (i.e., best linear unbiased predictors). The variance of the *Pearson-type residuals* is about $2.712014^2 = 7.36$, which is much larger than 1.0. This reinforces that over-dispersion in the count data needs to be handled.

A simple adjustment for over-dispersion is to add an extra dispersion parameter (i.e., a scale parameter ϕ) in the count variable's variance function to allow variance greater than the variance expected from the Poisson distribution (see eq. 5.6.2.1). The *RANDOM _RESIDUAL_* statement in *PROC GLIMMIX* allows us to add this scale parameter or residual variance to deal with over-dispersion problem. Alternatively, we may apply the negative binomial (NB) model that allows the count variable to have a larger variance than its mean. We will first demonstrate how to fit the multilevel Poisson regression with adjustment for over-dispersion, and then we will fit the BN model to the count data.

***SAS Program 5.6.4.3**;

```
...
PROC GLIMMIX noitprint noclprint;
class site;
MODEL Jailyear = Hinj_Region|Gender Ethnic Gdmc_age Highsch
 /S DIST=POISSON LINK = LOG DDFM = BW;
RANDOM int Gender/subject = site type=chol;
RANDOM _RESIDUAL_ /Subject = site;
nloptions tech = nrridg;
covtest 'H: var[random int] = 0' 0;
covtest 'H: var[random cov] = 0' . 0;
covtest 'H: var[random slope of gender] = 0' . . 0;
OUTPUT out=gout(keep=res2) pearson(blup) = res2;
Run;
PROC Means data=gout mean std; var res2;
Run;
```

Note that the key word *_RESIDUAL_* in the *RANDOM* statement adds an over-dispersion parameter to the model.

SAS Output 5.6.4.3 Selected results of multilevel poisson model with over-dispersion adjustment using *PROC GLIMMIX*

Model Information

Data Set	WORK.DATA3
Response Variable	Jailyear
Response Distribution	Poisson
Link Function	Log
Variance Function	Default
Variance Matrix Blocked By	site
Estimation Technique	Residual PL
Degrees of Freedom Method	Between-Within

Number of Observations Read	9824
Number of Observations Used	9685

Dimensions

G-side Cov. Parameters	3
R-side Cov. Parameters	1
Columns in X	7
Columns in Z per Subject	2
Subjects (Blocks in V)	20

Max Obs per Subject	1305

Optimization Information

Optimization Technique	Newton-Raphson with Ridging
Parameters in Optimization	3
Lower Boundaries	2
Upper Boundaries	0
Fixed-effects	Profiled
Residual Variance	Profiled
Starting From	Data

Convergence criterion (PCONV=1.11022E-8) satisfied.

Fit Statistics

-2 Res Log Pseudo-Likelihood	35184.58
Generalized Chi-Square	71267.01
Gener. Chi-Square / DF	7.36

Covariance Parameter Estimates

Cov Parm	Subject	Estimate	Standard Error
CHOL(1,1)	site	0.2949	0.08649
CHOL(2,1)	site	-0.1754	0.09441
CHOL(2,2)	site	0.1188	0.03766
Residual (VC)		7.3638	0.1061

Solutions for Fixed-effects

Effect	Estimate	Standard Error	DF	t Value	Pr > |t|
Intercept	0.1766	0.1424	18	1.24	0.2308
Hinj_Region	0.3180	0.1756	18	1.81	0.0869
Gender	0.9425	0.1238	9660	7.62	<.0001
Hinj_Region*Gender	-0.07803	0.1518	9660	-0.51	0.6072
Ethnic	0.2023	0.03813	9660	5.31	<.0001
Gdmc_age	0.03928	0.001800	9660	21.82	<.0001
Highsch	-0.00581	0.02869	9660	-0.20	0.8395

Tests of Covariance Parameters
Based on the Residual Pseudo-Likelihood

Label	DF	-2 Res Log P-Like	ChiSq	Pr > ChiSq	Note
H: var[random int]=0	1	35211	26.31	<.0001	MI
H: var[random cov]=0	1	35190	5.08	0.0242	DF
H: var[random slope of gender]=0	1	35191	6.00	0.0072	MI

```
DF: P-value based on a chi-square with DF degrees of freedom.
            MI: P-value based on a mixture of chi-squares.

                    Analysis Variable : res2 Pearson Residual

                              Mean           Std Dev
                         ---------------------------
                          -0.0045605        0.9985563
                         ---------------------------
```

In the above SAS output, the statement "*R-side Cov. Parameters*" in the *Dimension* panel shows there is one parameter in the *R* matrix (i.e., the level-1 residual variance/covariance matrix). This indicates that a scale parameter was included in the model. The *Generalized Chi-square /DF* statistic (7.36) in *Fit Statistics* panel remains the same as that in SAS Output 5.6.4.1. The variance of the *Pearson-type residuals* for the conditional Poisson distribution is $0.9986^2 \approx 1.0$. This indicates that simple adjustment by adding the scale parameter ϕ is effective.

Nevertheless, the result is still not conclusive for two reasons. First, in a model with random effects, two mechanisms, random effects and changing the variance function, are in competition for variability in the data. If we compare the current model with the previous model, the estimated dispersion parameter (i.e., *Generalized Chi-Square /DF*) is the same (i.e., 7.36). However, the variances and covariance of the random coefficients have been substantially reduced from 0.5124, 0.1747, and −0.4467 in the model without adjustment for over-dispersion to 0.2949, 0.1188, and −0.1754 in the model with adjustment for over-dispersion respectively (SAS Output 5.6.4.2 and SAS Output 5.6.4.3). These variance parameters became less significant and the Chi-square values for each component were reduced substantially.

A large estimate of residual dispersion is not necessarily indicative of an over-dispersion problem. The observed over-dispersion may be attributed to many other causes, such as model misspecification (e.g., some important fixed-effects covariates are omitted in the model), outliers in the data, or a distribution that is simply not a Poisson distribution (Littell et al., 2006).

Now, we turn to the multilevel negative binomial model that naturally allows over-dispersion. The SAS program follows:

***SAS Program 5.6.4.4;**

```
...
PROC GLIMMIX noitprint noclprint;
Class site;
MODEL Jailyear = Hinj_Region|Gender Ethnic Gdmc_age Highsch
 /S DIST = NB LINK = LOG DDFM = BW;
RANDOM int Gender/subject = site type = chol;
NLOPTIONS TECH = NRRIDG;
covtest 'H: var[random int] = 0' 0;
covtest 'H: var[random cov] = 0' . 0;
covtest 'H: var[random slope of gender] = 0' . . 0;
Run;
```

The SAS syntax for running multilevel negative binomial regression is almost the same as for multilevel Poisson regression. The only change is in the *MODEL* statement where we specify *DIST = NB rather than DIST = POISSON.*

SAS Output 5.6.4.4 Selected results of the multilevel negative binomial model using *PROC GLIMMIX*

Fit Statistics

-2 Res Log Pseudo-Likelihood	36910.56
Generalized Chi-Square	10360.10
Gener. Chi-Square / DF	1.07

Covariance Parameter Estimates

Cov Parm	Subject	Estimate	Standard Error
CHOL(1,1)	site	0.3585	0.08821
CHOL(2,1)	site	-0.2700	0.09795
CHOL(2,2)	site	0.1401	0.03713
Scale		2.0698	0.03683

Solutions for Fixed-effects

Effect	Estimate	Standard Error	DF	t Value	Pr > |t|
Intercept	0.02956	0.1449	18	0.20	0.8406
Hinj_Region	0.4610	0.1853	18	2.49	0.0229
Gender	1.0792	0.1299	9660	8.31	<.0001
Hinj_Region*gender	-0.2136	0.1689	9660	-1.26	0.2062
Ethnic	0.2190	0.04078	9660	5.37	<.0001
Gdmc_age	0.04845	0.002204	9660	21.99	<.0001
Highsch	-0.02921	0.03276	9660	-0.89	0.3725

Tests of Covariance Parameters
Based on the Residual Pseudo-Likelihood

Label	DF	-2 Res Log P-Like	ChiSq	Pr > ChiSq	Note
H: var[random int]=0	1	36952	41.19	<.0001	MI
H: var[random cov]=0	1	36922	11.18	0.0008	DF
H: var[random slope of gender]=0	1	36922	11.30	0.0004	MI

DF: P-value based on a chi-square with DF degrees of freedom.
MI: P-value based on a mixture of chi-squares.

The *Fit Statistics* panel in SAS output 5.6.4.4 shows that the ratio of the generalized chi-square statistic to its d.f. is very close to 1.0 (*Generalized Chi-square /DF* = 1.07). This indicates that there is almost no residual over-dispersion, thus the variability in these data has been properly modeled. A scale or dispersion parameter of 2.0698 was included in the model by default (see the *Covariance Parameter Estimates* section in the SAS output).

To compare the three models, we graph the predicted probability from each model against the distribution of the original data. Figure 5.6.4.2 shows that the negative binomial model appears to fit the observed data slightly better than either the standard Poisson or the over-dispersion adjusted Poisson model, but none of the models fits the data very well. This is because these models do not capture the zero-inflated part of the data, and thus may not be suitable. We will discuss how to deal with the extra zeros in count data in Chapter 6.

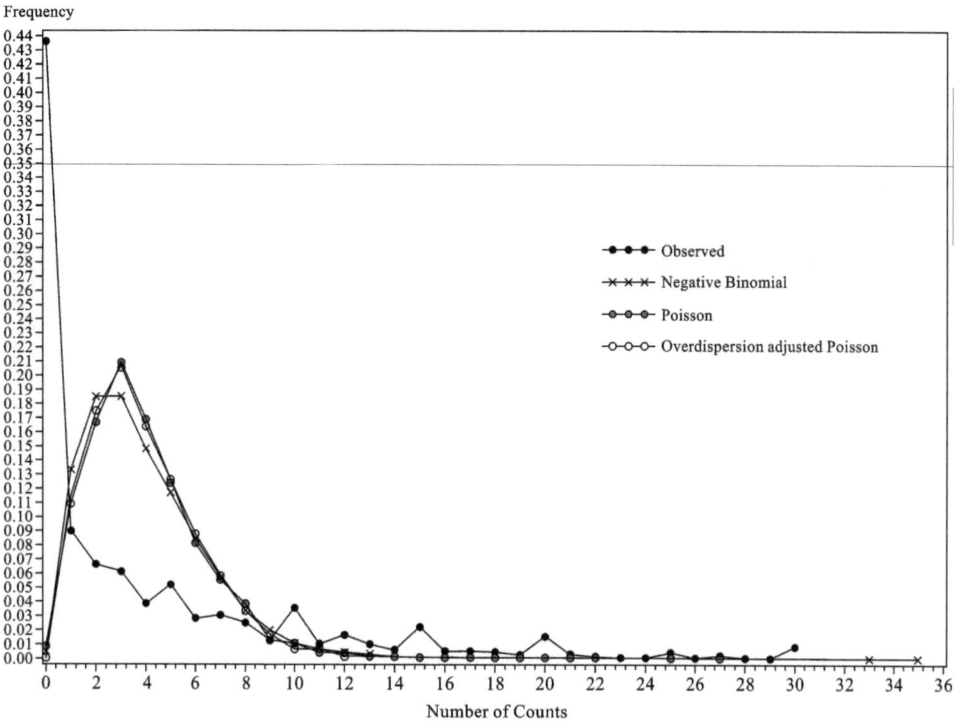

Figure 5.6.4.2 Distribution of observed and predicted number of years in jail in lifetime

The following SAS program generates Figure 5.6.4.2.

***SAS Program 5.6.4.5;**

```
...
PROC SQL;
Create table temp as select site, Heavy_inj,
 100*(mean(Heavy_inj)) as m_Heavy_inj, count(site) as Nj from data1
Group by site;
Create table data2 as select distinct site, m_Heavy_inj, Nj,
m_Heavy_inj>15 as Hinj_Region from temp;
Quit;

Data data3;
Merge data1 data2; by site;

*(1) Poisson;
Proc GLIMMIX data=data3 noitprint noclprint;
Class site;
```

```
ID site Jailyear;
Model Jailyear=Hinj_Region|Gender Ethnic Gdmc_age Highsch
 /S dist=POISSON link=LOG DDFM=BW;
RANDOM int Gender/subject=site g;
NLOPTIONS tech=nrridg;
OUTPUT out=psstats pred(ilink)=p;
Run;
Data psstats(keep=site rp dist);
 Set psstats;
rp=round(p);
dist='Poisson';
Run;
Title 'Poisson Predictions';
Proc Print data=psstats;
Run;

*(2) Overdispersed Possion;
Proc GLIMMIX data=data3 noitprint noclprint;
Class site;
Model Jailyear=Hinj_Region|Gender Ethnic Gdmc_age Highsch
 /S dist=POISSON link=LOG DDFM=BW;
RANDOM Int Gender/subject=site;
RANDOM _Residual_ /subject=site;
NLOPTIONS tech=nrridg;
OUTPUT out=psstats2 pred(ilink)=p;
Run;
Data psstats2(keep=site rp dist);
 Set psstats2;
rp=round(p);
dist='Poissonr';
Run;
title 'Overdispersed Poisson Predictions';
Proc print data=psstats2;
Run;

*(3) Negative Binomial;
Proc GLIMMIX data=data3 noitprint noclprint;
Class site;
Model Jailyear=Hinj_Region|Gender Ethnic Gdmc_age Highsch
 /S dist=NB link=LOG Ddfm=Bw;
RANDOM Int Gender/subject=site g;
NLOPTIONS tech=nrridg;
OUTPUT out=nbstats pred(ilink)=p;
Run;

Data nbstats(keep=rp dist);
 Set nbstats;
rp=round(p);
dist='NegBin';
Run;
Title 'Negative Binomial Predictions';
Proc Print data=nbstats;
Run;

* (4) Raw data;
Data rwstats(keep=rp dist);
 Set data3;
rename Jailyear=rp;
dist='Observed';
Run;
Title 'Raw Data';
```

```
Proc Print data=rwstats;
Run;

*Combined;
Data combined;
 Set rwstats psstats psstats2 nbstats;
Run;

Proc Freq data=combined;
by dist notsorted;
Tables rp / out=graphstats;
Run;

Data graphstats;
 Set graphstats;
prob=percent/100;
Label prob='Probability' rp='Number of Events';
Run;

Proc Print data=graphstats;
Run;

symbol1 i=j v=star r=1 c=blue;
symbol2 i=j v=dot r=1 c=black;
symbol3 i=j v=square r=1 c=cyan;
symbol4 i=j v=circle r=1 c=red;
axis1 label=('Number of Events') order=(0 to 36 by 1);
axis2 label=(angle=90 h=1 f=swiss 'Probability');
Title 'Poisson, Overdispersed Poisson and Negative Binomial Model Fit Com-
parison';
Proc Gplot data=graphstats;
Plot prob*rp=dist / haxis=axis1 vaxis=axis2;
Run;
Quit;
```

We now turn to fitting the model using *PROC NLMIXED*. Since *PROC NLMIXED* does not have an option to include a dispersion parameter, we will only fit the multilevel Poisson regression model without adjusting for over-dispersion; and then we will fit the negative binomial models to adjust for over-dispersion.

***SAS Program 5.6.4.6**;

```
...
PROC SQL;
Create table temp as select site, Heavy_inj,
 100*(mean(Heavy_inj)) as m_Heavy_inj from data1
 Group by site;
 Create table data2 as select distinct site, m_Heavy_inj,
 m_Heavy_inj>15 as Hinj_Region from temp;
Quit;

Data data3;
Merge data1 data2; by site;

Hinj_gender=Hinj_Region*Gender;
PROC NLMIXED;
PARMS   B0=0.03   B_Hinj_Region=0.46   B_Gender=1.08   B_Hinj_gender=-0.21
B_Ethnic=0.22  B_Gdmc_age=0.05 B_Highsch=-0.03  V_u0=0.35  V_u1=0.14  Cov_u01=0;
eta=B0+B_Hinj_Region*Hinj_Region+B_Gender*Gender+B_Hinj_gender*Hinj_gender+
```

```
    B_Ethnic*Ethnic+B_Gdmc_age*Gdmc_age+B_Highsch*Highsch+u0+u1*Gender;
  lambda=exp(eta);
  MODEL Jailyear ~ POISSON(lambda);
  RANDOM u0 u1 ~ NORMAL([0,0], [V_u0,Cov_u01,V_u1]) SUBJECT=SITE;
  RUN;
```

The parameters of the multilevel Poisson model are specified in the *PARMS* statement, and their starting values are from results of the model fitted by *PROC GLIMMIX*. The level-1 intercept and the slope of variable *Gender* are treated as random coefficients; and these two random effects are assumed to follow a bivariate normal distribution with zero means, free variances (i.e., *V_u*0, and *V_u*1), and covariance (i.e., *Cov_u*01) in the *G* matrix (see the *RANDOM* statement in the above SAS program) to be estimated.

SAS Output 5.6.4.6 Selected results of multilevel poisson model using PROC NLMIXED

Fit Statistics

-2 Log Likelihood	78244
AIC (smaller is better)	78264
AICC (smaller is better)	78264
BIC (smaller is better)	78274

Parameter Estimates

Parameter	Estimate	Standard Error	DF	t Value	Pr > |t|	Alpha	Lower	Upper	Gradient
B0	0.07040	0.1679	18	0.42	0.6799	0.05	-0.2823	0.4231	-0.01097
B_Hinj_Region	0.4520	0.2237	18	2.02	0.0585	0.05	-0.01808	0.9220	0.077375
B_Gender	1.0525	0.1577	18	6.67	<.0001	0.05	0.7212	1.3839	0.013015
B_Hinj_gender	-0.2394	0.2104	18	-1.14	0.2701	0.05	-0.6813	0.2026	-0.01665
B_Ethnic	0.1978	0.01433	18	13.80	<.0001	0.05	0.1676	0.2279	0.165664
B_Gdmc_age	0.03921	0.000664	18	59.02	<.0001	0.05	0.03782	0.04061	-1.33893
B_Highsch	-0.00528	0.01059	18	-0.50	0.6244	0.05	-0.02752	0.01697	-0.09941
V_u0	0.2333	0.08268	18	2.82	0.0113	0.05	0.05961	0.4070	-0.33143
V_u1	0.2033	0.07476	18	2.72	0.0141	0.05	0.04625	0.3604	-0.42482
Cov_u01	-0.2027	0.07627	18	-2.66	0.0160	0.05	-0.3629	-0.04246	-0.68785

The estimates of the fixed-effects are close to those estimated from *PROC GLIMMIX* (see SAS Output 5.6.4.2), but the residual variance/covariance estimates are somewhat different.

As noted earlier, *PROC NLMIXED* does not have an option for including a dispersion parameter to adjust for over-dispersion in count data. Although over-dispersion can be adjusted in fitting the multilevel negative binomial model using *PROC NLMIXED*, it is difficult to fit a multilevel negative binomial model with more than one random effect. Involving multiple random coefficients is very time consuming and often encounters model convergence problems. For the purpose of demonstration, we only include one random intercept in the following SAS program for multilevel negative binomial models.

```
*SAS Program 5.6.4.7;
...
PROC SQL;
Create table temp as select site, Heavy_inj,
 100*(mean(Heavy_inj)) as m_Heavy_inj from data1
Group by site;
Create table data2 as select distinct site, m_Heavy_inj,
 m_Heavy_inj>15 as Hinj_Region from temp;
Quit;

Data data3;
Merge data1 data2; by site;
Hinj_gender=Hinj_Region*Gender;

PROC NLMIXED tech = nrridg;
PARMS B0 = 0.07 B_Hinj_Region=0.46 B_Gender=1.05 B_Hinj_gender=-0.21
   B_Ethnic = 0.20 B_Gdmc_age=0.05 B_Highsch=0 STD=0.4 k=2;
eta =B0+B_Hinj_Region*Hinj_Region+B_Gender*Gender+B_Hinj_gender*Hinj_gender
   +B_Ethnic*Ethnic+B_Gdmc_age*Gdmc_age+B_Highsch*Highsch+u0;
lambda=exp(eta);
*pdf;
pdf=gamma(Jailyear+1/k)/(gamma(Jailyear+1) * (gamma(1/k))) * (((k*lambda)
**Jailyear)/(1+k*lambda) ** (Jailyear+1/k));
*loglikelihood;
if pdf > 1e-8 then LL = log(pdf); else LL = -1e20;
Model Jailyear ~ GENERAL(LL);
RANDOM u0 ~ NORMAL(0, STD*STD) SUBJECT=SITE;
Estimate 'intercept variance s2v' std*std;
Run;
```

In SAS program 5.6.4.7 we specified a probability density and a log-likelihood for the negative binomial distribution directly. The probability density function (PDF) is defined based on eq. 5.6.2.3. The *IF-THEN-ELSE* statement defines the log-likelihood function, and protects it from taking the log of zero, which is undefined. If the value of pdf is close to 0 (e.g., e^{-8}), we define the log-likelihood to be a very large negative number[9].

The variance of the random intercept is specified in terms of the standard deviation; i.e., *STD* STD*, where *STD* is the standard deviation of the random intercept. Coding in this way helps to stabilize the numerical process (Agresti et al., 2000).

SAS Output 5.6.4.7 Selected results of multilevel negative binomial model using *PROC NLMIXED*

```
                         Specifications

         Data Set                                      WORK.DATA3
         Dependent Variable                            jailyear
```

[9] In the IF statement of the SAS program, the SAS code –1e20 generates a very large negative log likelihood value, indicating that the corresponding likelihood value is very close to zero.

Distribution for Dependent Variable	General
Random Effects	u0
Distribution for Random Effects	Normal
Subject Variable	site
Optimization Technique	Newton-Raphson Ridge
Integration Method	Adaptive Gaussian Quadrature

Dimensions

Observations Used	9685
Observations Not Used	139
Total Observations	9824
Subjects	20
Max Obs Per Subject	1305
Parameters	9
Quadrature Points	21

Parameters

B0	B_Hinj_ Region	B_gender	B_Hinj_ gender	B_ethnic	B_gdmc_ age	B_highsch
0.07	0.46	1.05	-0.21	0.2	0.05	0

std	k	NegLogLike
0.4	2	21969.2293

Iteration History

Iter	Calls	NegLogLike	Diff	MaxGrad	Rho
1	23	21941.7677	27.46158	88.33897	1.7308
2	35	21933.5953	8.172406	43.60732	1
3	46	21929.8868	3.708557	63.58834	1.01129
4	57	21927.8764	2.01038	15.94016	2.00177
5	68	21927.7506	0.125819	2.077178	1.09058
6	79	21927.748	0.002531	0.051537	1.01786
7	90	21927.748	1.696E-6	0.000047	1.0013

NOTE: GCONV convergence criterion satisfied.

Fit Statistics

-2 Log Likelihood	43855
AIC (smaller is better)	43873
AICC (smaller is better)	43874
BIC (smaller is better)	43882

Parameter Estimates

Parameter	Estimate	Standard Error	DF	t Value	Pr > |t|	Alpha	Lower	Upper	Gradient
B0	0.09584	0.09545	19	1.00	0.3280	0.05	-0.1039	0.2956	-0.00002
B_Hinj_Region	0.4067	0.1105	19	3.68	0.0016	0.05	0.1754	0.6379	0.000011
B_Gender	1.0136	0.07567	19	13.40	<.0001	0.05	0.8553	1.1720	1.3E-6
B_Hinj_gender	-0.1594	0.08840	19	-1.80	0.0872	0.05	-0.3444	0.02561	-0.00001
B_Ethnic	0.2184	0.04304	19	5.08	<.0001	0.05	0.1284	0.3085	-5.45E-6
B_Gdmc_age	0.04912	0.002457	19	19.99	<.0001	0.05	0.04398	0.05426	-0.00001
B_Highsch	-0.02800	0.03460	19	-0.81	0.4284	0.05	-0.1004	0.04442	-3.79E-6
std	0.1630	0.03358	19	4.86	0.0001	0.05	0.09277	0.2333	-0.00005
k	2.3187	0.04629	19	50.09	<.0001	0.05	2.2218	2.4156	-4.24E-7

Additional Estimates

Label	Estimate	Standard Error	DF	t Value	Pr > |t|	Alpha	Lower	Upper
intercept variance s2v	0.02658	0.01095	19	2.43	0.0253	0.05	0.003666	0.04950

Because we use the *general*() distribution to specify the negative binomial distribution whose likelihood functions are defined using programming statements, the distribution for the response variable in the *Specifications* panel in the above SAS output is listed as "*General*" rather than negative binomial. The model converged after 7 iterations. The SAS output reports a dispersion parameter $k = 2.3187$ ($p < 0.0001$). The significant dispersion parameter indicates that an over-dispersion problem exists for the data, and the model is properly specified otherwise. All of the fit statistics are smaller than those of the multilevel Poisson model, indicating that the multilevel negative binomial model has a relatively better fit.

In this section, we have discussed and demonstrated how to fit count data with Poisson regression and negative binomial models. Up to this point, our attention has been placed on model formulation and model fit using different SAS procedures. The fixed-effects parameters of both the Poisson regression model and the negative binomial model are interpreted in the same way, as the effects on log of expected count, because the two models use the same log link function.

Let's interpret the parameter estimates shown in SAS Program 5.6.4.7 Output as an example. The intercept ($B_0 = 0.096$, $p = 0.3280$) of the multilevel negative binomial model can be interpreted as the average log of expected total years ever spent in jail for the reference group who are female IDUs (*Gender* = 0), White (*Ethnic* = 0) with less than high school education (*Highsch* = 0), with an average age among the IDUs (i.e., *Gdmc_age* = 0), and in a lower prevalence of heavy injection region (*Hinj_Region* = 0). In other words, the average total years ever spent in jail for the reference group was exp(0.096) = 1.10 years.[10]

The slope coefficient of the contextual variable, *Hinj_Region* is positive and statistically significant (i.e., *B_Hinj_Region* = 0.41, $p = 0.0016$), indicating that the total expected number of years ever spent in jail for IDUs in the region of higher prevalence of heavy injection (i.e., *Hinj_Region* = 1) was about exp(0.41) = 1.51 times longer than in the region of lower preva-

[10] The intercept B_0 is not statistically significant indicating that the log of expected years ever spent in jail for the reference group is not significantly different from 0. In other words, the expected number of years in jail is not significantly different from 1 year.

lence heavy injection (i.e., *Hinj_Region* = 0), controlling for other factors. The ratio of expected number of years for male IDUs to the expected number of years in jail for female IDUs was exp(*B_Gender*) = exp(1.01) = 2.75 ($p < 0.0001$), indicating that male IDUs were likely to be in jail for a much longer time than female IDUs.

The cross-level interaction between the contextual variable *Hinj_Region* and the level-1 covariate *Gender* is negative (*B_Hinj_gender* = – 0.16), indicating that the gender difference in regard to total expected number of years ever spent in jail was smaller in the region of higher prevalence of heavy injection, compared with the region of lower prevalence of heavy injection (e.g., the total expected number of years ever spent in jail for male IUDs was about exp(1.01 − 0.16) = 2.34 times longer than female IDUs in the region of higher prevalence of heavy injection). However, this moderating effect of region on gender effect was only marginally significant ($p = 0.0872$).

Being a Black IDU is positively associated with a longer expected years in jail (i.e., exp(0.22) = 1.25), about 25% longer than White IDUs. A one year increase in age would increase the total expected number of years ever spent in jail by about 1.05 times (exp(0.05) =1.05), other things being equal. However, education did not seem to have a significant effect on the expected number of years in jail (*B_Highsch* = −0.03, $p = 0.4284$).

In regard to the random effect, the variance of the random intercept is statistically significant ($s2v = 0.03$, $p = 0.0253$). As interpreted in other multilevel models, this means that the total expected number of years ever spent in jail among IDUs significantly varies across the project sites controlling for the covariates in the model.

Both the Poisson and the negative binomial models are useful for multilevel modeling of count data. The negative binomial model may be preferred for data with over-dispersion However, as formulated, both the Poisson and negative binomial models may not be appropriate to fit the count data when there are extra zeros (see Figure 5.6.4.1). In next chapter, we will discuss a special model to fit count data with extra zeros.

Chapter 6
Other applications of multilevel modeling and related issues

In Chapters 3—5 we have discussed various multilevel models to fit continuous or discrete outcome measures with cross-sectional and longitudinal data. In this chapter we will discuss methods that extend the range of multilevel applications and other considerations for applying multilevel modeling.

6.1 Multilevel zero-inflated models for count data with extra zeros

In analysis of count data, the standard Poisson regression model is limited by the assumption that the mean of the outcome equals the variance. This assumption is often violated in real world data. In Chapter 5 we introduced the Poisson regression with adjustment for over-dispersion and the negative binomial regression model that allows for over-dispersion or under-dispersion. However, when count data have a large number of zeros, Poisson regression and negative binomial models are not appropriate.

Under many circumstances, count data contain more zeros than that would be expected in the Poisson distribution. Zero outcomes in count data may arise from two different processes. For example, suppose we study the number of years that subjects have ever been incarcerated. There may be two different types of zero count measures. Among those who have ever been arrested, the number of years incarcerated could be zero or non-zero integers. Zero counts in this case obey the Poisson process and can be called sampling zeros. On the other hand, among those who have never been arrested, the outcome should always to be zero. This type of zero outcomes is structural, and is called "extra zeros" or "excess zeros" in count data.

Count data with extra zeros can be better analyzed using zero-inflated models. In recent years a number of zero-inflated models have been developed, including the zero-inflated Poisson (ZIP) model (Lambert, 1992), the zero-inflated negative binomial (ZINB) model (Green, 1994), the zero-inflated hurdle model (King, 1989; Mullahy, 1986), and the zero-altered Poisson model (ZAP) (Heilbron, 1994). These models have also been expanded to multilevel and longitudinal data to accommodate both observation dependence and the extra-zeros (Hall, 2000; Hur et al., 2002; Lee & Wang, 2006; Min & Agresit, 2005; Yau et al., 2003).

In this section we will introduce some basic concepts of the ZIP and ZINB models, discuss expansion of these models to multilevel data and demonstrate how to fit these models to clustered count data with extra zeros using SAS.

6.1.1 Fixed-effect zero-inflated Poisson (ZIP) model

The zero-inflated Poisson (ZIP) model employs two components that correspond to two zero generating processes. The first process governed by a binary distribution generates structural zeros, and the second process governed by the Poisson distribution generates Poisson counts that may also include zero values. The two model components are described as follows (Lambert, 1992; Hur et al., 2002; Hedeker & Gibbons, 2006):

$$\Pr(y_i = 0) = \pi_i + (1-\pi_i)\exp(-\lambda_i) \tag{6.1.1.1}$$

$$\Pr(y_i = h_i) = (1-\pi_i)\frac{\exp(-\lambda_i)\lambda_i^{h_i}}{h_i!},\ h_i > 0 \tag{6.1.1.2}$$

where the outcome variable y_i has any non-negative integer value; λ_i is the expected Poisson count for the ith individual; π_i is the probability of extra zeros. Given no additional information, zero counts in eq. 6.1.1.1 could be generated either from the process governed by a binary distribution or from a process governed by a Poisson process. The overall probability of zero counts (i.e., $y_i = 0$) is defined as a combination of the probability of structural zeros and the probability of expected Poisson zeros[1] (Hur et al., 2002). Equation 6.1.1.2 defines the probability of positive Poisson counts (i.e., $y_i = h_i > 0$, where h_i is the event count). Equations 6.1.1.1 and 6.1.1.2 can be combined into one equation (Hur et al., 2002):

$$\Pr(y_i) = \pi_i I(y_i) + (1-\pi_i)f(y_i) \tag{6.1.1.3}$$

When $\pi_i > 0$, we have a zero-inflated distribution; when $\pi_i < 0$, we have a zero-deflated distribution, which seldom arises in practice; and when $\pi_i = 0$, the ZIP distribution reduces to the standard Poisson distribution. $I(y_i)$ is an indicator function taking a value of 1 if the observed response is zero ($y_i = 0$), and a value of 0 if the observed response is greater than zero ($y_i = h_i > 0$). $f(y_i)$ is the Poisson distribution function:

$$f(y_i) = \frac{\exp(-\lambda_i)\lambda_i^{y_i}}{y_i!} \tag{6.1.1.4}$$

The ZIP model simultaneously estimates the two parameters, π_i — the probability of extra zeros, and λ_i — the expected Poisson count. The *logit* link function is used to model the likelihood of extra/structural zeros:

$$\log\left(\frac{\pi}{1-\pi}\right) = \text{logit}(\pi) = Z\gamma \tag{6.1.1.5}$$

The log link function is used to model the counts that follow the Poisson distribution:

$$\log(\lambda) = X\beta \tag{6.1.1.6}$$

where X and Z are covariate matrices, which may or may not coincide depending on the theoretical model; and β and γ are vectors of regression coefficients in the two equations.

There are two types of fixed-effect ZIP models (Lambert, 1992): ZIP and ZIP(τ) models. The latter is a special case of the former. In the ZIP(τ) model, the parameters in the logit part of the model are functionally related to those in the Poisson part of the model. For example, the two

[1] When count equals zero, the Poisson distribution $e^{-\lambda}\lambda^{y}/y!$ becomes $e^{-\lambda}$.

parts of the model are formulated as: $\text{logit}(\pi) = -\ \tau\beta'x$ and $\log(\lambda) = \beta'x$, where the parameter in the logit part of the model is simply a scalar multiple of those in the Poisson part of the model. As a result, the number of parameter estimates in the ZIP(τ) model is greatly reduced; however, "this assumption of common covariate effects is rarely satisfied in practice" (Hur et al., 2002). In this book, we only discuss the ZIP model.

6.1.2 Random effect zero-inflated Poisson (RE-ZIP) models

To model count data with extra zeros in hierarchically structured data, the multilevel ZIP model or mixed-effect ZIP model (i.e., the ZIP Model with random effects) should be applied to account for both the extra zeros and the intraclass correlation coefficient (ICC). Hall (2000) incorporated random effects into the Poisson component of the ZIP model for repeated measures data. Yau and Lee (2001) included two distinct random effects for the Poisson and logistic parts of the model, but the parameter estimates were implemented separately. Hur (1999) and Hur et al. (2002) developed a mixed-effect ZIP model by including a single random effect for both the logistic and the Poisson parts of the model. Xie et al. (2000) illustrated the application of Hur's model by analyzing longitudinal hospital admission data. In this section, we will discuss and illustrate Hur's mixed-effect or random-effect ZIP (RE-ZIP) model.

Given $j = 1, \ldots, J$ level-2 groups and $i = 1, \ldots, n_j$ level-1 individuals for group j, the total number of observations is $N = \sum_{j=1}^{J} n_j$. The RE-ZIP model is:

$$\Pr(y_{ij} = 0) = \pi_{ij} + (1-\pi_{ij})\exp(-\lambda_{ij}) \tag{6.1.2.1}$$

$$\Pr(y_{ij} = h_{ij}) = (1-\pi_{ij})\frac{\exp(-\lambda_{ij})\lambda_{ij}^{h_{ij}}}{h_{ij}!},\ h_{ij} > 0 \tag{6.1.2.2}$$

The combined form is:

$$\Pr(y_{ij}) = \pi_{ij} I(y_{ij}) + (1-\pi_{ij}) f(y_{ij}) \tag{6.1.2.3}$$

where

$$f(y_{ij}) = \frac{\exp(-\lambda_{ij})\lambda_{ij}^{y_{ij}}}{y_{ij}!} \tag{6.1.2.4}$$

As for the fixed effect ZIP model, the indicator function, $I(y_{ij})$, takes a value of 1 if the observed response is zero ($y_{ij} = 0$) and a value of 0 if the observed response is positive ($y_{ij} > 0$). The two link functions that link the linear predictors to the probability of extra zeros and the expected Poisson counts are:

$$\log\left(\frac{\pi}{1-\pi}\right) = \text{logit}(\pi) = Z\gamma + u_1 = Z\gamma + \sigma_1\theta_1 \tag{6.1.2.5}$$

$$\log(\lambda) = X\beta + u_2 = X\beta + \sigma_2\theta_2 \tag{6.1.2.6}$$

where Z and X are covariate matrixes in the logistic and Poisson parts of the model, respectively, and Z and X are not necessarily the same. Random effects are incorporated into the model by adding residual terms u_1 and u_2 into the above equations to account for ICC in hierarchically structured data. In eqs. 6.1.2.5 and 6.1.2.6, u_1 and u_2 represent the unexplained variation in the probability of having extra zeros and variation in the expected Poisson counts across level-2 units, respectively. As in other random effect models, the random effects in eqs. 6.1.2.5 and 6.1.2.6 are assumed to be normally distributed. Further, u_1 and u_2 can be standardized: $\theta_1 = u_1 / \sigma_1$ and $\theta_2 = u_2 / \sigma_2$, so that $\theta \sim N(0, 1)$.

6.1.3 Random effect zero-inflated negative binomial (RE-ZINB) models

The ZIP model is suitable for count data with extra zeros only if there is no over-dispersion. In the context of standard Poisson regression, over-dispersion has little effect on parameter estimates, but leads to underestimation of standard errors (Cox, 1983). Over-dispersion will result in significant bias in parameter estimates for the ZIP model (Grogger & Carson, 1991; Ridout et al., 2001; Yau et al., 2003). For count data with extra zeros and over-dispersion, an alternative zero-inflated negative binomial (ZINB) model is appropriate.

The ZINB model is formulated like the ZIP model by replacing the Poisson distribution $f(y_{ij})$ in eq. 6.1.2. 3 with the negative binomial distribution shown in eq. 5.6.2.3 in Chapter 5. As the dispersion parameter k approaches 0, there is no over-dispersion in the count data, and the ZINB distribution converges to the ZIP distribution (Land et al., 1996). Although a score test has been developed for testing the ZIP models vs. ZINB models (Ridout et al., 2001), the models can also be compared by testing the hypothesis:

$$H_0 : k = 0 \text{ vs. } H_A : k > 0$$

If k is significantly greater than 0, then over-dispersion in the count data is significant, and the ZINB model is preferred.

When count data have over-dispersion and are hierarchically structured, random effect ZINB (RE-ZINB) models should be applied to accommodate for extra zeros, over-dispersion, as well as ICC, simultaneously (Mwalili et al., 2006; Ridout et al., 2001; Yau et al., 2003). The formulation of the RE-ZINB model is similar to the RE-ZIP model shown in eqs. 6.1.2.5 and 6.1.2.6, however, a dispersion parameter k is included in model estimation.

6.1.4 Application of RE-ZIP and RE-ZINB models

To demonstrate the applications of multilevel ZIP and ZINB models, we will use the same outcome measures that we used for the Poisson regression model in Chapter 5: the number of years in jail in lifetime (i.e. variable *Jailyear*). The level-one covariates are *Gender* (1-Male; 0-Female) *Ethnic* (1-Non-white; 0-White), *Gdmc_age* (grand mean centered age), and *Highsch* (1-High school or above; 0-Less than high school). The contextual variable *Hinj_Region* is the dummy variable (1- Region of higher prevalence of heavy injection; 0- Region of lower prevalence of heavy injection). The cross-level interaction (variable *Hinj_gender*) between *Hinj_Region* and *Gender* is included in the model. We will first fit the RE-ZIP model and then the RE-ZINB model.

The plot in Figure 5.6.4.1 in Chapter 5 shows that our outcome variable, *Jailyear*, has a clear spike (about 44%) of zeros representing jail-free IDUs. The non-zero values are skewed to the right. The data are also hierarchically structured with individuals nested within 20 project sites. Our model must take into account zero-inflation, skewness, and within-site dependence simultaneously. Multilevel or random-effect zero-inflated models (ZIP and ZINB) are more appropriate than the multilevel Poisson regression or multilevel negative binomial model for fitting such data. In addition to predicting the number of years in jail in lifetime among IDUs, it is also interesting to predict the existence of extra zeros or structural zeros. In this case, the odds of having an extra zero is the odds of never been arrested.

The RE-ZIP model can be implemented in different ways. Depending on theoretical interests and model fit, we may include a logistic component of the model without specifying linear predictors; we may specify the same predictors in both the logistic and Poisson parts of the model, or we may specify different predictors in each part of the model; We may specify a random effect in one part, either the logistic or the Poisson equations, or in both parts of the model. In addition, we may allow the random effects in the two parts of the model to be either independent or correlated with each other on the basis of our theory or empirical findings.

Currently *PROC GLIMMIX* does not have option for modeling zero inflated data, but *PROC NLMIXED* does. We have used *PROC NLMIXED* to fit a series of models with various model specifications. We included the same set of linear predictors (covariates) in the logistic and Poisson parts of the RE-ZIP model. Unique random effects were specified for each part of the model, and were also allowed to be correlated. The model can be fitted with the following SAS program:

***SAS Program 6.1.4.1**;

```
LIBNAME lib 'D:\MLM\data';

Data data1;
 set lib.ML_data_Chap5;
proc sort; by site;

PROC SQL;
Create table temp as select site, Heavy_inj,
 100*(mean(Heavy_inj)) as m_Heavy_inj from data1
 Group by site;
 Create table data2 as select distinct site, m_Heavy_inj,
 m_Heavy_inj>15 as Hinj_Region from temp;
Quit;

Data data3;
 Merge data1 data2; by site;

Hinj_gender=Hinj_Region*Gender;
Proc NLMIXED;
PARMS
A0=0.3 A_Hinj_Region=0.2 A_Gender=0.3 A_Hinj_Gender=0 A_Ethnic=-0.02
A_Gdmc_age=-0.2 A_Highsch=-0.2

B0=1.5   B_Hinj_Region=0.1   B_Gender=0.5   B_Hinj_gender=0   B_Ethnic=0.2
B_Gdmc_age=0.04 B_Highsch=0.02

V_u1=0.5 Cov_u12=0 V_u2=0.5;

*linpinfl = linear predictor in the logistic part of the model;
linpinfl= A0 + A_Hinj_Region*Hinj_Region + A_Gender*Gender + A_Hinj_gender
*Hinj_gender + A_Ethnic*Ethnic + A_Gdmc_age*Gdmc_age + A_Highsch*Highsch
```

```
+ u1;

*infprob = probability of inflated zeros, which is the logistic transform
of the linear predictor;
infprob  = 1/(1 + exp(-linpinfl));

* eta is the linear predictor for the Poisson;
eta=B0 + B_Hinj_Region*Hinj_Region + B_Gender*Gender + B_Hinj_gender*Hinj_gender
   + B_Ethnic*Ethnic + B_Gdmc_age*Gdmc_age + B_Highsch*Highsch + u2;
* lambda is the expected Poisson counts;
lambda=exp(eta);
* Build the ZIP log likelihood;
if Jailyear=0 then LL = log(infprob + (1-infprob) *exp(-lambda));
  else LL=log(1-infprob) + Jailyear*log(lambda)-lambda-lgamma(Jailyear + 1);
Model Jailyear ~ general(LL);
RANDOM u1 u2 ~ NORMAL([0,0], [V_u1,Cov_u12,V_u2]) SUBJECT=SITE;
Run;
```

In the above SAS program, the *PARMS* statement defines the starting values for model parameters. The parameters with prefix A are starting values for the intercept and slopes of the logistic part of the model. The parameters with prefix B are starting values for the intercept and slopes of the Poisson part of the model. The last line in the *PARMS* statement defines the starting values of the variances (i.e., *V_u*1 and *V_u*2) of the two level-2 residuals (i.e., *u*1 and *u*2), as well as their covariance (i.e., *Cov_u*12). The starting values for all the parameters defined in the *PARMS* statement are based on our preliminary modeling results.

The statement *linpinfl* defines the linear predictors in the logistic part of the model (see eq. 6.1.2.5). The statement *infprob* is the logistic transform of the linear predictor and defines the probability of extra zeros in the logistic part of the ZIP model. The statement *eta* (i.e., log(λ) in eq. 6.1.2.6) defines the linear predictors in the Poisson part of the model, and the *lambda* statement defines the expected Poisson counts, which is exp(*eta*).
The *IF-THEN-ELSE* statement defines the log-likelihood function for the ZIP model.

- When the outcome measure *Jailyear* = 0, the log likelihood function (i.e., LL) is defined based on eq. 6.1.2.1.
- When *Jailyear* > 0 (when $y > 0$, it is defined $y = h > 0$ in eq. 6.1.2.2), the LL is based on eq. 6.1.2.2. The SAS *lgamma* function is used to calculate the log of $h_{ij}!$ in eq. 6.1.2.2.

The *general* distribution is used to fit the model with the log likelihood function defined in the *IF-THEN-ELSE* statement. The *RANDOM* statement defines the distribution and variance/covariance structure for the two random effects (i.e., *u*1 and *u*2), which are assumed to have bivariate normal distribution with zero means and a full variance and covariance matrix. The selected SAS output of the model follows:

SAS Output 6.1.4.1 Selected results of the RE-ZIP model

```
The NLMIXED Procedure

                              Specifications

Data Set                                                WORK.DATA3
Dependent Variable                                      jailyear
```

```
Distribution for Dependent Variable        General
Random Effects                             u1 u2
Distribution for Random Effects            Normal
Subject Variable                           site
Optimization Technique                     Dual Quasi-Newton
Integration Method                         Adaptive Gaussian
                                           Quadrature
```

Dimensions

Observations Used	9685
Observations Not Used	139
Total Observations	9824
Subjects	20
Max Obs Per Subject	1305
Parameters	17
Quadrature Points	1

Iteration History

Iter	Calls	NegLogLike	Diff	MaxGrad	Slope
1	6	28888.4745	286.1967	10355.52	-1.142E7
2	8	28084.1145	804.36	2910.089	-1204.16
3	9	28043.9053	40.20927	992.2724	-110.492
...					
72	153	27520.9844	0.000501	3.161033	-0.00012

NOTE: GCONV convergence criterion satisfied.

Fit Statistics

-2 Log Likelihood	55042
AIC (smaller is better)	55076
AICC (smaller is better)	55076
BIC (smaller is better)	55093

Parameter Estimates

Parameter	Estimate	Standard Error	DF	t Value	Pr>|t|	Alpha	Lower	Upper	Gradient
A0	0.9682	0.1162	18	8.33	<.0001	0.05	0.7241	1.2124	1.5915
A_Hinj_Region	-0.4729	0.1319	18	-3.59	0.0021	0.05	-0.7499	-0.1959	1.22989
A_Gender	-1.2048	0.09702	18	-12.42	<.0001	0.05	-1.4087	-1.0010	0.204505
A_Hinj_gender	0.1178	0.1132	18	1.04	0.3120	0.05	-0.1201	0.3556	-0.64589
A_Ethnic	-0.1920	0.05409	18	-3.55	0.0023	0.05	-0.3056	-0.07834	-1.57317
A_Gdmc_age	-0.04227	0.003137	18	-13.48	<.0001	0.05	-0.04886	-0.03568	0.459022
A_Highsch	0.09263	0.04515	18	2.05	0.0550	0.05	-0.00222	0.1875	-0.32593
B0	1.4745	0.05480	18	26.91	<.0001	0.05	1.3593	1.5896	-0.62747
B_Hinj_Region	-0.01459	0.06727	18	-0.22	0.8308	0.05	-0.1559	0.1267	0.033461
B_Gender	0.2471	0.03657	18	6.76	<.0001	0.05	0.1703	0.3240	-0.16654

B_Hinj_gender	0.1240	0.04056	18	3.06	0.0068	0.05	0.03878	0.2092	-0.1173
B_Ethnic	0.1069	0.01441	18	7.41	<.0001	0.05	0.07658	0.1372	-0.06609
B_Gdmc_age	0.02724	0.000699	18	38.99	<.0001	0.05	0.02577	0.02870	-0.20575
B_Highsch	0.02282	0.01070	18	2.13	0.0469	0.05	0.000352	0.04530	-0.14764
V_u1	0.03131	0.01443	18	2.17	0.0436	0.05	0.000996	0.06162	-0.91833
Cov_u12	-0.01257	0.006435	18	-1.95	0.0665	0.05	-0.02609	0.000952	3.161033
V_u2	0.01449	0.004850	18	2.99	0.0079	0.05	0.004303	0.02468	0.460964

The upper panel of the *Parameter Estimates* section of the SAS Output presents parameter estimates of the logistic regression part of the RE-ZIP model. Interpretation of the parameters is the same as those in the regular logistic regression models, except that we are seeking to explain the probability of extra/structural zeros. For example, with respect to the effect on likelihood of having an extra zero, the contextual variable *Hinj_Region* has a significant negative effect (*A_Hinj_Region* = -0.47, $p = 0.0021$), and the individual-level explanatory variable *Gender* also has significant negative effect (*A_ gender* = $-$ 1.20, $p < 0.0001$). However, the cross-level interaction between *Hinj_ Region* and *Gender* is not statistically significant (*A_Hinj_ gender* = $-$ 0.12, $p = 0.3120$).

The middle panel of the *Parameter Estimates* section presents parameter estimates for the Poisson part of the model. Interpretation is the same as in the regular Poisson regression model. The intercept *B0* = 1.47 ($p < 0.0001$) indicates that the expected number of years in jail in lifetime was about exp (1.47) = 4.39 years for the reference group that has zero values for all covariates included in the model. Gender has a significant slope coefficient (*B_Gender* = 0.25, $p < 0.0001$), indicating that the expected number of years in jail is about exp (0.25) = 1.28 years or about 28% longer for male IDUs, controlling for other covariates. The slope coefficient of the contextual variable *Hinj_Region* is not statistically significant (*B0_Hinj_Region* = $-$ 0.02, $p = 0.8308$), but its interaction with *Gender* is (*B_Hinj_gender* = 0.12, $p = 0.0068$). This indicates that being a male in a higher prevalence of heavy injection region further contributes to expected number of years in jail in lifetime. Male IDUs (i.e., *Gender* = 1) in the region of higher prevalence of heavy injection (i.e., *Hinj_Region* = 1) spent about 42% longer in jail on average than the female IDUs in the region of lower prevalence of heavy injection (i.e., exp(0.25 − 0.02 + 0.12) = 1.42).

The random effects for both the *logistic* and the Poisson parts of the ZIP model are statistically significant (*V_u*1 = 0.03, $p = 0.0436$; *V_u*2 = 0.01, $p = 0.0079$). This indicates that both the probability of years in jail (extra zeros) and the expected number of years in jail vary significantly across project sites. However, the association between the two random effects was only marginally significant (*Cov_u*12 = -0.01, $p = 0.0665$).

The observed and predicted frequency distributions for the number of years spent in jail in lifetime estimated from the RE-ZIP model are plotted in Figure 6.1.4.1. This plot shows that the RE-ZIP model has substantially improved model fit when compared with the multilevel Poisson or the negative binomial models discussed in Chapter 5 (see Figure 5.6.4.2) [2].

The SAS program for generating Figure 6.1.4.1 follows:

***SAS Program 6.1.4.2;**

```
...
Hinj_gender = Hinj_Region*Gender;
```

[2] Alternatively, Vuong's test can be used to compare the non-nested ZIP, Poisson regression, and negative binomial models (Vuong, 1989).

```
Proc NLMIXED data = data3;
PARMS
A0 = .3 A_Hinj_Region = 0.20 A_Gender = .3 A_Hinj_gender = 0 A_Ethnic = -.02
A_Gdmc_age = -.2 A_Highsch = -.2
B0 = 1.5 B_Hinj_Region = 0.1 B_Gender = .5 B_Hinj_gender = 0 B_Ethnic = .2
B_Gdmc_age = .04 B_Highsch = .02
V_u1 = .5 Cov_u12 = 0 V_u2 = .5;

*linpinfl = linear predictor in the logistic part of the model;
linpinfl = A0 + A_Hinj_Region*Hinj_Region + A_Gender*Gender + A_Hinj_gender
*Hinj _gender + A_Ethnic*Ethnic + A_Gdmc_age*Gdmc_age + A_Highsch*Highsch + u1;

*infprob = probability of inflated zeros, which is the logistic transform
of the linear predictor;infprob = 1/(1 + exp(-linpinfl));

*eta is the linear predictor for the Poisson;
eta = B0 + B_Hinj_Region*Hinj_Region + B_Gender*Gender + B_Hinj_gender*Hinj_gender
+ B_Ethnic*Ethnic + B_Gdmc_age*Gdmc_age + B_Highsch*Highsch + u2;
lambda = exp(eta); * lambda is the Poisson mean;

* Build the ZIP log likelihood;
if Jailyear = 0 then LL = log(infprob + (1-infprob) *exp(-lambda));
else LL = log(1-infprob) + Jailyear*log(lambda)-lambda-lgamma(Jailyear+1);
Model Jailyear ~ general(LL);
RANDOM u1 u2 ~ NORMAL([0,0], [V_u1,Cov_u12,V_u2]) SUBJECT = SITE;

Predict infprob out = pred_zidata (keep = pred rename = pred = pred_zi);
Predict lambda out = preddata (keep = Jailyear pred);
Run;

Proc SQL;
    select max(Jailyear) as y_max into :y_max
    from data1;
Quit;
Run;
%let y_max = &y_max;

%include 'D:\MLM\Graphsmacro.sas';
%graphZIP(title = RE-ZIP Model Fit, dataset = data3, depvar = Jailyear);
```

The first part of the program is copied from SAS Program 6.1.4.1 to run the RE-ZIP model and output predicted probability of the inflated zeros and the Poisson mean. The SAS macro *Graphsmacro.sas* (SAS, 2007) was called to generate the plot.

Although the RE-ZIP model takes into account the extra zeros in the count data, it cannot handle over-dispersion in the data. The zero-inflated negative binomial (ZINB) model should be applied to count data with extra zeros and over-dispersion simultaneously. Accordingly, we now demonstrate how to fit the RE-ZINB model using *PROC NLMIXED*. The same set of covariates and random effects are used.

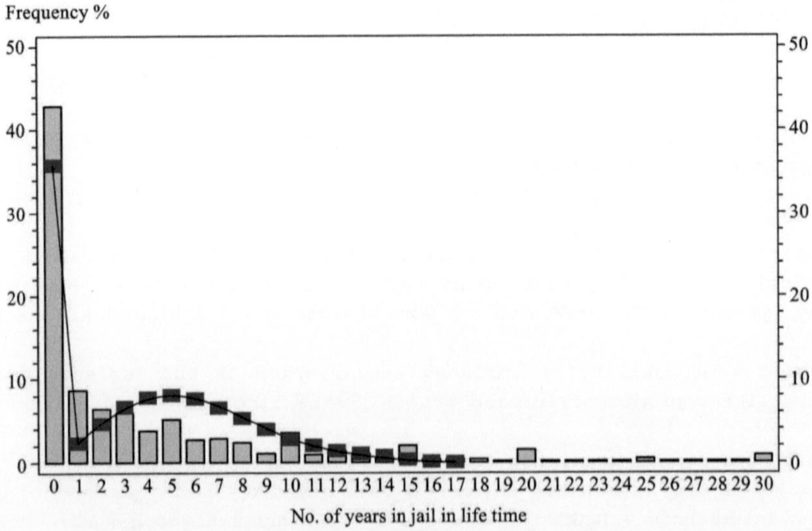

Figure 6.1.4.1 Distibution of observed and predicted number of years in jail in lifetime: Multilevel ZIP model

*SAS Program 6.1.4.3;

```
...
Hinj_gender = Hinj_Region*Gender;
Proc NLMIXED;
PARMS
A0 = .6 A_Hinj_Region = -.009 A_Gender = -1.12 A_Hinj_gender = 0 A_Ethnic = -.18
A_Gdmc_age = -.05 A_Highsch = .09
B0 = 1.5 B_ Hinj_Region = .002 B_Gender = .334 B_Hinj_gender = .002 B_Ethnic = .1
B_Gdmc_age = .027 B_Highsch = .022
V_u1 = .05 Cov_u12 = -.02 V_u2 = .01 k = 1;

*linpinfl--the linear predictor for logistic  part of the model;
*p -- the probability of being extra zeros;
linpinfl = A0 + A_Hinj_Region*Hinj_Region + A_Gender*Gender + A_Hinj_gender
*Hinj_gender + A_Ethnic*Ethnic + A_Gdmc_age*Gdmc_age + A_Highsch*Highsch
+ u1;
p = 1/(1 + exp(-linpinfl));

*eta -- the linear predictor for the negative binomial (NB) part of the model;
*lambda -- the mean of the NB distribution;
eta = B0 + B_Hinj_Region*Hinj_Region + B_Gender*Gender + B_Hinj_gender
*Hinj_gender + B_Ethnic*Ethnic + B_Gdmc_age*Gdmc_age + B_Highsch*Highsch
+ u2;
lambda = exp(eta);

*define NB pdf;
pdf = gamma(Jailyear + 1/k)/(gamma(Jailyear + 1)*(gamma(1/k)))*(((k*lambda)
**Jailyear)/(1 + k*lambda)**(Jailyear + 1/k));
if jailyear = 0 then pdf = p + (1-p)*pdf;
```

```
else pdf = (1-p)*pdf;

*define loglikelihood for the ZINB;
if pdf > 1e-8 then LL = log(pdf);
else LL = -1e20;
Model Jailyear ~ general(LL);
RANDOM u1 u2 ~ NORMAL([0,0], [V_u1,Cov_u12,V_u2]) SUBJECT = SITE;
Run;
```

The SAS codes for fitting RE-ZINB model are almost the same as those for fitting the RE-ZIP model. The key difference is that the probability density function (pdf) used to define the likelihood function for the RE-ZINB model is the pdf of the NB distribution (see eq. 5.6.2.3 in Chapter 5), instead of the pdf of the Poisson distribution (see eq. 6.1.2.2). In addition, a dispersion parameter k with a starting value of 1 is included in the model. The SAS output of the RE-ZINB Model follows:

SAS Output 6.1.4.3 Selected Results of the RE-ZINB Model

The NLMIXED Procedure

Specifications

Data Set	WORK.DATA3
Dependent Variable	jailyear
Distribution for Dependent Variable	General
Random Effects	u1 u2
Distribution for Random Effects	Normal
Subject Variable	site
Optimization Technique	Dual Quasi-Newton
Integration Method	Adaptive Gaussian Quadrature

Dimensions

Observations Used	9685
Observations Not Used	139
Total Observations	9824
Subjects	20
Max Obs Per Subject	1305
Parameters	18
Quadrature Points	1

Iteration History

Iter	Calls	NegLogLike	Diff	MaxGrad	Slope
1	6	21577.8352	95.28015	1313.171	-3372271
2	8	21573.2026	4.632656	315.4175	-19.3664
3	10	21568.0989	5.103729	376.3362	-2.79175
			...		
63	125	21521.4949	0.005628	11.59124	-0.00012

NOTE: GCONV convergence criterion satisfied.

Fit Statistics

-2 Log Likelihood	43043
AIC (smaller is better)	43079
AICC (smaller is better)	43079
BIC (smaller is better)	43097

Parameter Estimates

Parameter	Estimate	Standard Error	DF	t Value	Pr > |t|	Alpha	Lower	Upper	Gradient
A0	0.6150	0.1348	18	4.56	0.0002	0.05	0.3318	0.8982	-0.71741
A_Hinj_Region	-0.5107	0.1487	18	-3.44	0.0030	0.05	-0.8230	-0.1983	0.267128
A_Gender	-1.2851	0.1144	18	-11.24	<.0001	0.05	-1.5254	-1.0449	0.70077
A_Hinj_gender	0.1055	0.1361	18	0.77	0.4484	0.05	-0.1805	0.3915	0.3667
A_Ethnic	-0.1601	0.06984	18	-2.29	0.0342	0.05	-0.3068	-0.01333	-0.17694
A_gdmc_age	-0.04100	0.004272	18	-9.60	<.0001	0.05	-0.04998	-0.03203	-0.52899
A_Highsch	0.1241	0.05924	18	2.09	0.0506	0.05	-0.00038	0.2485	0.003834
B0	1.1933	0.09607	18	12.42	<.0001	0.05	0.9915	1.3952	-1.12143
B_Hinj_Region	0.05356	0.1067	18	0.50	0.6218	0.05	-0.1706	0.2778	0.771146
B_gender	0.3526	0.08417	18	4.19	0.0006	0.05	0.1757	0.5294	0.507975
B_Hinj_gender	0.07727	0.09444	18	0.82	0.4239	0.05	-0.1211	0.2757	-0.27943
B_Ethnic	0.1545	0.03649	18	4.24	0.0005	0.05	0.07789	0.2312	0.037654
B_Gdmc_age	0.03356	0.002063	18	16.26	<.0001	0.05	0.02922	0.03789	-11.5912
B_Highsch	0.01367	0.02889	18	0.47	0.6418	0.05	-0.04702	0.07436	-1.47916
V_u1	0.03518	0.01829	18	1.92	0.0704	0.05	-0.00325	0.07360	-1.58503
Cov_u12	-0.01251	0.007995	18	-1.56	0.1350	0.05	-0.02931	0.004285	-0.55211
V_u2	0.01550	0.006413	18	2.42	0.0265	0.05	0.002027	0.02897	-5.14135
k	0.8557	0.03201	18	26.73	<.0001	0.05	0.7884	0.9229	-0.19516

Running the program is very time consuming, taking a total number of 63 iterations for convergence. As expected, the parameter estimates in the RE-ZINB model are somewhat different from those of the RE-ZIP model, and the standard errors for all the parameter estimates in the RE-ZIP model are smaller (see SAS Output 6.1.4.1), compared with those in the RE-ZINB model. This is because over-dispersion in count data not only biases the standard errors of parameter estimates, but also biases the parameter estimates themselves.

Four of the fixed effects estimates, *A_Hinj_Region, A_Gender, A_Ethnic,* and *A_Gdmc_age,* in the *logit* part or the model remain statistically significant, leading to the same conclusions as those in the RE-ZIP model. However, in the fixed effects in the negative binomial part of the model, the cross-level interaction effect (*B_Hinj_gender* = 0.08, $p = 0.4239$) became statistically insignificant; and the main effect of education on the count outcome is no longer significant (*B_highsch* = 0.01, $p = 0.6418$).

The dispersion parameter is highly significant ($k = 0.86$, $p < 0.0001$), indicating a considerable over-dispersion in the count data. This confirms the appropriateness of the RE-ZINB model as opposed to the RE-ZIP for the count portion of the data.

The variance of the random effect (*V_u*1 = 0.04, $p = 0.0704$) in the *logistic* part of the model has become statistically insignificant in the RE-ZINB model. This implies that the individual heterogeneity in the *logistic* part of the model disappeared after the over-dispersion in the data was taken into account.

Figure 6.1.4.2 shows that the RE-ZINB model fits the data remarkably well. To compare the ZIP and ZINB models, several authors recommended the score test (Lee & Wang, 2006; Ridout et al., 2001); However, based on the model fit statistics provided by *PROC NLMIXED* and the predicted probability, we have sufficient evidence to conclude that the RE-ZINB model fits the data much better than all other models we have used to fit the count data in both Chapters 5 and 6. The following are the SAS program for generating Figure 6.1.4.2:

***SAS Program 6.1.4.4;**

```
...
Hinj_gender = Hinj_Region*Gender;
Proc NLMIXED;
PARMS
A0 = .6 A_Hinj_Region = -.009 A_gender = -1.12 A_Hinj_gender = 0 A_Ethnic = -.18
 A_Gdmc_age = -.05 A_Highsch = .09
B0 = 1.5 B_Hinj_Region = .002 B_Gender = .334 B_Hinj_gender = .002 B_Ethnic
 = .1 B_Gdmc_age = .027 B_Highsch = .022 V_u1 = .05 Cov_u12 = -.02 V_u2
 = .01 k = 1;

*linpinfl is the linear predictor for logit part of the model;
*p is the probability of being extra zeros;
linpinfl = A0 + A_Hinj_Region*Hinj_Region + A_Gender*gender + A_Hinj_gender
*Hinj_gender + A_Ethnic*Ethnic + A_Gdmc_age*Gdmc_age + A_Highsch*Highsch
+ u1;
p = 1/(1 + exp(-linpinfl));

*eta is the linear predictor for the negative binomial (NB)part of the model;
*lambda is the mean of the NB distribution;
eta = B0 + B_Hinj_Region*Hinj_Region + B_Gender*Gender + B_Hinj_gender
*Hinj_gender + B_Ethnic*Ethnic + B_Gdmc_age*Gdmc_age + B_Highsch*Highsch
+ u2;
lambda = exp(eta);
*define NB PDF;
pdf = gamma(Jailyear + 1/k)/(gamma(Jailyear + 1)*(gamma(1/k)))*(((k*lambda)
**Jailyear)/(1 + k*lambda)**(Jailyear + 1/k));
if Jailyear = 0 then pdf = p + (1-p)*pdf;
else pdf = (1-p)*pdf;
*define loglikelihood for the ZINB;
if pdf >1e-8 then LL = log(pdf);
else LL = -1e20;
Model Jailyear ~ general(LL);
RANDOM u1 u2 ~ NORMAL([0,0], [V_u1,Cov_u12,V_u2]) SUBJECT = SITE;

Predict p out = pred_zidata (keep = pred rename = pred = pred_zi);
Predict lambda out = preddata (keep = Jailyear pred);
Predict k out = pred_k (keep = pred rename = pred = k);
Run;

Proc SQL;
select max(Jailyear) as y_max into :y_max
from data1;
Quit;
Run;
%let y_max = &y_max;
%include 'D:\MLM\Graphsmacro.sas';
%graphZINB(title = RE-ZINB Model Fit, dataset = data3, depvar = Jailyear);
```

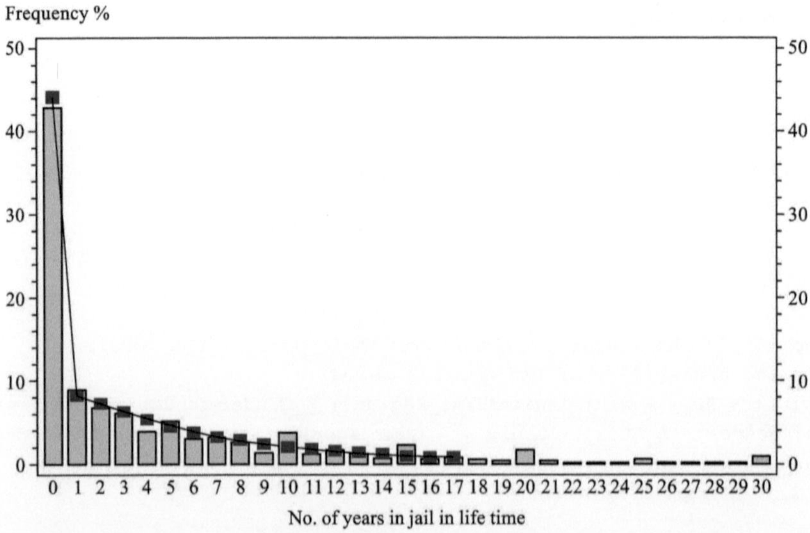

Figure 6.1.4.2 Distribution of observed and predicted number of years in jail in lifetime: Multilevel ZINB model

6.2 Mixed-effect mixed-distribution models for semi-continuous outcomes

In Section 6.1 we have discussed special models for multilevel count data with extra zeros. Researchers also encounter continuous outcome measures with a large number of observed values clustered at zero called a semi-continuous outcome, and this semi-continuous outcome may also have high level of skewness among the nonzero values. Highly skewed data with observation dependence and values clustering at zero have long been a challenge in statistical analysis. The assumption that this kind of data is normally distributed is inappropriate. Log-transformation can be used to normalize positively skewed, non-zero values, but excess zeros remains a problem. Recoding data into a dichotomous categorical variable (zeros vs. ones) for analysis with logistic regression is one common approach, but this practice reduces the question of "how much" something occurred to "whether or not" it occurred and thereby discards important information.

Alternatively, the two-part model developed in econometrics in the early 1980s is often used to analyze the data with a concentration of zero values (Duan et al., 1983). This approach uses one equation (usually logistic or probit regression) to model the probability of having a non-zero value, and a second equation (linear regression) to model the non-zero values. The two-part model is conceptually attractive, but the original model was developed for cross-sectional data and does not deal with dependency among observations arising from longitudinal or clustered data. In addition, the econometric two-part model assumes two separate or unconnected models (Manning et al., 1987; Wanning et al., 1987). In most instances the *likelihood* of having a non-zero value (versus zero value) and the *amount* or frequency of non-zero values observed in an outcome are correlated. For example, an individual with a low probability of using health services may opt to use less of that service, if any. By modeling the *likelihood* and *amount* separately, the econometric two-part model ignores an important piece of information and may introduce biases into parameter estimates (Olsen & Schafer, 2001).

In this section we introduce the *Mixed-Effect Mixed Distribution Model* that handles semi-continuous longitudinal outcome data. In such a model, random coefficients are allowed for both the logistic and linear regression parts of the model. The two parts of the model can be linked by allowing the two random coefficients to be correlated. Olsen and Schafer (2001) and Tooze et al. (2002) developed this model independently. Both approaches are based on maximum likelihood estimation, but the authors use different numerical methods to obtain model parameter estimates. Tooze et al. (2002) also developed a SAS Macro, *MIXCORR*, to facilitate the implementation of the model. In this section we focus on the model developed by Tooze et al. (2002) and demonstrate how to fit the model using SAS macro *MIXCORR*.

6.2.1 Mixed-effect mixed distribution model

Based on the work of Lachenburch (1976, 1992) for cross-sectional data and Grunward and Jones (2000) for time series data, Tooze et al. (2002) proposed the *Mixed-Effect Mixed Distribution Model* to fit semi-continuous outcomes for longitudinal data. The term "*mixed-distribution*" refers to a mixture of two distributions representing *occurrence* and *intensity*. The former models whether the outcome measure has a non-zero value and the latter models the *amount* or values of the observed non-zero measure. For example, suppose the outcome measure y_{ij} is the number of days a subject is hospitalized in a given time period, let R_{ij} represent the occurrence variable:

$$R_{ij} = \begin{cases} 0, & \text{if } y_{ij} = 0 \\ 1, & \text{if } y_{ij} > 0 \end{cases} \tag{6.2.1.1}$$

Correspondingly, we have a logistic regression model:

$$\text{logit}(p_{ij}) = \beta_{10} + \beta_{11}t_{ij} + \sum_{k=1}^{K} \beta_{1j}x_{1kj} + u_{1j} \tag{6.2.1.2}$$

where p_{ij} is the probability of having non-zero outcome values (i.e., $R_{ij} = 1$); t is time score; x_1's are time-invariant covariates that predict Pr ($R_{ij} = 1$). Regression coefficients β_{10}, β_{11}, and β_{1k}'s are intercept, fixed time effect, and fixed effects of covariates, respectively, on the log-odds of $R_{ij} = 1$; the term u_{1j} is the random effect of individuals on the log-odds, accounting for unobserved heterogeneity among individual subjects. In multilevel modeling terms, ($\beta_{10} + u_{1j}$) is considered the random intercept in the logit model, which allows the probabilities of having a non-zero outcome value to vary across individuals.

For non-zero outcome measures (i.e., $y_{ij} > 0$), define $S_{ij} \equiv (y_{ij} \mid R_{ij} = 1)$ to be the *intensity* variable that is a measure of the *amount* or values of the outcome $y_{ij} > 0$. A linear regression is used to model the *intensity*:

$$\log(S_{ij}) = \beta_{20} + \beta_{21}t_{ij} + \sum_{k=1}^{K} \beta_{2k}x_{2kj} + u_{2j} + e_{ij} \tag{6.2.1.3}$$

where the dependent variable is a log transformation of non-zero outcome measures (i.e., $y_{ij} > 0$); x_2's are time-invariant covariates that predict y_{ij}; e_{ij} is the level-1 residuals; and u_{2j} is the random effect on the initial level of *intensity*. The random intercept ($\beta_{20} + u_{2j}$) in the intensity model accounts for unobserved heterogeneity of mean non-zero outcome values among subjects.

Like the RE-ZIP model discussed in Section 6.1, the *Mixed-Effect Mixed Distribution Model* also has two parts: the logistic regression that models the probability of *occurrence* of non-zero outcome values (eq. 6.2.1.2); and the linear regression model (eq. 6.2.1.3) that predicts *amount* of the outcome as a function of linear predictors. The two parts of the model are integrated and simultaneously estimated. This model is also called the *Logistic-Lognormal-Normal Model*, where *logistic* refers to modeling *occurrence*, *lognormal* to modeling *intensity*, [3] and *normal* to the assumption of the bivariate normal distribution of the random effects. The covariates used in the *occurrence* and *intensity* parts of the model may or may not be the same.

As in other multilevel models, the two random effects are assumed to be jointly normally distributed.

$$\begin{pmatrix} u_{1j} \\ u_{2j} \end{pmatrix} \sim \text{BVN}\left[\begin{pmatrix} 0 \\ 0 \end{pmatrix}, \begin{pmatrix} \sigma_1^2 & \rho\sigma_1\sigma_2 \\ \rho\sigma_1\sigma_2 & \sigma_2^2 \end{pmatrix}\right] \tag{6.2.1.4}$$

where σ_1^2 and σ_2^2 are the variance of the random effects u_{1j} and u_{2j}, respectively, and ρ is the correlation between u_{1j} and u_{2j}.

The theoretical details about the *Mixed-Effect Mixed Distribution Model* are beyond the scope of this book. Interested readers are referred to Tooze et al. (2002). In this section we focus on application of the *Mixed-Effect Mixed Distribution Model* using the SAS macro *MIXCORR* developed by Tooze et al. (2002). The SAS macro works in the following interactive way. First, it calls *PROC GENMOD* to fit the logistic and linear regression parts of the model without random effects. The parameter estimates are then used as starting values in estimating the separate *occurrence* and *intensity* models with uncorrelated random effects using *PROC NLMIXED.* Next, the parameter estimates from the two uncorrelated random effects models are used as the starting values for the final integrated model.

6.2.2 Application of the mixed-effect mixed distribution model

The New Hampshire Dual Disorder Study Study project data used in the longitudinal models of Chapter 4 are used here to demonstrate the *Mixed-Effect Mixed Distribution Model.* The variables used for modeling are:

Outcome measure:

- *Hospdays*—number of days staying in psychiatric hospitals in the 6 months prior to each follow-up interview. As the measure has a wide range of values (i.e., from 0 to 182 days), we treat it as a continuous measure for the purpose of demonstration.

Micro and time-varying explanatory variables:

- *Time*—time score coded as 0, 1, 2, 3, 4, 5, 6 representing a 6-month time period from baseline to 6 month, 6 to 12 month, …, and 30 to 36 month follow-up interviews, respectively.

Individual level time-invariant covariate:

- *Trt*—dummy variable (1-Treatment group; 0-Control group).
- *Gdmc_Age*—continuous measure, grand-mean centered age at baseline interview;
- *Gender*—dummy variable (1-Male; 0-Female);
- *Edu*—education, dummy variable (1-High school or above; 0-Less than high school);

[3] If the non-zero outcome measures are normally distributed, the "logistic-normal-normal" model may be fit, where the dependent variable in the intensity part of the model is assumed to have a normal distribution.

- *Diagnosis*—primary psychiatric diagnosis, dummy variable (1-Schizophrenia/ schizoaffective disorder; 0-Bipolar disorder).
- *Hospdays0*—number of days staying in psychiatric hospitals in the past 6 months prior to the baseline interview.

Figure 6.2.2.1 shows the frequency of hospitalized days in each 6-month observation period after the baseline interview. As many participants reported not being hospitalized, the outcome measure (i.e., variable *Hospdays*) has a large proportion of zeros in each 6-month observation period. The number of days in hospital ranged from 0 to 184, and the number of zero outcomes ranged from 59% to 74%.

As described above, the mixed distribution model consists of two parts: the logistic regression part models the likelihood of being hospitalized during each of the six month period; and the linear regression part models variations in log-transformed non-zero outcomes, that is, the number of days staying in a hospital given that the respondent was hospitalized during each of the six month interval.

The SAS Macro *MIXCORR* is used to fit the model. The *%INCLUDE* statement is used to call the Macro. The "*DISTIN* = " statement specifies the lognormal distribution for non-zero part of the data. In theory, other distributions such as gamma distribution can also be assumed for the model. However, the current version of *MIXCORR* only allows lognormal or normal distributions, while a binomial distribution is assumed for the binary part of the data. The covariates included in the two parts of the model do not have to be the same. The SAS program follows.

***SAS Program 6.2.2.1**;

```
LIBNAME lib 'D:\MLM\Data';

Data data1;
Keep id Time Hospdays Trt Gdmc_age Gender Edu Diagnosis Hospdays0;
Set lib.Chap4_rawdata;

Array T(6) t2—t7;
Array A(6) Hospdays1 Hospdays2 Hospdays3 Hospdays4 Hospdays5 Hospdays6;
Do i = 1 to 6;
 Time = T(i)-1;
 Hospdays = A(i);
 Gdmc_age = (age-33.7); *create grand-mean centered age variable;
Output;
End;
%INCLUDE 'D:\MLM\Mixcorr.sas';
%MIXCORR(data = data1, distn = lognormal, response = Hospdays,
COVARSB = Time Trt Gdmc_age Gender Edu Hospdays0 Diagnosis,
COVARSL = Time Trt Gdmc_age Gender Edu Hospdays0 Diagnosis,
subject = ID);
Run;
```

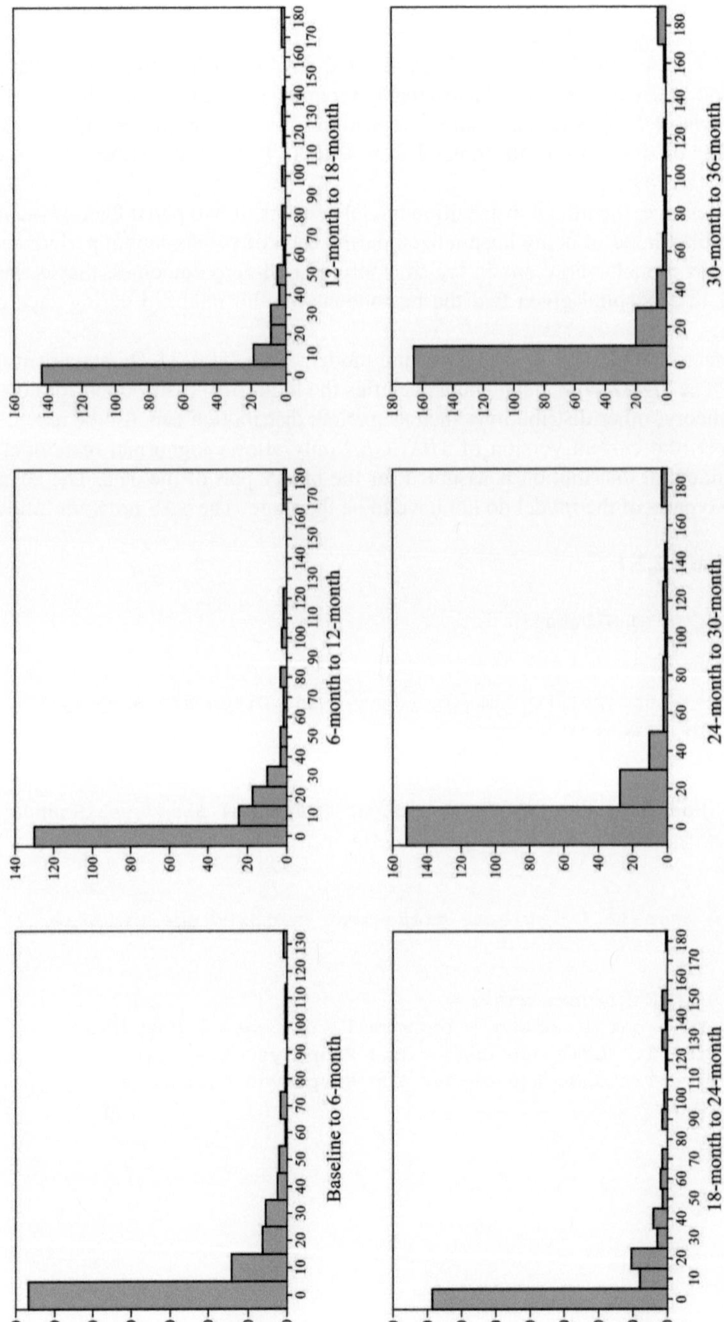

Figure 6.2.2.1 Distribution of number of hospitalidzed days in the past 6 months over a 3-year study period

The data used for longitudinal studies in Chapter 4 (SAS data file *Chap4_rawdata*) are retrieved, and converted into person-period format and saved in the temporary SAS data set *Data1*. The number of days in hospital observed in each 6-month period after the baseline were used to create the outcome measure (i.e., *Hospdays*), while the number of days in the past 6 months prior to the baseline is treated as a predictor (i.e., *Hospdays0*) of the outcome measure. The time-point indicator for 6-month, 12-month, … , and 36-month follow-up (i.e., t2 — t7) are used to create the time score (i.e., *Time*) in which the score of 0 represents the time-point of 6-month follow-up. The Grand-mean centered age (i.e., *Gdmc-age*) is also created by subtracting the baseline age from the overall mean age (i.e., 33.7 years old).

The SAS *%MIXCORR* macro includes the input parameters for data set, outcome distribution, outcome/response variable, covariates used for the logistic regression (i.e., eq. 6.2.1.2) and linear regression models (i.e., eq. 6.2.1.3), and subject ID in parenthesis. In this study, the *lognormal* distribution is specified. If the non-zero outcome measures follow a normal distribution, *DISTN = lognormal* option can be replaced with *DISTN = normal* option. The covariates for the binary logistic regression and the linear regression models are specified in the *COVARSB* and *COVARSL* statements, respectively.

The *%MIXCORR* macro produces two sets of model parameter estimates: one for the binary logistic regression model and another for the linear regression model (see SAS Output 6.2.2.1). The SAS macro fits the model twice without and with correlation between the two random effects (i.e., u_{1j} in eq. 6.2.1.2 and u_{2j} in eq. 6.2.1.3); and a likelihood ratio (LR) test is conducted for model comparison.

The upper panel of SAS Output 6.2.2.1 presents the results of the model without correlation between the two random effects. The lower panel presents the model with the correlation. The difference in −2LL between the two models is 4614.45−4593.28 = 21.17 with d.f. = 1, which has a Chi-square distribution. The corresponding *p*-value is less than 0.0001, indicating that the model with correlated random effects is preferred.

SAS Output 6.2.2.1 Selected results of the mixed-effect mixed-distribution model
Model with uncorrelated random effects

Parameter	Name	Estimate	Std Err	Prob>|t|
a1	Intercept--bi	-1.3922	0.3814	0.0003
b1	Time--bi	-0.1696	0.0436	0.0001
c1	Trt--bi	0.4558	0.2390	0.0579
d1	Gdmc_age--bi	-0.0129	0.0151	0.3935
e1	Gender--bi	0.0241	0.2702	0.9290
f1	Edu--bi	-0.0215	0.2414	0.9292
g1	Hospdays0--bi	0.0090	0.0033	0.0071
h1	Diagnosis--bi	0.6556	0.2841	0.0221
u1v	Var(Rndm Effect)--bi	1.5327	0.3397	0.0000
a2	Intercept--in	2.6345	0.2231	0.0000
b2	Time--in	0.0248	0.0318	0.4351
c2	Trt--in	-0.1593	0.1362	0.2441
d2	Gdmc_age--in	-0.0051	0.0084	0.5506
e2	Gender--in	-0.0886	0.1544	0.5667
f2	Edu--in	-0.0879	0.1389	0.5280
g2	Hospdays0--in	0.0074	0.0018	0.0001
h2	Diagnosis--in	0.1809	0.1706	0.2906

Continued

s2	Residual--in	0.9039	0.0807	0.0000
u2v	Var(Rndm Effect)--in	0.2554	0.0755	0.0009

Name	Value	Sum
AIC--bi	1333.98	.
AIC--lognormal	3318.47	4652.45
-2 Log Likelihood--bi	1315.98	.

Model with correlated random effects

Parameter Name		Estimate	Std Err	Prob>|t|
a1	Intercept--bi	-1.3572	0.3797	0.0004
b1	Time--bi	-0.1700	0.0436	0.0001
c1	Trt--bi	0.4736	0.2391	0.0489
d1	Gdmc_age--bi	-0.0122	0.0151	0.4197
e1	Gender--bi	0.0249	0.2701	0.9266
f1	Edu--bi	-0.0575	0.2422	0.8125
g1	Hospdays0--bi	0.0090	0.0033	0.0072
h1	Diagnosis--bi	0.6433	0.2830	0.0241
u1v	Var(Rndm Effect)--bi	1.5384	0.3381	<.0001
a2	Intercept--in	2.4237	0.2326	<.0001
b2	Time--in	0.0109	0.0312	0.7280
c2	Trt--in	-0.1712	0.1405	0.2243
d2	Gdmc_age--in	-0.0063	0.0087	0.4701
e2	Gender--in	-0.1434	0.1593	0.3692
f2	Edu--in	-0.0898	0.1427	0.5302
g2	Hospdays0--in	0.0079	0.0018	<.0001
h2	Diagnosis--in	0.2136	0.1744	0.2221
s2	Residual--in	0.8939	0.0789	<.0001
u2v	Var(Rndm Effect)--in	0.3391	0.0946	0.0004
u12v	Covariance	0.6291	0.1553	<.0001

Name	Value	Diff in -2ll	p-value
AIC	4633.28	.	.
-2 Log Likelihood	4593.28	21.17	0.0000

The data indicate that the likelihood of being hospitalized and number of days staying in hospital are significantly correlated ($u12v = 0.6291$, $p < 0.0001$). As such, the results from the model with correlated random effects will be interpreted. The intercept in the logit part of the model (*Intercept--bi* $= -1.3572$, $p = 0.0004$) indicates that the probability of being hospitalized for the reference group (i.e., the first 6-month after the baseline interview (*Time* = 0), female (*Gender* = 0), with education less than high school (*Edu* = 0) and mean age (*Gdmc_age* = 0)) is: $\exp(-1.3572)/(1 + \exp(-1.3572)) = 0.20$. In contrast, the intercept in the linear regression part of the model (*In-*

tercept--bi = 2.4237, $p < 0.0001$) indicates that the overall mean number of days in hospital for the reference group was about exp(2.4237) = 11.28 days.

The probability of being hospitalized significantly declined over time (*Time--bi* = −0.1700, p = 0.0001) after the first 6-month period; the number of days staying in hospital, however, remains unchanged over time once a participant was hospitalized (*Time--in* = 0.0109, p = 0.7280). The variable *Trt* has a significant positive slope coefficient in the *logit* model (*Trt--bi* = 0.4736, p = 0.0489), indicating that participants in the treatment group were more likely to be hospitalized after the baseline interview; however, there was no significant difference between the treatment and control groups in the number of days staying in hospital (*Trt--in* = −0.1712, p = 0.2243) among those who were hospitalized.

For covariates, the length of hospitalization prior to the baseline interview was a significant predictor of being hospitalized after the baseline. The variable *Hospdays*0, which is a measure of days in hospital in the past 6 months prior to the baseline interview, is positively associated with the likelihood of being hospitalized over time (*Hospdays*0*--bi* = 0.0090, p = 0.0072), and it also positively influences the length of hospitalization (*Hospdays*0*--in* = 0.0079, $p < 0.0001$).

The psychiatric disorder diagnosis (schizophrenia vs. bipolar) has a significant positive slope coefficient in the *logistic* part of the model (*Diagnosis--bi* = 0.6433, p = 0.0241), indicating that a participant being diagnosed with schizophrenia was more likely to be hospitalized than those who were diagnosed with bipolar disorder. Once hospitalized, however, the number of days staying in hospital did not vary significantly by diagnosis (*Diagnosis--in* = 0.2136, p = 0.2221). Individual characteristics such as gender, age, and education did not have significant effects in the *logistic* or the linear regression part of the model.

The variances of the random effects (i.e., u_{j1} and u_{j2}) in both the logit and the linear regression parts of the correlated mixed-effect mixed-distribution model were statistically significant ($u1v$ = σ_1^2 =1.5384, $p < 0.0001$; $u2v$ = σ_2^2 = 0.3391, p = 0.0004), indicating there is significant heterogeneity among the participants with respect to the likelihood of being hospitalized and the number of days being hospitalized. The covariance between the two random effects was positive and statistically significant ($u12v$ = $\rho_{\sigma1\sigma2}$ = 0.6291, $p < 0.0001$) indicating that participants who had a greater likelihood of being hospitalized would also stay in hospital longer on average once they were hospitalized. The parameter estimate S^2 = 0.8939 ($p < 0.0001$) is the estimated variance of residuals in the linear regression model shown in eq. 6.2.1.3, representing unexplained variation at level of the person-period observations.

6.3 Bootstrap multilevel modeling

In multilevel modeling, researchers often encounter data with non-normality of residuals and/or a relatively small number of units at the higher levels (e.g., small number of groups). The maximum likelihood methods used for multilevel modeling assume: 1) the micro and macro residuals are normally distributed; and 2) the sample size is large, i.e., both the number of "groups" (e.g., level-2 units) and the size of the groups (i.e., number of level-1 units per group) are large. In general, having a large number of groups is more important than having a large group size or a large number of individuals per group (Van der Leeden et al., 1997; Browne & Draper, 2000). If these assumptions are violated, the level-1 parameter estimates are not necessarily biased; but the variance components and standard errors of parameter estimates at the group level tend to be biased downward (Van der Leeden et al., 1997; Busing, 1993; Hox, 1998).

When the number of groups is small and/or the residuals are not normally distributed, the restricted maximum likelihood (REML) estimatoris usually recommended for model estimation, as it is considered superior to the maximum likelihood (ML)estimator (Kreft, 1996; Littell et al., 1996). However, the results of simulation studies show that REML estimates of variance components may be also biased when the number of groups is small and/or the residuals are not normally distributed (Busing, 1993; Carpenter et al., 2003).

Bootstrap multilevel modeling can be an effective method to address non-normality and/or a small number of groups in multilevel modeling (Carpenter et al., 2003; Maas & Hox, 2004). Simulation studies have shown that bootstrapping can correct bias in parameter estimates and improve the accuracy of inferences about parameter values (Carpenter et al., 2003; Maas & Hox, 2004; Meijer et al., 1994).

Bootstrapping is a re-sampling method for statistical inference (Miller, 1974; Efron, 1979, 1981, 1982; Efron & Tibshirani, 1986; Mooney & Duval, 1993; Davison & Hinckley, 1997). Bootstrapping uses the computer to draw random "resamples" from the original sample repeatedly with replacement and with the same size of the original sample. Once the pool of "bootstrap resamples" has been established, the sampling distribution of the bootstrap estimates of a parameter of interest can be created. In this way the bias, biased-corrected estimate, standard error, and confidence interval of the parameter can be estimated. Suppose $\hat{\theta}$ is the estimate of the parameter of interest, θ, computed from the original sample. The parameter estimate from a bootstrap resample is $\hat{\theta}_b$, and the mean and the variance of the bootstrap $\hat{\theta}_b$ s are $\hat{\theta}^*_b$ and $\sigma^2{}_b$, respectively. Then, the bias of $\hat{\theta}$ is estimated as Bias($\hat{\theta}$) = ($\hat{\theta}^*_b - \hat{\theta}$), a bias-corrected estimate is $\hat{\theta}_{bc}$ = [$\hat{\theta}$ − Bias($\hat{\theta}$)]. The standard error of $\hat{\theta}$ is estimated by the standard deviation of the empirical distribution of $\hat{\theta}_b$ (Efron, 1982).

Bootstrapping cannot be simply implemented with hierarchically structured data for multilevel modeling. This is because conducting bootstrapping from individual units in multilevel data will destroy the natural hierarchy of the data. In general, there are three kinds of bootstrap methods for multilevel modeling: 1) case-resampling; 2) parametric residual bootstrap; and 3) nonparametric residual bootstrap (Carpenter et al., 2003).

Case-resampling bootstrap is a non-model-based approach. Samples are randomly drawn before modeling. With a case-resampling bootstrap, each sample would have different design matrices because the explanatory variables in the design matrix are randomly sampled. This is undesirable because inference is conditioned upon explanatory variables. Variation in the design matrix would cause variation in information although the difference is often negligible in large data sets (Davison & Hinkley, 1997, p. 265; Carpenter et al., 2003). The most commonly used bootstrap methods for multilevel modeling are the model-based residual bootstrap methods — nonparametric residual bootstrap or parametric residual bootstrap—which are model-based resampling techniques (Carpenter et al., 2003).

6.3.1 Nonparametric residual bootstrap multilevel modeling

The steps for conducting nonparametric residual bootstrap multilevel modeling follow:

1) Run the theoretical multilevel model with the original sample and save the model estimated level-1 and level-2 crude residuals.

2) Create new level-1 and level-2 residuals, transformed from the corresponding crude residuals estimated from the multilevel model with the original sample.

3) Randomly draw samples with replacement independently from the two sets of new level-1 and level-2 residuals, separately.

4) Create a bootstrap sample by first generating the new level-1 random coefficients, and then generate the new response variable y^*_{ij} Using the model specified in eqs. 2.2.1—2.2.3 as an example, the new level-1 random coefficients $\hat{\beta}^*_{0j}$ and $\hat{\beta}^*_{1j}$ are generated using eqs. 6.3.1.1 and 6.3.1. 2. The new response variable y^*_{ij} is generated using eq. 6.3.1.3.

$$\hat{\beta}^*_{0j} = \hat{\gamma}_{00} + \hat{\gamma}_{01} w_{1j} + \hat{u}^*_{0j} \quad (6.3.1.1)$$

$$\hat{\beta}^*_{1j} = \hat{\gamma}_{10} + \hat{\gamma}_{11} w_{1j} + \hat{u}^*_{1j} \quad (6.3.1.2)$$

$$y^*_{ij} = \hat{\beta}^*_{0j} + \hat{\alpha}_1 X_{1ij} + \hat{\beta}^*_{1j} z_{1ij} + \hat{e}^*_{ij} \quad (6.3.1.3)$$

where the regression coefficients ($\hat{\gamma}_{00}$, $\hat{\gamma}_{01}$, $\hat{\gamma}_{10}$, $\hat{\gamma}_{11}$, and $\hat{\alpha}_1$) were estimated from the multilevel model using the original sample; residuals $\hat{u}^*_{0j}$, $\hat{u}^*_{1j}$, and $\hat{e}^*_{ij}$ are the transformed new level-2 and level-1 residuals, respectively. The new response variable y^*_{ij} will be merged together with the original level-1 and level-2 independent variables to generate a bootstrap sample.

5) Refit the model specified in Step 1 to the bootstrap sample and save the model parameter estimates.

6) Repeat steps 3—5 ($B-1$) times[4] and append the B (e.g., 1000) sets of model parameter estimates. The mean and standard deviation of the empirical distribution of the bootstrap estimates for a particular model parameter are the bootstrap-estimated parameter and its standard error, respectively.

The crucial part of nonparametric bootstrap multilevel modeling is the residual transformation as described in Step 2. Because the crude residuals in both the micro- and macro-level equations of the multilevel model do not necessarily have a zero mean, and their variance/covariance matrices do not match the model-estimated residual variance/covariance matrices, both the crude level-1 and level-2 residuals must be rescaled and then transformed. Rescaling is accomplished by centering the residuals to avoid biased estimates of the intercept parameters caused by the nonzero means of the residuals. Then, the rescaled residuals need to be transformed, at each level of the hierarchy to generate new residuals independently across levels. The variance/covariance matrix of the new residuals at each level should be identical to the corresponding model-estimated residual variance/covariance matrix. *Endnote* EN6.1 provides a proof of the validity of this technique.

6.3.2 Parametric residual bootstrap multilevel modeling

The steps for conducting parametric residual bootstrap multilevel modeling are similar to those discussed above. The only difference is that the parametric bootstrap uses the parametrically estimated distribution function of the data, instead of model estimated crude residuals, to generate the new residuals at each level. Subsequently, the bootstrap resamples are created. More Specifically, the new residuals (e.g., e^*_{ij}, u^*_{0j} and u^*_{1j}) are generated from the normal distributions of $N\sim(0, \hat{\sigma}^2)$, $N\sim(0, \hat{\sigma}^2_{u0})$ and $N\sim(0, \hat{\sigma}^2_{u1})$, where $\hat{\sigma}^2$, $\hat{\sigma}^2_{u0}$ and $\hat{\sigma}^2_{u1}$ are the model estimated level-1 and level-2 residual variances, respectively, using the original sample. For conducting a parametric bootstrap, residuals, e^*_{ij}, u^*_{0j} and u^*_{1j}, are randomly drawn from these estimated normal distribu-

[4] When B is large, the standard deviation of a parameter's bootstrap sampling distribution approaches the standard error of the parameter estimate. According to statisticians, $B \geqslant 800$ is usually required for bootstrap variance estimation (Booth & Sarkar, 1998).

tion functions, respectively, and then used to generate bootstrap resamples using eqs. 6.3.1.1—6.3.1.3. Once the bootstrap resamples are generated, then do Steps 5 and 6.

Usually, the nonparametric residual bootstrap is preferable to the parametric residual bootstrap, because the former takes into account non-normality in the distribution of the residuals, and gives more accurate inferences (Carpenter et al., 2003). In this section, we focus on application of nonparametric residual bootstrap linear multilevel modeling.

6.3.3 Application of nonparametric residual bootstrap multilevel modeling

Although the technique of bootstrap linear multilevel modeling is well-developed, there have been very few applications of the method. This is due, in part, to the lack of computer programs for bootstrap multilevel modeling. Wang, Carpenter & Kepler (2006) addressed this problem by developing SAS programs for nonparametric bootstrap linear multilevel modeling. In this section, we demonstrate how to use these SAS programs to conduct nonparametric bootstrap multilevel model using the data set that we used in Chapter 3.

The model specified in SAS Program 3.7.2 of Chapter 3 is used for this demonstration. In nonparametric bootstrap multilevel modeling, a new response variable will be generated from a linear function of the explanatory variables and transformed residuals. The 3-level categorical explanatory variable *Edu_gp* is replaced with two dummy variables : *Edu_gp*1 representing less than high school education, and *Edu_gp*2 representing high school or above education. The omitted category for education is college or above. The modified model is shown below:

$$\begin{aligned}\text{Lninj}_{ij} &= \beta_{0j} + \beta_{1j}\text{Ethnic}_{ij} + \beta_2\text{Gender}_{ij} + \beta_3\text{Gdmc\_age}_{ij} \\ &\quad + \beta_4\text{Edu\_gp1}_{ij} + \beta_4\text{Edu\_gp2}_{ij} + e_{ij}\end{aligned} \tag{6.3.3.1}$$

$$\beta_{0j} = \gamma_{00} + \gamma_{01}\text{HIV\_Region}_j + u_{0j} \tag{6.3.3.2}$$

$$\beta_{1j} = \gamma_{10} + \gamma_{11}\text{HIV\_Region}_j + u_{1j} \tag{6.3.3.3}$$

$$\begin{aligned}\text{Lninj}_{ij} &= \gamma_{00} + \gamma_{01}\text{HIV\_Region}_j + \gamma_{10}\text{Ethnic}_{ij} + \gamma_{11}\text{HIV\_Region}_j * \text{Ethnic}_{ij} \\ &\quad + \beta_2\text{Gender}_{ij} + \beta_3\text{Gdmc\_age}_{ij} + \beta_4\text{Edu\_gp1}_{ij} + \beta_4\text{Edu\_gp2}_{ij} \\ &\quad + (u_{0j} + \text{Ethnic}_{ij} * u_{1j} + e_{ij})\end{aligned} \tag{6.3.3.4}$$

First, analysts should check the distributions of the level-1 and level-2 residuals. Recall that the response variable (*Lninj*) used in the model is a log-transformation of the original response variable (*Inject*). As such, we expect the level-1 residual e_{ij} would be approximately normally distributed. However, the level-2 residual u_{0j} and u_{1j} may not be normally distributed as the number of groups (i.e., level-2 units in this study) is relatively small with only 20 project sites. The following SAS program is used to check residual normality:

***SAS Program 6.3.3.1**;

```
OPTIONS NOCENTER PS = 500 ls = 150;
LIBNAME Lib 'D:\MLM\data';

Data data1;
  Set Lib.ML_data_Chap3;
```

```
Proc Mixed Method = REML noclprint COVTEST;
  Class Site Edu_gp;
  Model Lninj = HIV_Region|Ethnic Gender Gdmc_age Edu_gp1 Edu_gp2
    /Solution ddfm = bw notest OUTP = Predicted;
  Random INT Ethnic /Solution type = UN Subject = Site;
  ODS Output SOLUTIONR = Level_2_resd_out;
Run;

Proc Transpose Data = Level_2_resd_out Out = lib.Level_2_resd
  (keep = site intercept Ethnic rename = (intercept = u0) rename = (Ethnic
  = u1));
  By site;
Var estimate;
ID effect;

*Split level-1 residual from the composite residual;
Data data1;
 Set Predicted;
Proc sort;
 By site;
Data Lib.Level_1_resd (keep = site e_ij);
 Merge data1 Lib.Level_2_resd;
 By site;
u_j = u0 + u1*Ethnic;

*The variable name of composite residual internally signed by SAS is Resid;
e_ij = Resid-u_j;*Split level-1 residuals from the composite residuals;
Run;

*Check normality of level-1 residuals;
Proc Univariate Data = lib.Level_1_resd Noprint;
VAR e_ij;

*The normal option superimposes a normal distribution curve on the histo-
gram;
Histogram /Normal midpoints = -5 to 5 by .3;

*The midpoints = option specifies the range and bin width.;
QQPlot / Normal;
Run;

*Check normality of level-2 residuals;
Proc Univariate Data = lib.Level_2_resd Noprint;
Var u0;
Histogram / Normal midpoints = -2 to 2 by .1;
QQPlot / Normal;
Run;

Proc Univariate Data = Lib.Level_2_resd Noprint;
```

```
Var u1;
Histogram / Normal midpoints = -.7 to .7 by .1;
QQPlot /normal;
Run;
```

The *OUTP* = *Predicted* option in the *MODEL* statement creates a new SAS data file *Predicted* containing the composite residuals[5]. The option *Solution* in the *Random* statement and the SAS *Output Delivery System* (ODS) option "*ODS Output SOLUTIONR*" were used to save crude level-2 residuals in the new SAS file *Level_2_resd_out*. This output dataset was reformatted using *PROC TRANSPOSE* procedure and saved in SAS file *Level_2_resd*. Readers with access to these datasets are encouraged to view the formatting differences between *Level_2_resd_out* and *Level_2_resd*. To check distribution of the level-1 residuals, we need to separate e_{ij} from the composite residual *Resid* (i.e., ($u_{0j} + u_{1j} * Ethnic + e_{ij}$) in this model) that was saved in the SAS file *Predicted* using option *OUTP*. Level-1 residual e_{ij} was calculated after merging the level-1 and level-2 residuals files.

Once the residuals were estimated, *PROC UNIVARIATE* was used to check normality of residual distributions. The *Histogram* statement produced a histogram of the residuals with equal bin widths. The *Normal* option placed a normal curve on top of the histogram. The *Midpoints* option allowed us to specify the minimum, maximum, and incremental values in the distribution, thus to change the resolution of the underlying histogram.

The *QQPLOT* statement creates a quantile-quantile plot of the quantiles of the residuals against the quantiles of a specified theoretical distribution. The normal distribution is specified by using the *Normal* option. The Q-Q plot tells how much the residual distributions deviate from the normal distribution. If residuals are normally distributed, the plot would approximately form a 45-degree straight line. The greater is the departure from the line, the greater the evidence that the residuals are not normally distributed.

The histograms and Q-Q plots shown in Figures 6.3.3.1—6.3.3.6 indicate that the level-1 residuals are approximately normally distributed. This is due to the log-transformation of the outcome measure. However, the level-2 residuals (i.e., u_{0j} and u_{1j}) are clearly not normally distributed. This is very likely due to the small number of project sites in this study. With this violation of the normality assumption, parameter estimates using REML may not be biased, but their standard errors are very likely to be biased. We will now demonstrate how to use nonparametric residual bootstrap multilevel modeling to address the problem.

Following the steps discussed in the beginning of this section, we first estimate the multilevel model specified in eqs. 6.3.3.1—6.3.3.4 using *PROC MIXED* with the original sample (data file *ML_data_Chap3*). Next, transform the model estimated residuals; create a bootstrap sample with the new response variable; refit the model to the bootstrap sample and save model parameter estimates; and repeat these steps by *B* times (e.g., 1000). The crucial step of conducting nonparametric residual bootstrap multilevel modeling is the model estimated residual transformation.

---

[5] The variable name of the composite residuals in the output SAS dataset is internally assigned as *Resid*, representing the observed minus predicted outcome measures.

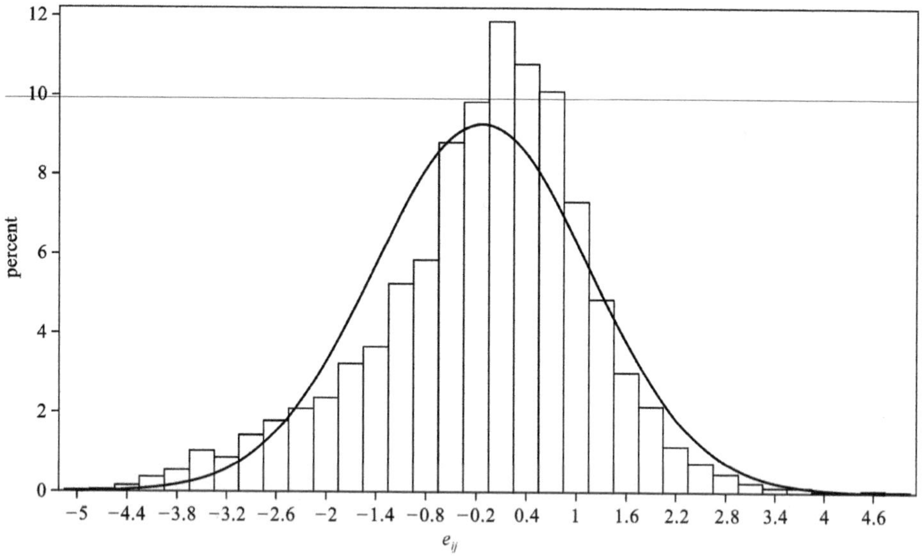

Figure 6.3.3.1 Distribution of level residuals

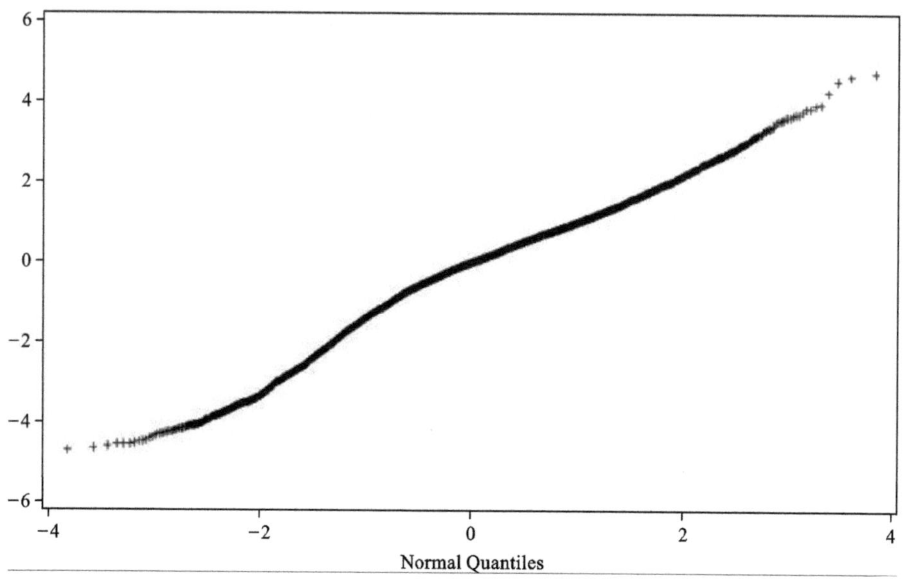

Figure 6.3.3.2 Level 1 residual Q-Q plot

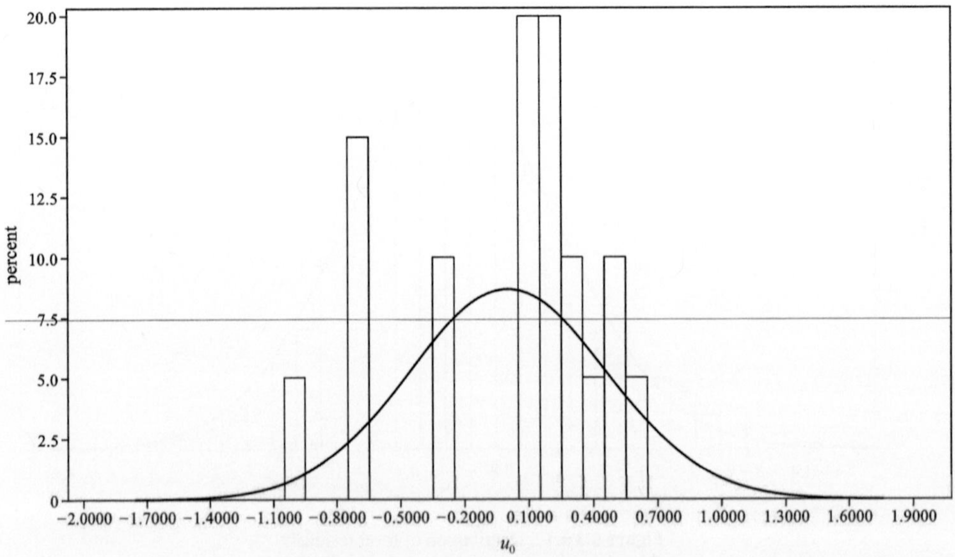

Figure 6.3.3.3 Level 2 residual u_0 distribution

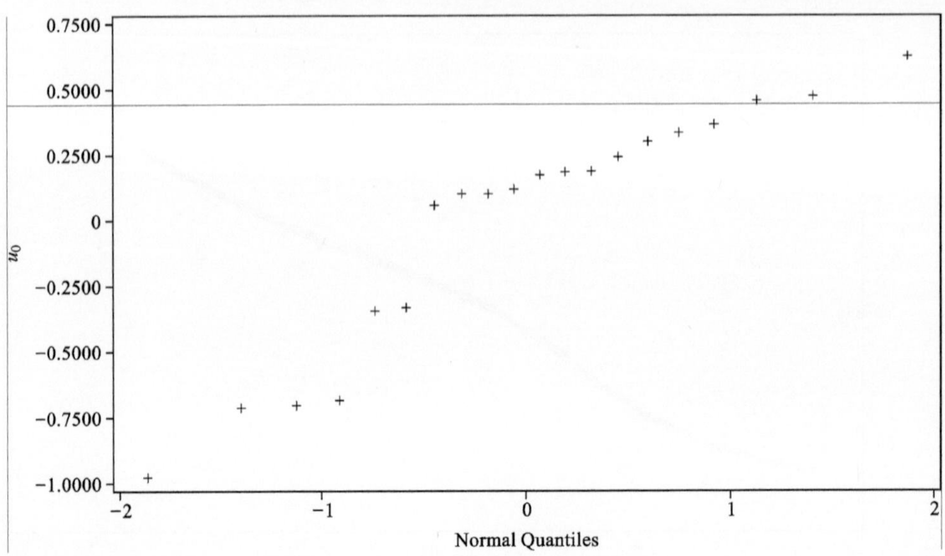

Figure 6.3.3.4 Level 2 residual u_0 Q-Q plot

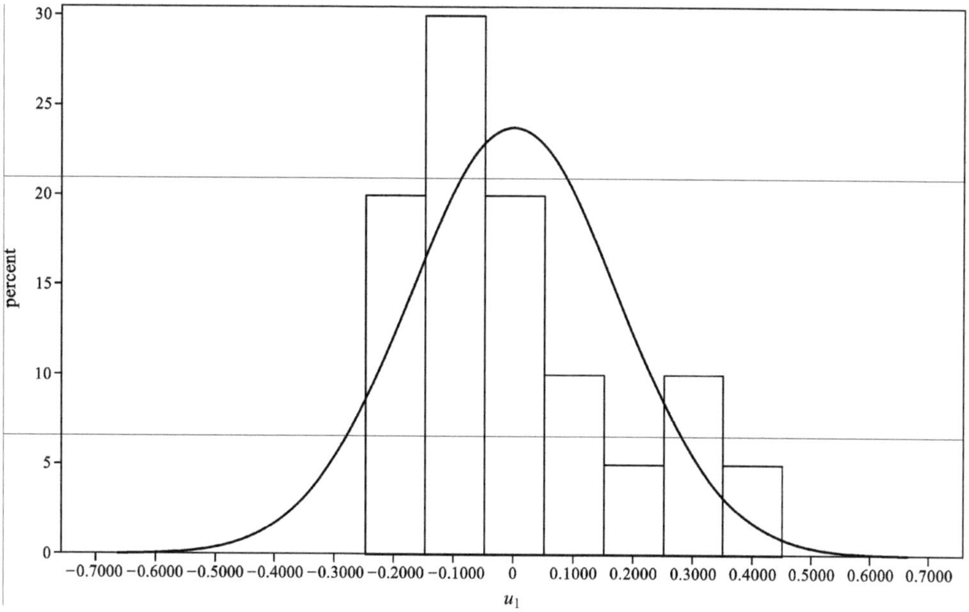

Figure 6.3.3.5 Level 2 residual u_1 distribution

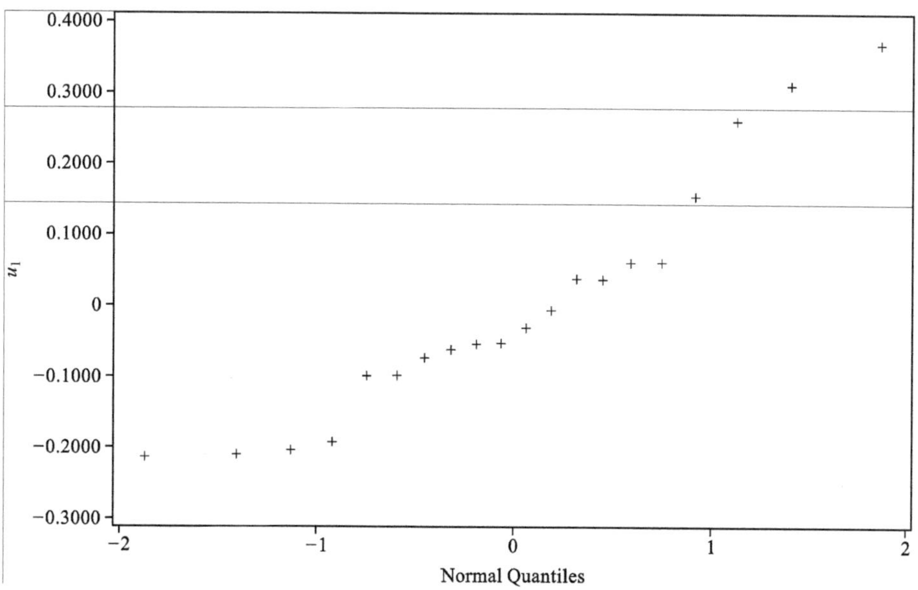

Figure 6.3.3.6 Level 2 residual u_1 Q-Q plot

Residual transformation: In SAS Program 6.3.3.1, the model estimated level-1 and level-2 residuals were created and were saved in permanent SAS files names *Level_1_resd* and *Level_2_resd*. We start with level-2 residual transformation as level-1 residual transformation is just a special case of the transformation. In SAS Program 6.3.3.2, the model estimated level-2 residuals are read from data file *Level_2_resd*, then rescaled or centered, and saved with new names (*c_u*0 and *c_u*1 for the random level-1 intercept β_{0j} and random slope β_{1j}, respectively) in *PROC SQL*. The rescaled level-2 residuals *c_u*0 and *c_u*1 were transformed using *PROC IML* following the formulas described in *Endnote* EN6.1. In this SAS program, matrix *S* is the empirical variance/covariance matrix of the rescaled level-2 residuals, which is the sum of squares of the residuals divided by the number of level-2 units (J = 20 in this study), rather than (J-1) (Carpenter et al., 2003). *Sigma* is the model-estimated level-2 residual variance/covariance matrix[6] using the original sample (see Table 6.3.3.1). The lower triangular Cholesky decomposition of *S* and *Sigma* were calculated using the *root* function of SAS PROC IML,[7] and are used to create matrix *A*. Then matrix *A* is used to transform the level-2 residuals as shown in *Endnote* EN6.1. The transformed level-2 residuals were saved under names *new_u*0 and *new_u*2 in matrix *NewU* and exported to the SAS file *New_Level_2_Resd.* To insure that the transformed level-2 residuals are created correctly, we recommend creating the variance/covariance matrix of the new level-2 residuals (i.e., matrix *S*1 in the SAS program) and checking whether it matches the model-estimated level-2 residual variance/covariance matrix *Sigma.*

***SAS Program 6.3.3.2;**

```
OPTIONS NOCENTER PS = 500 ls = 150;
LIBNAME lib 'D:\MLM\data';

*Rescale level-2 residuals;
Proc SQL;
Create Table Rescaled_U as select *, u0-mean(u0) as c_u0, u1-mean(u1) as
c_u1
from Lib.Level_2_resd;
Quit;
Proc Print Data = Rescaled_U;
Run;

*Define macro variables for model estimated level-2 residual var/cov matrix
elements;
%let c11 = 0.2369;
%let c12 = - 0.01372;
%let c21 = - 0.01372;
%let c22 = 0.04917;

*Transform rescaled level-2 residuals;
Proc IML;
Reset print;
USE Rescaled_U;
READ all var {c_u0 c_u1} into C_U; *Read c_u0 c_u1 into matrix C_U;
S = (C_U'*C_U)/20;
Ls = (root(S))'; *Create lower triangular of the Cholesky decomposition of S;
```

[6] Note, to avoid hard coding in programming, the values of the elements of the model estimated level-2 residual variance/covariance matrix (i.e., $\hat{\Sigma}$) (see Table 6.3.3.1) were set in SAS macro variables c11—c22 in this example.

[7] Note that the *ROOT* function in SAS *PROC IML* procedure generates the upper triangular of the Cholesky decomposition. The lower triangular is generated by transposing the upper triangular of the Cholesky decomposition.

```
*Input the model estimated level-2 residual variance/covariance matrix;
Sigma = {&c11 &c12,
         &c21 &c22};
Lsigma = (root(Sigma))';   *Create lower triangular of the Cholesky decompo-
                            sition of Sigma;
Sigma2 = Lsigma*Lsigma';   *Sigma2 should equal to Sigma;
A = (Lsigma*inv(Ls))';     *Inv(Ls_lt) is the inverse of Ls;
NewU = C_U*A;
S1 = (NewU'*NewU)/20;      *S1 should equal to Sigma;
col = {"new_u0" "new_u1"};

*Save the variables in a SAS data file;
CREATE lib.New_Level_2_resd FROM NewU [colname = col];
APPEND FROM NewU;
Quit;
Proc means Data = lib.New_Level_2_resd;
Run;
```

*Transforming Level-*1 *Residuals*: In the last paragraph of *Endnote* EN6.1 we show that transformation of level-1 residuals is a special case of transformation of level-2 residuals. This transformation was conducted in SAS Program 6.3.3.3. Firstly, the crude level-1 residuals that were created in SAS Program 6.3.3.1 were read from SAS data file *Level_1_resd* and centered at zero using the SAS *PROC SQL* procedure. The rescaled level-1 residuals *c_e_ij* were saved in the SAS active file *Resd_e_ij*. Following the formulas describe in the last paragraph of *Endnote* EN6.1, level-1 residual transformation was conducted using *PROC IML*. The new transformed level-1 residuals were saved as *New_resd_*1 in a permanent SAS data file *New_level_1_resd.*

***SAS Program 6.3.3.3**;

```
OPTIONS NOCENTER PS = 500 ls = 150;
LIBNAME lib 'D:\MLM\data';

*Rescale level-1 residuals;
Proc SQL;
 Create table Resd_e_ij as select site, e_ij-mean(e_ij) as c_e_ij from
 Lib.Level_1_resd;
Quit;

*Transform rescaled leve-1 residuals;
Proc IML;
Reset Print;
Use Resd_e_ij;
Read all var {c_e_ij} into E;     *Read the centered level-1 residuals into
                                   vector E;
Ls1 = sqrt((E`)*E/9824);          *Empirical standard deviation of level-1 re-
                                   siduals N = 9824;
Lsigma1 = sqrt(1.5531);           *Standard deviation of model estimated level
                                   -1 residuals;
A1 = Lsigma1/Ls1;
```

```
New_e_ij = E*A1;                 *Create transformed new level-1 residuals;
S1 = (New_e_ij'*New_e_ij)/9824;  *Equal to model estimated level-1 residual
                                 variance;

*Save the new level-1 residuals;
col = {"New_resd_1"};

*Output the new level-1 residuals;
Create lib.New_level_1_resd from New_e_ij [colname  = col];
Append from New_e_ij;
Quit;
Run;
Proc Means data = Lib.New_level_1_resd;
Run;
```

Nonparametric Residual Bootstrap Multilevel Modeling: Once the level-1 and level-2 residuals have been transformed, the next step is to draw random samples with replacement from the new level-1 and level-2 residuals; merge them with the original sample; generate a new response variable; and run the same multilevel model with the new response variable. This process is repeated for B times (e.g., $B = 1\ 000$) to generate empirical distributions of the model parameter estimates. The mean and standard deviation of the empirical distribution of parameter estimates are the bootstrap estimates of the parameter and its standard error, respectively. This example of bootstrap multilevel modeling is conducted in SAS Program 6.3.3.4.

***SAS Program 6.3.3.4;**

```
...
ODS listing close;
ODS noresults;
%let samplesize = 9824;          *total sample size;
%let groups = 20;                *number of groups in the multilevel data;
%let gamma00_hat = 3.5955;       *overall mean;
%let gamma01_hat = 0.5310;       *main effect of HIV_Region;
%let gamma20_hat = - 0.5476;     *main effect of ethnic;
%let gamma21_hat = 0.3311;       *interaction between ethnic and HIV_Region;
%let beta2_hat= - 0.0452;        *main effect of gender;
%let beta3_hat = 0.01395;        *main effect of age;
%let beta4_hat = - 0.0852;       *main effect of edu_gp1;
%let beta5_hat = - 0.0497;       *main effect of edu_gp2;
%Macro ML_Bootstrap;
 %do sample = 1 %to 500;
  %put samp = &sample;
Proc Surveyselect out = Data1 data = lib.New_level_1_resd method = URS n =
&samplesize
outhits noprint;

Data Newdata1;
 Merge Data1 lib.data_level_1;
```

```
Proc Surveyselect out = Data2 data = lib.New_level_2_resd method = URS n =
&groups outhits
noprint;

Data Newdata2;
 Merge Data2 lib.Data_level_2;
new_alpha = &gamma00_hat + &gamma01_hat*HIV_region + new_u0;
new_beta1 = &gamma20_hat + &gamma21_hat*HIV_region + new_u1;

Data Bootstrap_data (keep = site new_y new_alpha new_beta1 Gender Ethnic
Gdmc_age
 Edu_gp1 Edu_gp2 HIV_Region New_resd_1);
Merge Newdata1 Newdata2; by site;
New_y = new_alpha + new_beta1*Ethnic + &beta2_hat*Gender + &beta3_hat*Gdmc_age
 + &beta4_hat*Edu_gp1 + &beta4_hat*Edu_gp2 + New_resd_1;

Proc Mixed Method = REML noinfo noitprint COVTEST;
 Class site;
Model New_y = HIV_Region|Ethnic Gender Ethnic Gdmc_age  Edu_gp1 Edu_gp2
/SOLUTION DDFM = BW notest;
Random int ethnic /SOLUTION SUBJECT = site type = UN;

*Output fixed coefficients and residual var/covariances;
 ODS Output Solutionf = Reg_coef Covparms = Resd_Cov;
 Title 'Nonparametric Residual Bootstrap Multilevel modeling';
 Run;

%Put &syslast;
/*The existence of output file REG_COEF as the most current data set indi-
cates convergence of modeling using the most current bootstrap resample.
The name of the SAS working data file is case sensitive. */
%if &syslast = WORK.REG_COEF %then %do;

*Transpose the output of fixed parameter estimates;
Proc Transpose Data = Reg_coef Out = Fixed_p; Var estimate; id effect;

*Transpose the output of residual var/covariances;
Proc Transpose Data = Resd_Cov Out = Random_p (keep = UN_1_1_ UN_2_1_
UN_2_2_
  Residual rename = (UN_1_1_ = u0_v) rename = (UN_2_1_ = u01_cv) rename =
(UN_2_2_ = u1_v)
  rename = (Residual = e_v));
Var estimate; id CovParm;

%if &sample = 1 %then %do;
Data fixed_effects;
 Set Fixed_p;
 %end;
```

```
%else %do;
Data fixed_effects;
 Set fixed_effects Fixed_p;
 %end;
 %if &sample = 1 %then %do;
Data random_effects;
 Set Random_p;
 %end;
 %else %do;
Data random_effects;
 Set random_effects Random_p;
 %end;
 %end;
 %end;
 %mend;
 %ML_Bootstrap;

*Save residual bootstrap MLM results;
Data lib.fixed_effect_all; set fixed_effects;
Data lib.random_effect_all; set random_effects;
Run;

ODS listing;
Proc means data = lib.fixed_effect_all; Run;
Proc means data = lib.random_effect_all; Run;
```

In the beginning of the SAS program, sample size, number of groups, and model regression coefficients estimated from the original sample are defined as SAS macro variables to avoid hard coding in SAS macro *%ML_Bootstrap*. The SAS program follows several steps of data manipulation and modeling.

SAS *PROC SURVEYSELECT*, which provides a variety of methods for probability sampling, was used to draw random samples with replacement from the new level-1 and level-2 residuals created from SAS Programs 6.3.3.2 and 6.3.3.3, respectively. The *METHOD = URS* option specifies unrestricted random sampling, which selects units with equal probability, with replacement. Note with replacement, a unit could be selected more than once or not at all in a sample. By default, the output data contain only one observation for each selected unit, even if the unit is selected more than once. The *OUTHITS* option tells the procedure to keep a separate observation in the output data set for each selection when the same unit is selected more than once. The option *n* = specifies sample size. A bootstrap resample, which is saved by the *OUT* option, is usually defined to have the same size of the original sample. In our example, the bootstrap samples of the level-1 and level-2 new residuals are named *Data*1 and *Data*2, respectively.

Then, *Data*1 and *Data*2 were merged with the original level-1 and level-2 data sets (i.e., *data_level_*1 and *data_level_*2), that were created in SAS Program 3.3.3.1. Finally, new level-1 and level-2 data sets (i.e., *newdata*1 and *newdata*2) were merged to create the final bootstrap data set named *Bootstrap_data* for modeling. In the process of generating the new data, a new response variable *New_y* was created. The same multilevel model was used to fit the bootstrap data with the new response. Estimates of the fixed parameters and variance components were output, using options *ODS OUTPUT SOLUTIONF* and *COVPARMS* in *PROC MIXED*.

The fixed parameter estimates, saved in SAS active data file *Reg_coef*, were transposed and saved in SAS active data file *Fixed_p*. The variance/covariance estimates of the level-1 and level-2 residuals, saved in SAS active data file *Resd_Cov*, were transposed and then saved in SAS active data file *Random_p*[8].

Finally, resampling, data generation, and modeling were repeated by looping through the core sequence of SAS statements 500 times[9]. The 500 sets of estimates of the fixed parameter, level-1 residual variance, and level-2 residual variances and covariance were appended in data sets *FIXED_EFFECTS* and *RANDOM_EFFECTS* and saved in separate SAS data files *FIXED_EFFECT_ALL* and *RANDOM_EFFECT_ALL*, respectively. The mean and standard deviation of the 500 bootstrap estimates of a particular model parameter were used to estimate the mean and standard error of the parameter.

Note that the SAS statements "*ODS LISTING CLOSE*" and "*ODS NORESULTS*" at the beginning of the program stop printing the model results to the output window to increase computing speed and prevent the program from running out of memory when a large number of bootstrap resamples is involved. The SAS Macro statement "*%PUT SAMP = &SAMPLE* " prints the current bootstrap sample number to the SAS log window, allowing the researcher to see how the computing progresses. The SAS macro statements "*%PUT &SYSLAST;*" and "*%IF &SYSLAST = WORK.FIXED %then %DO;*" check the SAS data file "*REG_COEF* " that was created from *PROC MIXED*, using the most current bootstrap sample. The program will not write to the appended data sets of the bootstrap parameter estimates unless modeling converges with the current bootstrap sample. The existence of the most currently created data set indicates convergence of the model using the most current bootstrap sample. This allows us to know at the end of modeling how many times the modeling did not converge using the $B = 500$ bootstrap samples.

The results of the nonparametric residual bootstrap multilevel model, as well as those of the same model with the original sample, are shown in Table 6.3.3.1. The sample-estimated and the bootstrap-estimated main effects and their standard errors were close to each other, while the sample-estimated standard errors of level 2 variance components were slightly smaller than those of the bootstrap procedure. However, the cross-level interaction effect was statistically significant when fitting the model to the original sample ($\hat{\gamma}_{11} = -0.3311$, SD = 0.1671), but became insignificant in the nonparametric residual bootstrap multilevel model ($\hat{\gamma}_{11} = -0.2461$, SD = 0.1774). This result indicates that although the total sample size is large ($n = 9\,824$ in this study), testing for cross-level interaction effect is biased when the number of groups is small (e.g., $J = 20$ in this study) even when REML was used for model estimation.

In this section, we have demonstrated how to use SAS to conduct nonparametric residual bootstrap multilevel modeling when the number of groups is small. This bootstrap method can also be applied to multilevel modeling when the outcome measure is not normally distributed. Instead of using a log-transformation, bootstrap multilevel modeling with the original measure makes the estimated regression coefficients more interpretable. The methods demonstrated here are readily applicable for this purpose (Carpenter et al., 2003). A user-friendly SAS macro, RBMLM, has been recently developed for non-parametric residual bootstrap multilevel modeling. The macro is available upon request (jiwang@gwu.edu). Residual bootstrap multilevel modeling for discrete outcome measures is complex and is beyond the scope of this book.

---

[8] Note that the variance component estimates (e.g., level-1 residual variance, level-2 residual variances and covariance) were internally named as *Residual*, *UN_1_1*, *UN_2_2*, and *UN_2_1*, respectively, in the SAS file *Resd_Cov*.

[9] When sample size is large (n=9,824 in this study), bootstrap multilevel modeling is very time consuming. For the purpose of demonstration, we chose B=500, and it took about 7 hours to finish on a 2.4 GH PC with 512 M RAM.

Table 6.3.3.1 Comparing results between the original sample multilevel model and the nonparametric residual bootstrap multilevel model

Parameter	Original Sample		Nonparametric Residual Bootstrap[1]	
	Estimate	S.E.	Estimate[2]	S.E.[2]
Fixed Parameter				
Intercept	3.5955*	0.1335	3.7251*	0.1305
HIV Prevalence				
Region				
Lower Level	-	-	-	-
Higher Level	0.5310	0.2742	0.4568	0.2635
Ethnicity				
Non-White	-	-	-	-
White	−0.5476*	0.0707	−0.5256*	0.0746
Interaction between HIV Prevalence Region & Ethnicity	0.3311*	0.1671	0.2461	0.1774
Gender				
Female	-	-	-	-
Male	−0.0452	0.0292	−0.0550	0.0288
Age[3]	0.0140*	0.0018	0.0150*	0.0017
Education				
Edu1	−0.0852*	0.0358	−0.0837*	0.0352
Edu2	−0.0497	0.0353	−0.0823*	0.0330
Edu3	-	-	-	-
Variance Component[4]				
$\hat{\text{Var}}(e_{ij}):\hat{\sigma}^2$	1.5531*	0.0222	1.4522*	0.0248
$\hat{\text{Var}}(u_{0j}):\hat{\sigma}_0^2$	0.2369 *	0.0885	0.2714*	0.10037
$\hat{\text{Var}}(u_{1j}):\hat{\sigma}_1^2$	0.0492*	0.0290	0.0677*	0.0328
$\hat{\text{Cov}}(u_{0j},u_{1j}):\hat{\sigma}_{01}^2$	−0.0137	0.0418	−0.0198	0.0409

Notes:
*-Statistically significant at 0.05 level.
1-Nonparametric residual bootstrap multilevel model based on 500samples.
2-Bootstrap estimated parameter and standard error.
3-Grand-mean centered age.
4-Residual variances were tested by one-tail test for H_0:variance > 0

6.4 Group-based models for longitudinal data analysis

In this section we introduce the group-based trajectory models that have been developed recently for longitudinal data analysis (Nagin & Land, 1993; Nagin, 1999, 2005; Jones et al., 2001). As mentioned previously, longitudinal data with repeated measures are considered to be hierarchically structured data with measurement occasions as level-1 units and individuals as level-2 units. Figure

6.4.1A illustrates hypothetical growth trajectories in which each line represents an individual's growth trajectory, and different individuals have different intercepts and slopes. The bold line represents the overall average growth trajectory. In Chapter 4, this kind of data is analyzed using growth models (GM). In GMs the intercept and slope coefficients of individual growth trajectories are assumed to vary randomly across individuals. Growth models assume that all individuals in the sample come from a single population. Individual variations around the mean growth trajectory are captured by the random intercept and slope coefficients.

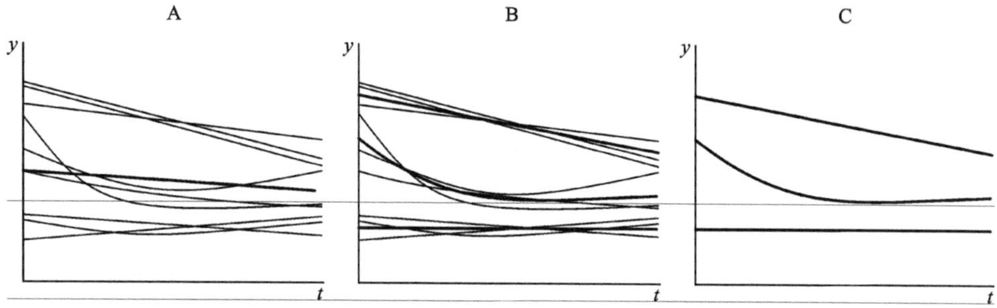

Figure 6. 4.1 Group-based trajectory model vs. growth model (GM) and latent growth mixture model (LGMM) in longitudinal data analysis

Growth models (GM) can also be studied within the framework of structural equation modeling (SEM), or as latent growth models (LGM) in which the random intercept and slope coefficients are called latent growth factors (Muthén, 2001). The LGM can be further generalized to estimate latent growth mixture models (LGMM) accommodating population heterogeneity in the outcome (Muthén, 2001; Muthén & Muthen, 2000; Muthén & Muthén, 1998—2010). Unlike GM and LGM, LGMM not only models intra-individual growth trajectories and inter-individual variations in the trajectories, but also identifies distinct trajectory groups within which individuals share similar, but not identical growth trajectories. Figure 6.4.1B illustrates three trajectory groups:

- Group 1: low initial level of outcome and no significant change in the outcome over time;
- Group 2: moderately high initial level of outcome and significant nonlinear change in the outcome;
- Group 3: high initial level of outcome and rapid linear change in the outcome.

The three bold lines in Figure 6.4.1 C represent three groups with different growth trajectories, assuming no within-group variation.

Like LGMM, the *group-based trajectory model* can be used to study population heterogeneity in growth trajectories. The group-based model identifies distinct growth trajectories and classifies them into a finite number of groups, but the group-based trajectory model assumes a homogenous growth trajectory within each group. For example, using the data shown in Figure 6.4.1A, group-based trajectory model may identify three distinct growth trajectory groups, but no individual variation within each group (see Figure 6.4.1C). Individual variation in growth trajectory in group-based models is captured by the distinct trajectory groups that are defined by their size and trajectory shape (Nagin, 2005).

Both LGMM and group-based trajectory models belong to the family of person-centered analysis. Unlike traditional variable-centered analysis that focus on relationships between variables, person-centered analysis focus on the structure of cases or individuals, identifying the heterogeneity in observed outcome that is believed to be the result of an unobserved finite number of sub-populations in the target population[10]. This method can be used to identify the typology of outcome responses. In longitudinal studies this method can be used to assess whether the population under study is composed of a mixture of identifiable groups based on their growth trajectories. LGMM is a more generalized model, which can be readily estimated using structural equation modeling software like Mplus (Muthén & Muthén, 1998—2010). However, group-based trajectory models are less complicated, and its data manipulation, modeling, and graphing can be conducted in SAS. In this book, we only discuss group-based trajectory models using the SAS procedure, *PROC TRAJ*, written by Jones, Nagin and Roeder (2001).[11]

PROC TRAJ supports different outcome distributions including censored normal, Bernoulli, zero-inflated Poisson (ZIP) and regular Poisson distributions. As such, group-based trajectory models with censored outcome measures, group-based logit trajectory models, group-based Poisson trajectory models, and group-based ZIP trajectory models can be estimated using *PROC TRAJ.*

6.4.1 Introduction to group-based trajectory model

The goal of group-based trajectory models is to identify groups of individuals with similar growth trajectories. Let $Y_i = \{y_{i1}, y_{i2}, \ldots, y_{iT}\}$ denote the longitudinal sequence of the ith individual's outcome measures over m time periods of measurements; π_g denotes group membership probability, i.e., the probability of individual i belonging to group g; and $P^g(Y_i)$ denotes the conditional probability of Y_i given membership in group g. Thus, the probability of observing individual i's longitudinal sequence of measurement is:

$$P(Y_i) = \sum_{g=1}^{G} \pi_g P^g(Y_i) \tag{6.4.1.1}$$

where $P^g(Y_i)$ is a conditional probability, defined as the following:

$$P^g(Y_i) = \prod_{t=1}^{T} p^g(y_{it}) \tag{6.4.1.2}$$

where $p^g(y_{it})$ is the probability distribution function (PDF) of y_{it}, given membership in group g. This equation implies an assumption that the sequential realizations of the elements of Y_i are independent, conditional on group membership. This conditional independence assumption is made in most applications of finite mixture modeling with longitudinal data for the purpose of simplicity. In group-based models, however, repeated outcome measures are assumed not to be serially correlated in the sense that individual deviations from the group average trajectory are uncorrelated. For further elaboration of this assumption, the readers are referred to Nagin (2005).

The distribution form of $p^g(Y_{it})$ corresponds to the type of data under study. For binary data, $p^g(Y_{it})$ is defined as the binary logistic distribution. For count data, it is defined as the Poisson distribution or zero-inflated Poisson (ZIP) distribution. For censored data, it is defined as the censored normal distribution (Nagin, 2005). In group-based trajectory models, $p^g(Y_{it})$ is specified as a polynomial function of time. A set of parameter estimated from the polynomial function determines the shape of growth trajectories

Another component in eq. 6.4.1.1, π_g is the group membership probability or proportion of the population that is in sub-population j, and is defined as (Nagin, 2005):

---

[10] Other person-centered analysis models include latent class analysis (LCA) and latent transition analysis (LTA) (Clogg, 1995; Collins & Wugalter, 1992; Reboussin et al., 1998). LTA, LGMM, and group-based models are extensions of the LCA.

[11] SAS PROC TRAJ can be downloaded from http://www.andrew.cmu.edu/user/bjones.

$$\pi_g = \frac{\exp(\theta_g)}{\sum_{g=1}^{G} \exp(\theta_g)} \tag{6.4.1.3}$$

where θ_g is a set of base parameters, which are estimated from a multinomial logit model without covariates. The values of π_g across the G groups sum to 1.

Identifying the optimal number of trajectory groups: To determine the model that best fits the data, the number of trajectory groups and the shape of the outcome growth trajectory within each group need to be determined. A series of models with different combinations of number of groups and shapes of trajectories needs to be run and the results compared, starting with a single group model (i.e., $G = 1$), followed by 2-group, 3-group, 4-group models, and so on (i.e., $G = 2, 3, 4, \ldots$).[12] For each specified number of groups, various polynomial functions need to be tested within each group.[13] When G is greater than one and a higher order of polynomial functions is involved, the process involves a large number of model estimates. For example, when $G = 3$ and a third order polynomial function is involved, a maximum of $3\times(3 + 1) = 12$ possible models would be estimated and compared. These models would include constant, linear, quadratic, or cubic polynomial functions in each of the three groups. Models with the same number of groups, but different polynomial functions were compared first, and then models with different numbers of groups were compared.

Note that the − 2LL difference between two models with different numbers of latent groups does not follow a Chi-square distribution. Thus the LR test is not appropriate for comparing models with different number of groups. Use of the Bayesian Information Criterion (BIC) for model comparison is widely recommended [14]. A value of BIC that is closer to zero indicates a better model fit.

In addition to BIC, the log form of the Bayes factor is used for model selection. The change in the BIC between the alternative and the null models can be used to approximate the log of the Bayes factor. More specifically, the log of the Bayes factor is approximated by 2 times the difference in BIC (2(ΔBIC)) between the two models being compared: $2\log(B_{10}) \approx 2(\Delta BIC)$ (Raftery, 1995; Kass & Wasserman, 1995; Jones et al., 2001). The log of the Bayes factor can be interpreted as the degree of evidence favoring the alternative model. The interpretation of $2\log(B_{10})$ is described as: 0-2—Not worth mentioning; 2-6—Positive; 6-10—Strong; and >10—Very strong (Jones et al., 2001, p.389). That is, if the log of the Bayes factor is 6 or above, the alternative model would be favored.

Finally, another statistic recommended by Nagin (1999) for selecting the number of groups in group-based models is the "*probability of correct model,*" which is the posterior probability that model g is the correct model. The"*probability of correct model*" is approximated as:

$$P_g = \frac{e^{(BICg - BIC_{max})}}{\sum_g e^{(BICg - BIC_{max})}} \tag{6.4.1.4}$$

---

[12] In case of G=1, all individuals share the same trajectory, leading to the traditional repeated measure analysis which assumes only one growth trajectory group.

[13] The current version of SAS *PROC TRAJ* allows estimation of up to a fourth order polynomial.

[14] BIC is defined as: $BIC = \log(L) - 0.5K\log(N)$ where L is the model estimated maximum likelihood, K is the number of parameters in the model, and N is the sample size. The first term measures the model fit, and the second term functions like a penalty for including more parameters in a model (Nagin, 2005).

where BIC_{max} is the BIC with the maximum absolute score among all the models under consideration, and g is greater than 2 in general (Nagin, 1999, p.148). The "correct" model has the largest P_g value.

Predict group membership: Once the number of groups is identified, relationships between the probability of group membership and covariates can be assessed by the following multinomial logit model:

$$\pi_g(X_i) = \frac{\exp(\theta'_g X_i)}{\sum_{g=1}^{G} \exp(\theta'_g X_i)} \tag{6.4.1.5}$$

where X is a vector of covariates that predict the probability of group membership. Note that only time-invariant covariates (e.g., Gender, Ethnicity) are allowed to be included in eq. 6.4.1.5. Time varying covariates (e.g., marital status, employment, and health status) are used to predict $p^g(y_{it})$ in the group-based models, thus influence the shape of growth trajectories.

Group membership assignment: In group-based modeling, it is important to estimate the probability that each individual in the sample belongs to specific groups. The growth trajectory groups are unobserved, thus it is impossible to determine an individual's group membership with certainty (Nagin, 1999). Instead, an individual's probability of being in any trajectory group is estimated. This probability is called posterior group membership probability, which can be estimated as:

$$\hat{p}(g \mid Y_i) = \frac{\hat{p}(Y_i \mid g)\hat{\pi}_g}{\sum_g \hat{p}(Y_i \mid g)\hat{\pi}_g} \tag{6.4.1.6}$$

where $\hat{p}(g \mid Y_i)$ is the estimated probability of individual i's membership in group g, given Y_i comprising i's outcome measures in each time period; and $\hat{\pi}_g$ is the estimated population proportion in group g. An individual is assigned to a specific group for which his/her posterior probability is the largest. If there were no misclassification, the average posterior probability in a group would equal the proportion of the population in that group (i.e., $\hat{\pi}_g$). However, misclassification is unavoidable in group-based modeling because $\hat{p}(g \mid Y_i)$ is unlikely to be 1.0 for a specific group and zero for all other groups. According to Nagin (2005), a rule of thumb for acceptable group classification is that the probability of correct group membership assignment is 0.70 or greater.

Corresponding to the distributions of $p^g(y_{it})$, SAS *PROC TRAJ* allows us to run different kinds of group-based trajectory models: group-based logit trajectory models, group-based Poisson or ZIP trajectory models, and group-based trajectory models with censored outcome measure. In the following sections, we demonstrate how to run these models.

6.4.2 Group-based logit trajectory model

We start with a group-based logit trajectory model with a third order polynomial function in which the conditional probability $p^g(y_{it})$ is specified as binary logit distribution.

$$p^g(y_{it}) = \frac{\exp(\beta_{g0} + \beta_{g1}T_{it} + \beta_{g2}T^2_{it} + \beta_{g3}T^3_{it})}{1 + \exp(\beta_{g0} + \beta_{g1}T_{it} + \beta_{g2}T^2_{it} + \beta_{g3}T^3_{it})} \tag{6.4.2.1}$$

or

$$\ln\left(\frac{p^g(y_{it})}{1-p^g(y_{it})}\right) = \beta_{g0} + \beta_{g1}T_{it} + \beta_{g2}T_{it}^2 + \beta_{g3}T_{it}^3 \qquad (6.4.2.2)$$

We will show later that both time-varying and time-in variant covariates can be readily included in the model to predict the outcome growth and growth trajectory group memberships, respectively. The intercept and slope coefficients in the model determine the shapes of the growth trajectories. Within the same trajectory group, the coefficients β_{g0}, β_{g1}, β_{g2}, and β_{g3} are the same for all the individuals, while they vary across the G trajectory groups. In other words, individuals within the group share the same growth trajectory, while different groups have different trajectories.

In the case of $G = 1$, the model reduces to the traditional repeated measure analysis which assumes only one growth trajectory group. In group-based trajectory models, G is a finite number and the multinomial model shown in eq. 6.4.1.5 is used to predict group membership.

Application of group-based logit trajectory models: A sample of crack cocaine users recruited for a natural history health study in Dayton, Ohio, the U.S. (Falck et al., 2003) will be used to demonstrate the group-based Logit model. Starting in July 1996, 430 crack cocaine users were recruited using targeted sampling (Watters & Bernacki 1989; Carlson et al., 1994). After completing a baseline interview, each participant completed follow-up interviews every 6 months for three years. As such, the observation period of the study was comprised of 6-month assessment periods: T_1 covering the time between baseline and 6 month follow-up; T_2 covering 6 to 12 month follow-up, through T_6 covering 30 to 36 month follow-up. The purpose of this analysis is to assess the trajectory of crack-cocaine use in the population over time. The outcome is a binary variable (1-stopped crack use; 0-used crack), measuring whether an individual had stopped using crack during each 6-month assessment period between T_t and T_{t+1}.

Among the 430 participants, those who had at least one follow-up interview (n = 409) after baseline interview were included in the analysis. A small example of the data used for the analysis is shown in SAS Output 6.4.2.0.

Each individual is assigned a unique identification number (*id* in this data set). The time variables (T_1-T_6 in this data set) are required to have constant integer values for all individuals who participated in follow-up interviews in the specific time periods (e.g., $T_1 = 0$, $T_2 = 1$, … , and $T_6 = 5$) [15]. Variables y_1-y_6 denote the binary outcome (i.e., stopped crack-cocaine use) in each 6-month observation period, respectively, after baseline interview. There are four time-invariant variables: *Ethnic* (1-Black; 0-Other) and *Gender* (1-Male; 0-Female) are measured a dummy variables; *Age_c* and *Educ_c* are grand-mean centered numeric variable. The variables *T1_Trt* —*T6_Trt* are time-varying variables (1-Being in drug treatment; 0-Not in drug treatment), which measure whether a respondent participated in drug treatment in each specific 6-month observation period. The time-invariant variables were used to predict latent growth trajectory group membership; and the time-varying covariates were used to predict the shape of the growth trajectory.

SAS Output 6.4.2.0 The first 20 cases selected from the data used for group-based logit model

id	t1	t2	t3	t4	t5	t6	y1	y2	y3	y4	y5	y6	Gender	Ethnic	Age_c	Educ_c	t1_ Trt	t2_ Trt	t3_ Trt	t4_ Trt	t5_ Trt	t6_ Trt
1	0	1	2	3	4	5	0	0	0	0	0	0	0	1	-2.5	2.46	1	1	1	0	0	0
2	0	1	2	3	4	5	0	0	0	0	0	0	0	1	-2.5	0.46	0	0	0	0	0	0
3	0	1	2	3	4	5	0	0	0	0	0	0	1	1	11.5	3.46	0	0	0	0	0	0
4	0	1	.	.	.	5	0	0	.	.	.	1	1	1	-6.5	-1.54	0	0	.	.	.	0
5	0	1	2	.	4	5	0	0	0	.	0	0	1	1	0.5	2.46	1	0	0	.	0	0

[15] T_1 was set to 0, thus the intercept β_{g0} represents the log odds of stopping crack use in the first 6-month period after baseline interview in group g.

6	0	.	.	.	.	5	0	.	.	.	.	0	1	0	4.5	2.46	0	.	.	.	.	1
7	0	1	2	3	4	5	0	0	0	0	0	0	1	1	-5.5	0.46	0	0	0	0	0	0
8	.	1	2	3	4	5	.	0	0	1	0	0	1	0	17.5	-2.54	.	0	1	0	0	0
9	.	1	2	3	4	5	.	0	0	0	0	0	1	1	5.5	0.46	.	0	0	0	0	0
10	.	1	2	3	.	5	.	0	0	0	.	0	0	0	1.5	-1.54	.	0	1	0	.	0
11	0	1	2	3	4	.	0	0	0	0	0	.	.	.	.	-2.54	0	0	0	0	1	.
12	0	.	2	3	4	5	0	.	0	0	0	0	1	1	7.5	-1.54	0	.	0	0	0	0
13	0	1	2	3	4	5	0	0	0	0	0	0	1	1	6.5	-1.54	0	0	0	0	0	0
14	0	1	2	3	4	5	0	0	0	0	0	0	0	1	-13.5	-1.54	1	1	0	0	0	0
15	0	1	2	3	4	5	0	0	0	0	0	0	0	1	-11.5	-1.54	1	1	1	0	0	0
16	0	1	2	3	4	5	0	0	0	0	0	0	0	0	3.5	-1.54	0	0	1	1	0	0
18	0	1	2	3	4	5	0	0	0	0	1	0	0	1	-9.5	-1.54	1	0	0	0	1	0
19	.	1	2	3	4	5	.	0	1	1	1	1	0	0	10.5	2.46	.	1	0	0	0	0
21	.	1	2	3	4	5	.	0	0	0	0	0	0	1	6.5	3.46	.	0	0	0	0	0

The group-based logit trajectory model was conducted in the following sequence. First, we tested models with various polynomials with various numbers of trajectory groups. The best polynomial functions for models with different numbers of trajectory groups in this example are: linear for a 1-group model; linear in each group for a 2-group model; and linear in each group for a 3-group model. Since Bayesian Information Criterion (BIC) substantially increased after $G = 2$, we stopped modeling at $G = 3$. Second, we added the four time-invariant covariates into the program to predict trajectory group membership. Third, we classified individual crack cocaine users into appropriate trajectory groups. Fourth, we included the time-varying covariates into the model. And finally, we created individual profiles based on the estimated posterior group membership.

Table 6.4.2.1 Model comparison

No. of Groups	BIC	$2\log(B_{10})$ [1]	P_j [2]
1	–389.09	0	0
2	–350.61	76.96	0.9999
3	–361.83	–22.44	0

Notes:

[1]: The log of the Bayes factor is approximated by 2 times the difference in BIC (2(ΔBIC)) between the alternative and null models. $2\log(B_{10})$ is interpreted as: 0-2—Not worth mentioning; 2-6—Positive; 6-10—Strong; and >10—Very strong (Jones et al., 2001, p.389).

[2]: P_g is the posterior probability that model with g groups is the correct model (see eq. 6.4.1.4).

Model fit statistics for models with 1, 2 and 3 groups are shown in Table 6.4.2.1. The value of BIC improved (getting closer to zero) when the number of group increased from $g = 1$ to $g = 2$, then decreased when from $G = 2$ to $G = 3$. The BIC of the 2-group model is closest to zero, indicating it has a better fit than the other two models. The log of the Bayes factor $2\log(B_{10})$ and the probability of having a correct model P_g also favors the 2-group model.

The following SAS program is used to run the group-based logit trajectory model with 2 trajectory groups:

***SAS Program 6.4.2.1;**

```
LIBNAME lib 'D:\MLM\data';
```

```
Data GB_Logit_model;
 Set Lib.GB_Logit_Dat;

PROC TRAJ OUTPLOT = OP OUTSTAT = OS OUT = OF OUTEST = OE ITDETAIL;
 ID ID; VAR y1—y6; INDEP T1—T6;
MODEL LOGIT; *specify Logit model;
NGROUPS 2;   *specify number of trajectory groups;
ORDER 1 1;        *specify trajectory polynomial function for each group;
Run;

%TRAJPLOT(OP,OS, 'Stopping Crack vs. Time', 'Group-based Logit Model',
'Prob of no use', 'Time Period');
```

Data used for the group-based logit trajectory model were retrieved from SAS data file *GB_Logit_dat*. In *PROC TRAJ* the ID variable is specified so that observations can be identified in output data sets. Repeated measures of the outcome variables are specified in the *VAR* statement and independent variables (time score variable T_1-T_6 in this example) are specified in the INDEP statement. The model type (group-based logit model in this example) is specified in the *MODEL* statement. Another two statements, *NGROUPS* and *ORDER* specify the number of latent trajectory groups and polynomial function within each group. In this example, "2" is specified for two-groups and "1 1" call for a linear function of time in groups 1 and 2.

Five additional options can be specified in *PROC TRAJ*:

- *OUTPLOT*—produces data for plotting growth trajectories.
- *OUTSTAT*—produces parameter estimates used by SAS macro *TRAJPLOT*, including estimates of population proportions ($\hat{\pi}_j$).
- *OUT*—produces group membership and membership probabilities at individual level, which can be used to calculate average group membership probabilities by group.
- *OUTEST*—produces parameter estimates and their variance/covariance matrix.
- *ITDETAIL*—produces iteration information model estimation in SAS output.

Readers can visit the SAS *PROC TRAJ* website for detailed instructions and file downloads (http://www.andrew.cmu.edu/user/bjones).

The upper panel of SAS Output 6.4.2.1 presents linear Logit model parameter estimates for the two trajectory groups. The intercepts represent the log odds of stopping crack use in the first 6-month period after the baseline interview, while the slope coefficients represent the rate of change in the log odds in the trajectory groups. The *Group membership* panel presents model estimated population proportions (i.e., $\hat{\pi}_j$), that are also saved to a SAS active file OS using the option *OUTSTAT*.

SAS Output 6.4.2.1 Selected results of group-based logit trajectory model with 2 trajectory groups without covariates

Group	Parameter	Estimate	Standard Error	T for H0: Parameter = 0	Prob > |T|
1	Intercept	-5.19725	0.46508	-11.175	0.0000
	Linear	0.43898	0.12202	3.598	0.0003
2	Intercept	-2.31133	0.44560	-5.187	0.0000

```
        Linear            0.78845          0.15968                4.938          0.0000
        Group membership
 1                (%)        92.51277          1.96936               46.976          0.0000
 2                (%)         7.48723          1.96936                3.802          0.0001
BIC = -350.60834 (N = 2072)      BIC = -346.55196 (N = 409)   AIC = -336.51767
```

SAS macro *TRAJPLOT* [16] was used to plot the growth trajectories by group (last line of SAS Program 6.4.2.1). The macro requires specification of the output data sets (*OP* and *OS* in this example), the first and second titles (e.g., *'Stopping Crack vs. Time'* and *'Group-based Logit Model'*), as well as the name of the *Y* and *X* axes in the plot (*'Prob of no use'* and *'Time Period'*).

The growth trajectories of the two groups and the observed overall growth curve are depicted in Figure 6.4.2.1. The solid lines are the observed growth trajectories of crack-cocaine use, the dotted lines are the model estimated growth trajectories, and the bold line represents the observed sample overall growth trajectory. The corresponding numbers used to generate Figure 6.4.2.1 are shown in Table 6.4.2.2. The two groups can be defined as:

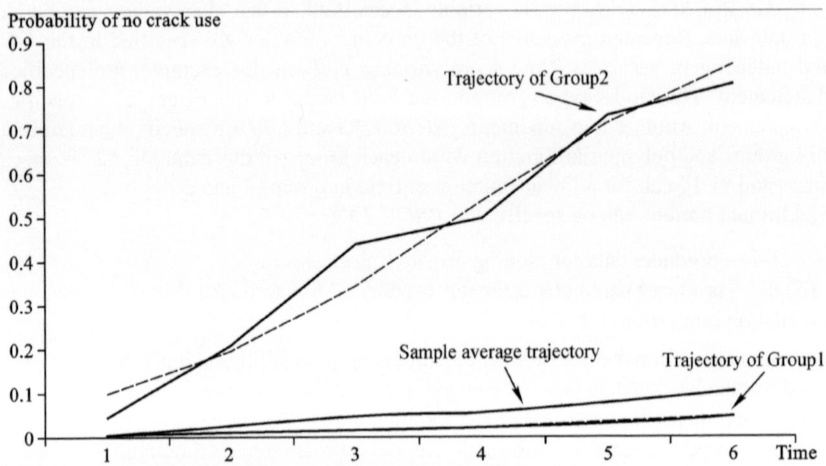

Figure 6.4.2.1 Group-based logit trajectory model: solid lines observed trajectories; dotted lines represents Predicted trajectories

- Group 1: Almost nobody had stopped crack use during the first 6-month period after the baseline interview; and then there was a significant, but small increase in stopping crack-cocaine use over time.
- Group 2: The probability of stopping crack use during the first 6-month period after the baseline interview was low; however there was a remarkable increase in stopping crack-cocaine use over time.

Individual group memberships are shown in Table 6.4.2.3. About 93.9% (n = 384) and 6.1% (n = 25) of the sample were classified into group 1 and 2, respectively. The percentages of correct group membership assignment are high (98% for group 1 and 89% for group 2). As misclassifica-

[16] SAS/GRAPH is required in order to run *TRAJPLOT*. SAS macro files *TRAJPLOT*, *TRAJPLOTNEW*, and *TRAJTEST* that are extracted from the compressed file *trajv9.exe* must be copied into the /CORE/SASMACRO folder. *TRAJPLOT* is helpful for viewing group trajectories. However, if unequal time intervals are involved, other computer software (e.g., MS Excel) are recommended for trajectory graphs using the data that the option *OUTPLOT* of PROC TRAJ produces.

tion was small in both groups, the estimated population group proportions (92.5% for group 1 and 7.5% for group 2) are very close to the sample membership proportions.

Table 6.4.2.2 Observed proportion and model estimated probability of stopping crack use by time period and trajectory group

	Observed			Estimated	
T_t	Overall[1]	AVG1[2]	AVG2[3]	PRED1[4]	PRED2[5]
1	0.0052	0.00246	0.04338	0.00592	0.09916
2	0.0243	0.01164	0.20709	0.00869	0.1922
3	0.0474	0.01526	0.43693	0.01359	0.33969
4	0.0526	0.02225	0.49409	0.01996	0.53454
5	0.0762	0.02869	0.72996	0.03211	0.71238
6	0.1027	0.04795	0.80311	0.04804	0.8436

Note:
[1]: Overall sample proportion of no crack use by time period.
[2]: Sample proportion of no crack use in group 1 by time period.
[3]: Sample proportion of no crack use in group 2 by time period.
[4]: Model estimated probability of no crack use in group 1 by time period.
[5]: Model estimated probability of no crack use in group 2 by time period.
[*] *AVG1—AVG2* and *PRED1—PRED2* are estimated and saved by SAS *PROC TRAJ*, using option *OUTPLOT*.

Table 6.4.2.3 Probability of trajectory group membership, sample membership classification, and estimated population proportions ($n = 409$)[1]

	Group 1 ($n = 384$)	Group 2 ($n = 25$)
Group 1	0.98	0.02
Group 2	0.11	0.89
Sample membership classification	93.9%	6.1%
Estimated population proportion	92.5%	7.5%

Note:
[1]: Participants who had never come back for follow-up interviews after baseline were excluded from the analysis.

Notice that the size of group 2 is small in this example, thus the model results may not be valid for inferences. However, for illustration purpose, we still present this model.

The estimated sample membership classification in Table 6.4.2.3 can be obtained by adding the following SAS codes into SAS Program 6.4.2.1:

```
Data group_inf;
 set OF;
Proc sort; by group;

Proc Freq;
Tables group;
Run;

Proc Means;
by group;
```

```
Var  GRP1PRB GRP2PRB;
Run;
```

To predict group membership, the time-invariant covariates (e.g., *Ethnic*, *Gender*, *Age_c*, and *Educ_c*) are specified in the SAS *RISK* statement[17] (see SAS Program 6.4.2.2a). In addition, time-varying covariates, such as *T1_trt—T6_trt* in this example, can be included to predict the outcome growth trajectories by specifying the covariates in the SAS TCOV statement (see SAS Program 6.4.2.2b).

***SAS Program 6.4.2.2a;**

```
...
Data GB_Logit_model;
 Set lib.GB_Logit_dat;

*with time-invariate covariate;
PROC TRAJ data = GB_Logit_model OUTPLOT = OP OUTSTAT = OS OUT = OF OUTEST
= OE ITDETAIL;
 ID id; VAR y1—y6; INDEP t1—t6;
 MODEL LOGIT;
NGROUPS 2;
ORDER 1 1;
    RISK Ethnic Gender Age_c Educ_c;
Run;
```

***SAS Program 6.4.2.2b;**

```
...
*with time-invariate and time-varying covariates;
PROC TRAJ data = GB_Logit_model OUTPLOT = OP OUTSTAT = OS OUT = OF OUTEST
= OE ITDETAIL;
 ID id; VAR y1—y6; INDEP t1—t6;
 MODEL LOGIT;
NGROUPS 2;
ORDER 1 1;
    RISK Ethnic Gender Age_c Educ_c;
    TCOV  T1_Trt  T2_Trt  T3_Trt  T4_Trt  T5_Trt  T6_Trt;
Run;
```

SAS Output 6.4.2.2a & 6.4.2.2b Selected results of group-based logit trajectory model

A. With time-invariant covariates

Group	Parameter	Estimate	Standard Error	T for H0: Parameter = 0	Prob > |T|
1	Intercept	-5.18395	0.47748	-10.857	0.0000
	Linear	0.42578	0.12439	3.423	0.0006

[17] In SAS Proc Traj procedure, time-invariant or time-stable covariates are called “risk factors”.

2	Intercept	-2.26014	0.44340	-5.097	0.0000
	Linear	0.77781	0.16621	4.680	0.0000
	Group membership				
1	Constant	(0.00000)	.	.	.
2	Constant	-2.97066	0.65937	-4.505	0.0000
	Ethnic	0.27385	0.54643	0.501	0.6163
	Gender	0.09188	0.48219	0.191	0.8489
	Age_c	-0.10807	0.03641	-2.969	0.0030
	Educ_c	0.07894	0.14543	0.543	0.5873

BIC = -348.75600 (N = 2016) BIC = -341.43245 (N = 396) AIC = -323.51608

B. With time-invariant and time-varying covariates

Group	Parameter	Estimate	Standard Error	T for H0: Parameter = 0	Prob > |T|
1	Intercept	-5.36888	0.49867	-10.766	0.0000
	Linear	0.43188	0.12209	3.537	0.0004
	t1_Trt	0.98204	0.40980	2.396	0.0166
2	Intercept	-2.18299	0.49290	-4.429	0.0000
	Linear	0.77688	0.18360	4.231	0.0000
	t1_Trt	-0.12099	0.65580	-0.184	0.8536
	Group membership				
1	Constant	(0.00000)	.	.	.
2	Constant	-3.00363	0.74143	-4.051	0.0001
	Ethnic	0.15511	0.58250	0.266	0.7900
	Gender	0.16130	0.51503	0.313	0.7542
	Age_c	-0.10953	0.03780	-2.897	0.0038
	Educ_c	0.09543	0.15057	0.634	0.5263

BIC = -353.96270 (N = 2016) BIC = -345.01169 (N = 396) AIC = -323.11392

T1_Trt—*T6_Trt*[18] had a significant positive effect on growth trajectory in group 1 (0.98204, $p = 0.0166$), but not in group 2 (−0.12099, $p = 0.8536$). This indicates that people who were in drug treatment in period T_t were more likely to stop crack use in Group 1. By contrast, being in treatment did not have a significant effect in Group 2.

Note that sample sizes associated with BIC in the SAS Output vary. The first N denotes the total number of occasions in the sample, and the second N is the total number of cases in the

[18] In the model, measures *T1_Trt-T6_Trt* actually represents the same variable measured at different time points. In the SAS Output, only the first measure (i.e., *T1_Trt*) is used to represent the time-varying covariate.

sample. The *N*'s in this SAS Output are different from those in SAS Output 6.4.2.1, due to missing values in the covariates.

When time-invariant covariates are included to predict group membership, SAS *PROC TRAJ* no longer provides population proportion estimates in SAS output. However, these estimates can be obtained from the SAS data file *OS* exported by the *OUTSTAT* option.

Finally, the estimated trajectory group membership can be saved and merged with the original data sets to create individual profiles. For example, in this study SAS data file *OF* containing group membership was merged with the 36-month follow-up interview data and with the observed individual characteristics at the end of the 3-year observation period. The individual profile by trajectory group from merged data is shown in Table 6.4.2.4. As the model predicts, older people were less likely to be in group 2, while the other time-invariant covariates were insignificant [19]. The trajectory group membership was not significantly associated with being in drug treatment at the end of the entire observation period. From Table 6.4.2.4, we can see that individual profile is useful to describe how the trajectory patterns are associated with other variables.

Table 6.4.2.4 Observed individual characteristics at the end of the 3-Year observation period by trajectory group ($n = 292$)

Variable[1]	Group 1 ($n = 272$)	Group 2 ($n = 20$)	*P*-value
Gender			
Female ($n = 117$)	108 (92.3%)	9 (7.7%)	
Male ($n = 172$)	161 (93.6%)	11 (6.4%)	0.6698
Ethnicity			
White ($n = 106$)	98 (92.5%)	8 (7.5%)	
Black ($n = 183$)	171 (93.4%)	12 (6.6%)	0.7493
Age Group			
< 25 ($n = 16$)	13 (81.3%)	3 (18.8%)	
25-34 ($n = 69$)	60 (87.0%)	9 (13.0%)	
35 + ($n = 203$)	195 (96.1%)	8 (3.9%)	0.0059
Education			
<High School ($n = 118$)	108 (91.5%)	10 (8.5%)	
High School ($n = 108$)	104 (96.3%)	4 (3.7%)	
College ($n = 66$)	60 (90.9%)	6 (9.1%)	0.2614
In Drug Treatment[2]			
No ($n = 247$)	230 (93.1%)	17 (6.9%)	
Yes ($n = 45$)	42 (93.3%)	3 (6.7%)	0.9579

Note:

[1]: Number of cases in categories of a variable may not sum up to the total sample size due to missing values.

[2]: Self-reported being in drug treatment in the past 6 months prior to the 36-month follow-up interview.

6.4.3 Group-based zero-inflated Poisson (ZIP) trajectory model

In the last section of Chapter 5 we discussed how to use multilevel Poisson regression to model count data. We introduced the multilevel ZIP model in Section 6.1 of this chapter for instances where there is a large number of 0's in the count data. When distinctive sub-groups in a popula-

[19] Note that the size of group 2 is small in this example. In real research this may be a concern.

tion exist in longitudinal count data, the group-based zero-inflated ZIP trajectory model should be used. This model accommodates extra-zeros in the count data, and correctly accounts for heterogeneity in individual outcome growth trajectories by identifying a small number of latent growth trajectory groups. Individual growth trajectories are similar within the same group, but different across groups (Nagin, 1999, 2005; Nagin & Land, 1993). In this section we demonstrate how to run the group-based ZIP model using SAS *PROC TRAJ.*

In the group-based ZIP trajectory model the conditional probability $p^g(y_{it})$ is defined as a zero-inflated Poisson (ZIP) distribution:

$$p^g(y_{it}=0)=\rho_{itg}+(1-\rho_{itg})e^{-\lambda_{itg}} \qquad (6.4.3.1)$$

$$p^g(y_{it}=h_{it})=(1-\rho_{itg})\frac{e^{-\lambda_{itg}}\lambda_{itg}^{h_{it}}}{h_{it}!},\ \text{if } h_{it}>0 \qquad (6.4.3.2)$$

where random variable y_{it} is any non-negative integer value; λ_{itg} is the expected number of event occurrences for individual *i* at time *t*, given membership in group *g*. ρ_{itg} is the probability of having extra zeros[20]. The ZIP model simultaneously estimates the two parameters ρ_{itg} and λ_{itg}. In group-based ZIP trajectory models, the link function that connects the likelihood of having extra zeros with time is:

$$\ln\left(\frac{\rho_{itg}}{1-\rho_{itg}}\right)=\alpha_{g0}+\alpha_{g1}T_{it}+\alpha_{g2}T_{it}^2+\alpha_{g3}T_{it}^3 \qquad (6.4.3.3)$$

The link function that connects the event count to time is:

$$\ln(\lambda_{itg})=\beta_{g0}+\beta_{g1}T_{it}+\beta_{g2}T_{it}^2+\beta_{g3}T_{it}^3 \qquad (6.4.3.4)$$

A third-order polynomial function of time is specified in both equations in this example. The two models are simultaneously estimated in SAS *PROC TRAJ.*

Application of the group-based ZIP trajectory model: Data used for the demonstration were from a natural history study of stimulant users in the U.S. The data format is similar to that shown in SAS Output 6.4.2.1. The outcome measure is self-reported number of days of methamphetamine (meth) use in the past 30 days prior to an interview. The data are saved in SAS file *GB_ZIP_dat* with variable names: *id* (individual identification number), *T0*-*T5* (time measured as 0, 1, 2, 3, 4, and 5 for baseline, 6-month, 12-month, … , and 30-month follow-ups, respectively), *meth_t0*—*meth_t5* (outcome measured as number of days of meth use in the past 30 days prior to *T0*, *T1*, … , *T5*, respectively), *Gender* (1-Male; 0-Female), *Ethnic* (1-Black; 0-Otherwise), *Age_c* (overall mean centered age), *Highsch* (1-High school education; 0-Otherwise), and college (1-College education; 0-Otherwise). No time-varying covariates are included.

Histograms of the self-reported number of days of meth use in the past 30 days in each 6-month time period are shown in Figure 6.4.3.1. The outcome distribution involves a substantial number of zeros indicating a zero-inflated Poisson distribution. More zeros were present than would be expected in a pure Poisson distribution. The group-based ZIP trajectory model appears to be appropriate for these data. A statistical test can be conducted to determine whether the group-based ZIP trajectory model or the group-based Poisson trajectory model would be favored for analysis with a given dataset. We will discuss this issue later in this section.

[20] Note that the notation used for extra zero probability here is different from that used in the equations of Section 6.1 as π is usually used to denote group membership probability in group-based trajectory models (see eq. 6.4.1.3).

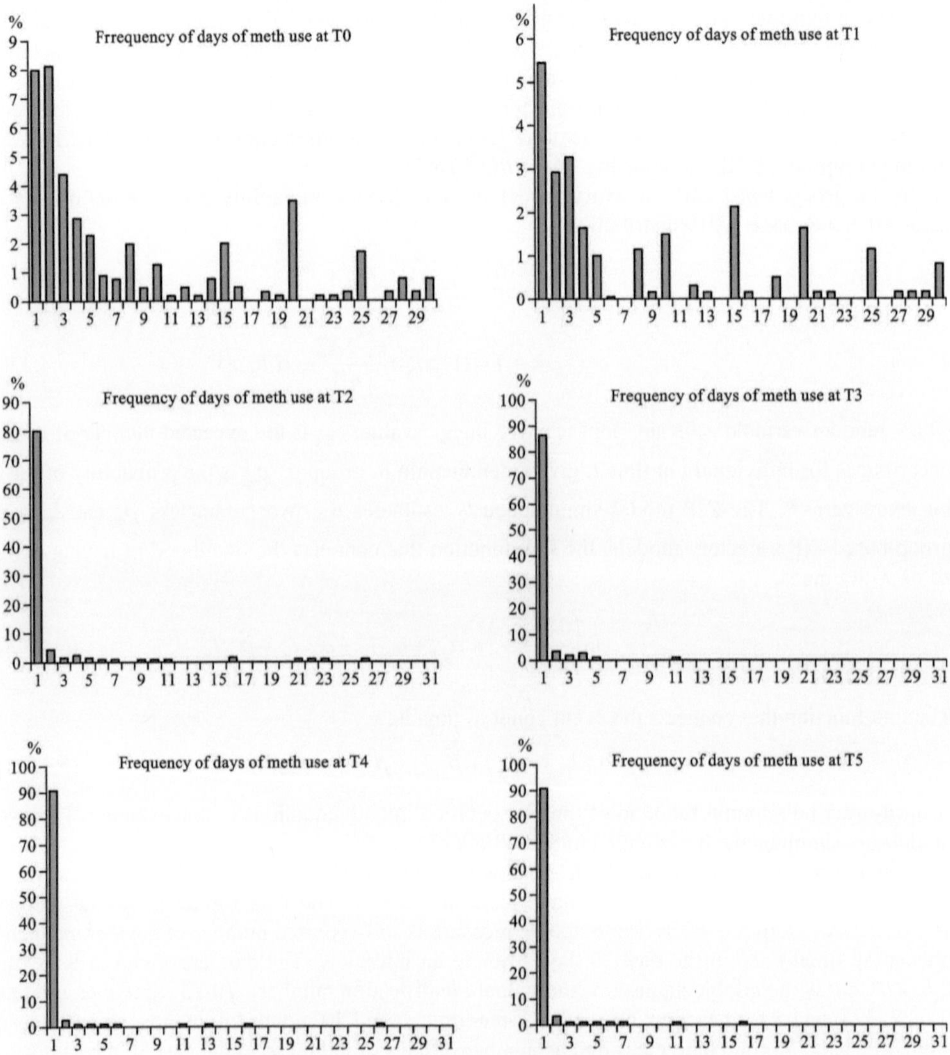

Figure 6.4.3.1 Distributions of count data over time

A series of models was tested. The best polynomial functions for models with different numbers of trajectories groups are:

Poisson model (eq. 6.4.3.4):
1-group model: quadratic
2-group model: linear, linear
3-group model: linear, linear, linear
4-group model: constant, linear, quadratic, quadratic
5-group model: constant, linear, linear, quadratic, quadratic

Extra zero logit model (eq. 6.4.3.3):
1-group model: linear

2-group model: linear, linear
3-group model: linear, linear, linear
4-group model: constant, linear, linear, quadratic
5-group model: constant, linear, linear, linear, quadratic

The absolute values of the BIC statistic are plotted against their respective number of groups in Figure 6.4.3.2. The BIC drops until the number of groups equals 4, then increases somewhat when the number of groups equals 5. The BIC of the 4-group model is closer to zero and thus the 4-group model is favored. The trajectories of the 4 groups are displayed in Figure 6.4.3.3. These groups can be characterized as follows:

- Group 1 - almost no meth use
- Group 2 - light meth use with no change over time
- Group 3 - initial heavy meth use with a dramatic decline during the first year and was sustained lower level over time
- Group 4 - initial heavy use of meth that increases somewhat in the first year, then declines significantly over time.

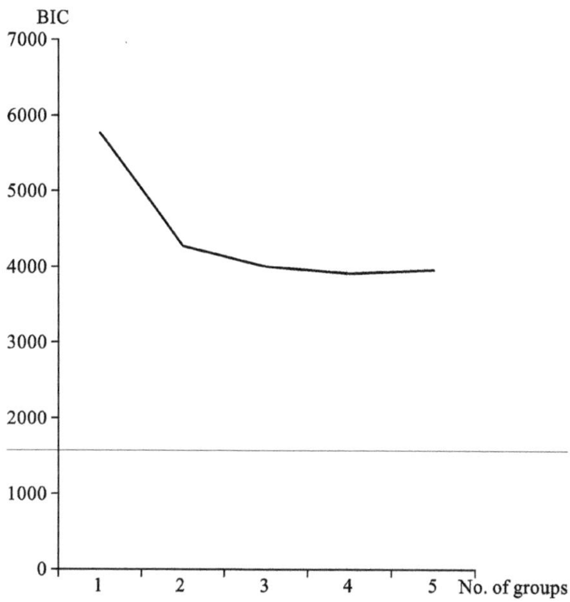

Figure 6.4.3.2 BIC vs. group numbers Group-based ZIP trajectory model

Individual group memberships are shown in Table 6.4.3.1. About 60.3% (n = 400), 23.2% (n = 154), 9.0% (n = 60), and 7.4%(n = 49) of the sample were assigned to Groups 1, 2, 3, and 4, respectively. The percentage of correct group membership assignment was 94% for group 1, 91% for group 2, 96% for group 3, and 98% for group 4. Misclassification was small in all groups. As a result, the estimated population group proportions are close to the sample group proportions (see Table 6.4.3.1).

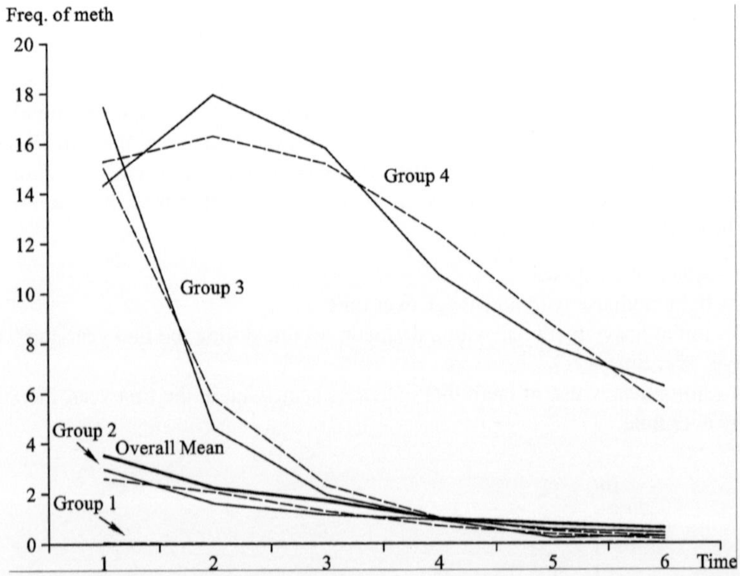

Figure 6.4.3.3 Observed and predicted trajctories of methamphatamine use Groupbased ZIP Model

Table 6.4.3.1 Average trajectory group membership probability based on maximum posterior probability ($n = 663$)[1]

	Group 1 ($n = 400$)	Group 2 ($n = 154$)	Group 3 ($n = 60$)	Group 4 ($n = 49$)
Group 1	0.94	0.06	0.00	0.00
Group 2	0.07	0.91	0.02	0.00
Group 3	0.00	0.02	0.96	0.02
Group 4	0.00	0.01	0.01	0.98
Sample Membership Classification	60.3%	23.2%	9.0%	7.4%
Estimated Population Proportion	58.3%	24.7%	9.5%	7.5%

Note:
[1]: Among the 710 participants in the study, 47 participants who did not come back for any follow-up interviews were excluded from analysis.

The following SAS program was used to estimate the 4-group ZIP trajectory model:

***SAS Program 6.4.3.1**;

```
...
Data GB_Zip_model;
 Set lib.GB_ZIP_Dat;

%let g = 4;        *define the number of trajectory groups;
%let ord1 = 0 1 2 2;     *define polynomial order of the Poisson model;
%let ord2 = 0 1 1 2;     *define polynomial order of the extra zero model;

PROC TRAJ OUTPLOT = OP OUTSTAT = OS OUT = OF OUTEST = OE;
```

```
PROC TRAJ OUTPLOT = OP OUTSTAT = OS OUT = OF OUTEST = OE;
   ID id; VAR meth_t0—meth_t5; INDEP t0—t5;
 MODEL ZIP;     *specify ZIP model;
 NGROUPS &g;
 ORDER &ord1;
 IORDER &ord2;
Run;
%TRAJPLOT(OP, OS, 'Meth use trajectory', 'Zero Inflated Poisson Model',
'Freq. of meth use', 'Time');
```

The number of trajectory groups, the polynomial function for the Poisson trajectory model, and the polynomial function for the extra zero logit trajectory model are defined as SAS macro variables (i.e., *% let g* = 4; *% let ord*1 = 0 1 2 2.*% let ord2* = 0 1 1 2;). As such, it is easier for programmers to specify different values for model testing.

SAS Output 6.4.3.1 Selected results of 4-group ZIP trajectory model without covariates

Group	Parameter	Estimate	Standard Error	T for H0: Parameter = 0	Prob > |T|
1	Intercept	−0.72088	0.47080	-1.531	0.1258
2	Intercept	1.18735	0.07238	16.404	0.0000
	Linear	0.00702	0.02779	0.253	0.8006
3	Intercept	2.92413	0.04145	70.554	0.0000
	Linear	-0.88625	0.09371	-9.458	0.0000
	Quadratic	0.09214	0.02447	3.765	0.0002
4	Intercept	2.76027	0.03638	75.876	0.0000
	Linear	0.15531	0.03828	4.058	0.0001
	Quadratic	-0.05136	0.00798	-6.440	0.0000
1	Alpha0	2.47031	0.31908	7.742	0.0000
2	Alpha0	-1.33027	0.22484	-5.916	0.0000
	Alpha1	0.87712	0.07469	11.743	0.0000
3	Alpha0	-1.38916	0.26275	-5.287	0.0000
	Alpha1	0.67812	0.09290	7.300	0.0000
4	Alpha0	-3.26636	0.62638	-5.215	0.0000
	Alpha1	0.81761	0.46553	1.756	0.0791
	Alpha2	-0.04455	0.07888	-0.565	0.5723
	Group membership				
1	(%)	58.28485	3.10096	18.796	0.0000
2	(%)	24.69978	2.81770	8.766	0.0000
3	(%)	9.51749	1.26529	7.522	0.0000
4	(%)	7.49788	1.05079	7.136	0.0000

```
BIC = -3903.37673 (N = 3465)  BIC = -3886.83980 (N = 663)  AIC = -3841.87205
```

In SAS Output 6.4.3.1, the first set of parameter estimates (e.g., *Intercept*, *Linear*, and *Quadratic*) are regression coefficient estimates of the Poisson regression part of the model, representing the effect of time on the outcome counts (i.e., times of meth use in the past 30 days) in the four trajectory groups. The second set of parameter estimates (e.g., *Alpha*0, *Alpha*1, *Alpha*2) are regression coefficient estimates of the extra-zero logit part of the model, representing the effect of time on the likelihood of extra zeros in each trajectory group.

Time-invariant covariates and time-varying covariates can be included into the model. In this example, only time-invariant covariates are included to predict trajectory group membership. The SAS program for the 4-group ZIP trajectory model follows:

SAS Program 6.4.3.2;

```
...
%let g = 4;           *define the number of trajectory groups;
%let ord1 = 0 1 2 2;      *define the polynomial order of the Poisson model;
%let ord2 = 0 1 1 2;      *define the polynomial order of the extra zero model;

PROC TRAJ OUTPLOT = OP OUTSTAT = OS OUT = OF OUTEST = OE;
    ID id; VAR meth_t0-meth_t5; INDEP t0-t5;
    MODEL ZIP;   *specify the ZIP model;
    NGROUPS &g;
    ORDER &ord1;
    IORDER &ord2;
    RISK Gender Ethnic Age_c Highsch College; *Time-invariant covariate or
    risk factors;
Run;
```

SAS Output 6.4.3.2 Selected results of 4-group ZIP trajectory model with covariates

Group	Parameter	Estimate	Standard Error	T for H0: Parameter = 0	Prob > |T|
1	Intercept	1.70040	0.04215	40.340	0.0000
2	Intercept	0.86906	0.06338	13.711	0.0000
	Linear	-0.31308	0.06699	-4.674	0.0000
3	Intercept	2.96097	0.03488	84.888	0.0000
	Linear	-1.13876	0.08436	-13.499	0.0000
	Quadratic	0.15130	0.02258	6.701	0.0000
4	Intercept	2.91335	0.03971	73.369	0.0000
	Linear	0.08619	0.04040	2.134	0.0330
	Quadratic	-0.04182	0.00842	-4.965	0.0000
1	Alpha0	-0.84158	0.15685	-5.366	0.0000
2	Alpha0	0.94563	0.09855	9.595	0.0000

	Alpha1	0.67820	0.07976	8.503	0.0000
3	Alpha0	-1.39223	0.28517	-4.882	0.0000
	Alpha1	0.78009	0.11413	6.835	0.0000
4	Alpha0	-3.37697	0.70585	-4.784	0.0000
	Alpha1	0.92326	0.52407	1.762	0.0782
	Alpha2	-0.06347	0.08852	-0.717	0.4734
	Group membership				
1	Constant	(0.00000)	.	.	.
2	Constant	2.00242	0.34510	5.802	0.0000
	gender	-0.25132	0.34143	-0.736	0.4617
	ethnic	3.54304	1.03224	3.432	0.0006
	age_c	-0.02667	0.01609	-1.657	0.0975
	highsch	-0.20632	0.35645	-0.579	0.5628
	college	0.87120	0.69117	1.260	0.2076
3	Constant	-0.10316	0.47489	-0.217	0.8280
	gender	-0.38278	0.44762	-0.855	0.3925
	ethnic	1.73603	1.19891	1.448	0.1477
	age_c	-0.12893	0.02802	-4.601	0.0000
	highsch	-0.41402	0.46512	-0.890	0.3735
	college	0.59803	0.85997	0.695	0.4868
4	Constant	0.11577	0.43761	0.265	0.7914
	gender	-0.45296	0.45108	-1.004	0.3154
	ethnic	0.97178	1.26394	0.769	0.4420
	age_c	-0.00876	0.02144	-0.408	0.6829
	highsch	-0.17007	0.46821	-0.363	0.7165
	college	0.10849	0.93212	0.116	0.9073

```
BIC = -3854.63483 (N = 3465)   BIC = -3825.69521 (N = 663)   AIC = -3747.00164
```

The upper panel of SAS Output 6.4.3.2 presents parameter estimates for the Poisson regression and extra zero logit parts of the model. The lower panel in the SAS Output presents results of the multinomial logit model to predict group membership. Group 1 is treated as the reference group, and three Logit models were simultaneously run to compare the likelihood of being in Groups 2, 3, and 4 over Group 1. The variable *Ethnic* (1-Black; 0-White) has a significant positive effect in the first Logit comparing trajectory group 2 vs. group 1 (3.54304, $p = 0.0006$), indicating that black, rural drug users were more likely than whites to be in trajectory group 2 over group 1. Age has a significant negative effect in the second Logit, indicating that the younger people were more likely than the older people to be in trajectory group 3. Other covariates show no significant effects on the group membership.

Both group-based trajectory ZIP and Poisson models can be estimated using *PROC TRAJ*. By specifying *%let Iord* $= -1\ -1\ -1\ -1$; in Program 6.4.3.2, we produce a 4-group Poisson trajectory model for which BIC $= -4770$, indicating that group-based ZIP trajectory model provides a better fit with the data.

6.4.4 Group-based censored normal trajectory models

In social science studies some outcome measures are censored. For example, when the exam questions are not sufficiently difficult a substantial number of students would receive the total possible score, and there would be no variation among the "top" students. Censoring on the high end creates a "ceiling" or maximum of the applicable scale. Another example is the amount of money that people spend on expensive durable goods (e.g., computers, furniture, cars, etc.) in a given year. There is "no" ceiling on what might be spent. However, many people buy no high end durables and spend nothing in the year. In this case, censoring arises on the low end or minimum of the scale, called the "floor." Among those who have a zero value of the outcome measure (i.e., spent nothing), some may never buy any high end durables (they have a real zero value); some may be ready to buy, but have not yet. Irrespective of the underlying propensity of the individual, the observed outcome values are "zero".

Censoring can also occur simultaneously at both low and high ends. For example, Likert scales typically provide a limited number of ordinal responses for an underlying continuous measure. For example, self-reported health status may be measured on a 5-point scale: 1-very bad, 2-bad, 3-neutral know, 4-good, and 5-very good. Under some circumstances responses might "pile up" at both ends of the spectrum.

When an outcome measure is censored, its distribution is distorted. Regression models that ignore the presence of censoring can produce biased coefficient estimates. The classical Tobit regression model (Tobin, 1958; Greene, 1990) is frequently used for modeling censored outcomes in econometric studies. In this section, we will discuss how to deal with censored data in group-based modeling. These models can be used for continuous outcome measures with and without censoring.

Let y_{it}^* denote a latent or unobserved outcome measure in a longitudinal study. Link y_{it}^* to its observed but censored counterpart, y_{it}:

$$y_{it} = \text{Min, if } y_{it}^* < \text{Min}$$

$$y_{it} = y_{it}^*, \text{ if Min} \leq y_{it}^* \leq \text{Max}$$

$$y_{it} = \text{Max, if } y_{it}^* > \text{Max}$$

where Min and Max denote the minimum and the maximum possible scores on a scale. If the latent variable is less than Min, the values of the observed measure y_{it} would cluster at the Min; if the latent variable is greater than Max, the values of the observed measure y_{it} would clustered at the max; and the latent variable y_{it}^* would equal its observed measure y_{it} between Min and Max.

For the censored normal distribution, the link between time and the latent outcome variable in the group-based model is defined as:

$$y_{it}^* = \beta_{g0} + \beta_{g1}T_{it} + \beta_{g2}T_{it}^2 + \beta_{g3}T_{it}^3 + \varepsilon_{it} \tag{6.4.4.1}$$

where the βs are parameters that determine the shape of growth trajectories, and the error term ε_{it} is assumed to be normally distributed with zero means and standard deviation σ. This is referred to as a "*censored normal*" model in which the distribution of a continuous, normally distributed latent variable is censored as observed in measure, y_{it}. Given membership in group g the probability distribution function (PDF) of y_{it} is defined as (Nagin, 2005):

$$p^g\left(y_{it} = \text{Min}\right) = \Phi\left(\frac{\text{Min} - \beta_g x_{it}}{\sigma}\right) \tag{6.4.4.2}$$

$$p^g(y_{it}) = \frac{1}{\sigma}\phi\left(\frac{y_{it} - \beta_g x_{it}}{\sigma}\right) \text{ for Min} \le y_{it} \le \text{Max} \tag{6.4.4.3}$$

$$p^g(y_{it} = \text{Max}) = 1 - \Phi\left(\frac{\text{Max} - \beta_g x_{it}}{\sigma}\right) \tag{6.4.4.4}$$

where ϕ is the probability density function (PDF) and Φ is the cumulative distribution function (CDF) of the normal random variable with mean $\beta_g x_{it}$ and standard deviation σ; $\beta_g x_{it}$ denotes the expected value of y_{it}^*, which is $\beta_{g0} + \beta_{g1}T_{it} + \beta_{g2}T_{it}^2 + \beta_{g3}T_{it}^3$; Min and Max are the scale minimum and maximum values.

Application of the group-based censored normal trajectory model: Sample data from a crack cocaine use study (Falck et al., 2003) were used for this demonstration. Phase I data collected between July 1996 and August 2001, as well as Phase II data collected between July 2001 and August 2005, were included.

The group-based censored normal trajectory model was used to assess the pattern of annual change in status of crack cocaine use. There are a total of 8 measurements from baseline (T_0), the first year follow-up (T_1), … , the third year follow-up of Phase I (T_3), the Phase II baseline (T_4) (about 2 years after T_3), the first year follow-up in Phase II (T_5), … , and the end of Phase II (T_7).

The outcome measure was a Likert scale measure of crack use status in the 6 months prior to an interview, with: 0—None; 1—Less than 4 times per month; 2—About 1 time a week; 3—2 to 6 times a week; 4—About 1 time a day almost every day; 5—About 2 to 3 times a day almost every day; and 6 — 4 or more times a day almost every day. The data in SAS file *GB_Censored_dat* include variables: *id* (individual identification number); *T0*-*T7* (time variables as described above);*Crack_t0*-*Crack_t7* (crack use status in the past 6 months measured on a 7-point Likert scale); *Gender* (1-Male; 0-Female), *Ethnic* (1-Black; 0-otherwise), *Age*, *Highsch* (1-Completed high school education; 0-Otherwise), and *College* (1-Completed college education; 0-Otherwise); and time varying covariates *Trt_t0*-*Trt_t7* (1-Ever being in drug treatment in the past year prior to an interview; 0-Otherwise).

Histograms of crack use status over time are shown in Figure 6.4.4.1. The outcome distribution shows that there are substantial high frequencies of the maximum values at earlier time points (e.g., T_0, T_1, and T_2) and minimum values at later time points (e.g., T_3-T_7). Suppose that the Likert scale, to some extent, measures unobserved addiction to crack-cocaine. The observed measures may have both high-end and low-end censoring. This outcome measure serves as an example for the purpose of demonstrating the group-based censored normal model. Models with and without censoring were compared.

After testing as series of models, four trajectory groups were identified with a quadratic polynomial function in Groups 1 and 2, and a cubic polynomial function in Groups 3 and 4 (see Figure 6.4.4.2). We can describe the trajectory groups as:

- Group 1— Starting with a relatively high initial level, crack use declined rapidly in the first 3 years and kept at a very low level over time with a little relapse in the end of the observation period.
- Group 2— Starting with a moderately high level of crack use, declined significantly but still remained with weekly use on average over time.
- Group-3— Starting with a moderately high level of crack use, decline rapidly in the first 4 years after baseline, and remained at a very low level.
- Group-4— Starting with a very high level of crack use, had some change over time, but maintained daily use during the observation period.

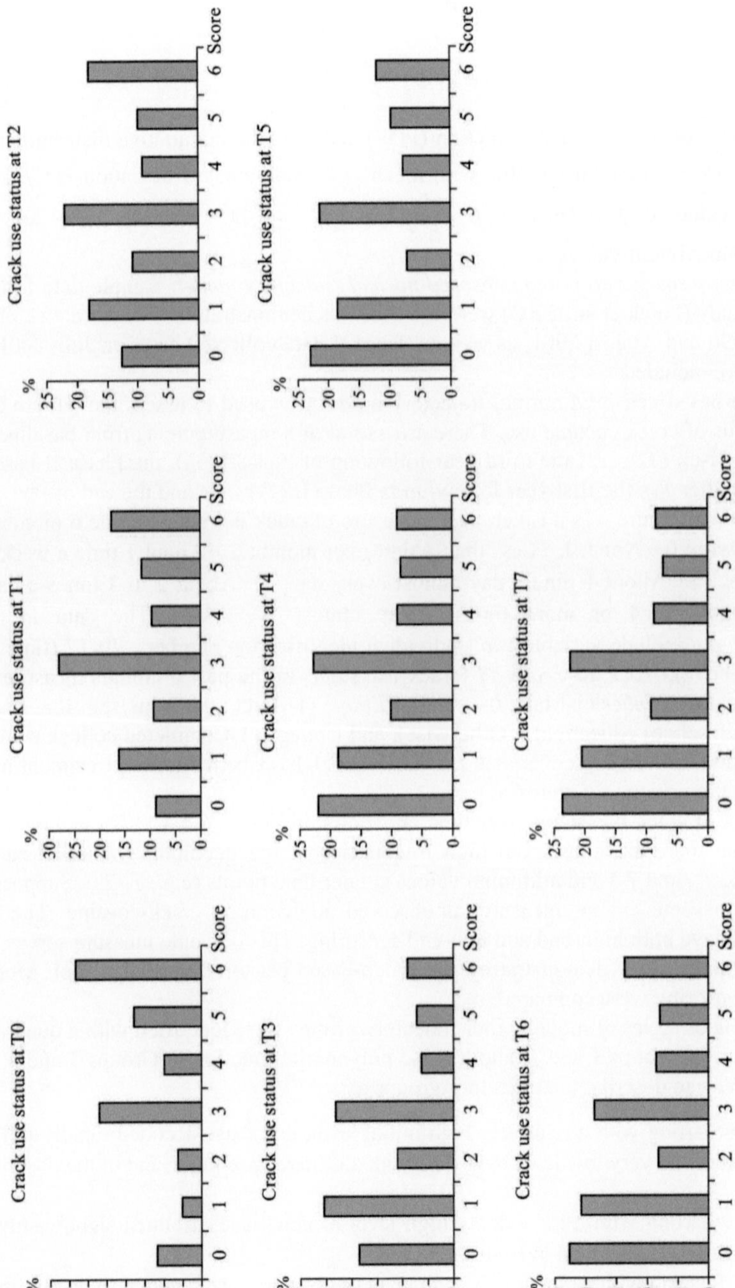

Figure 6.4.4.1 Distributions of crack-cocain use over time

Table 6.4.4.1 shows predicted group membership. The majority of the sample was classified into Group 2 (55.1%) and 29.4% of sample was classified into Group 4. Only 8.1% and 7.4% of the sample were classified into Groups 1 and 3, respectively. The percent of correct group membership assignment was 92% for group 1, 89% for group 2, 87% for group 3, and 90% for group 4. According to the rule of thumb (i.e., 70% correct assignment) proposed by Nagin (2005), the model with 4-group solution provides a good group classification.

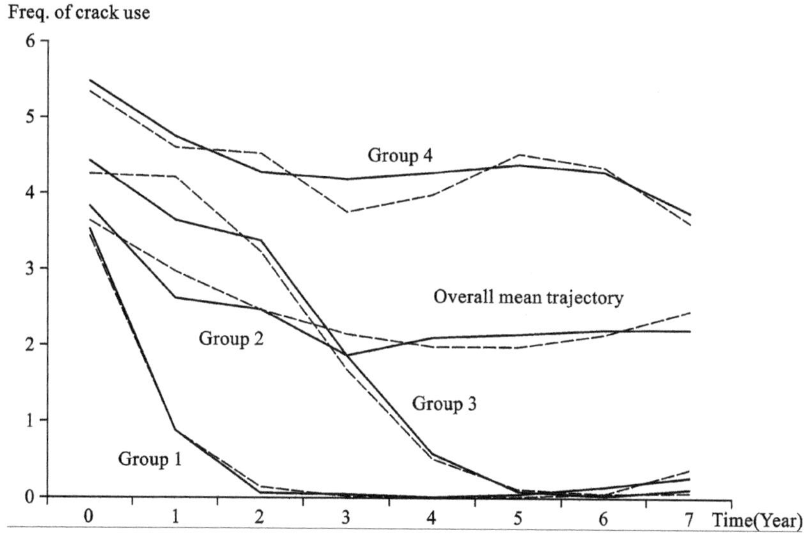

Figure 6.4.4.2 Observed and predicted trajectcnes Group-based censored normal trajectory model

Table 6.4.4.1 Average trajectory group membership probability based on maximum posterior probability (n = 394)[1]

	Group 1 (n = 32)	Group 2 (n = 217)	Group 3 (n = 29)	Group 4 (n = 116)
Group 1	0.92	0.04	0.03	0.00
Group 2	0.01	0.89	0.03	0.07
Group 3	0.02	0.10	0.87	0.00
Group 4	0.00	0.10	0.01	0.90
Sample membership classification	8.1%	55.1%	7.4%	29.4%
Estimated population proportion	8.4%	53.1%	8.5%	30.0%

Note:

[1]: Participants who never came back for follow-up interviews were excluded from analysis.

Model selection is both statistically-driven and theory-driven. In addition to model fit statistics, other factors, such as the number of individuals per group, similarity of trajectory shapes, and association with theoretically related external criteria should be considered in model selection (Muthén & Muthén, 2000). In this example, the numbers of individuals in Group 1 (n = 32) and in Group 3 (n = 29) estimated from the model are relatively small. Although the 4-group model fits the data better than alternative models, small group size is a concern in real research.

The following SAS program is used to estimate the 4-group censored normal trajectory model without covariates:

***SAS Program 6.4.4.1**;

```
...
Data GB_Censored_model;
 Set Lib.GB_Censored_Dat;

%let g = 4; *define number of trajectory groups;
%let ord = 2 2 3 3; *define trajectory shape;
%let scale_min = 0; *define scale minimum;
%let scale_max = 6; *define scale maximum;

Proc Traj out = of outplot = op outstat = os outest = oe;
Var Crack_t0—Crack_t7;
INDEP t0—t7;
MODEL CNORM; *specify censored normal model;
MIN &scale_min;
MAX &scale_max;
NGROUPS &g;
ORDER &ord;
Run;
%TRAJPLOT(OP,OS, 'Trajectory of crack use frequency', 'Group-base censored
normal model','Crack use freq.', 'Time (Year)');
```

In this SAS program, the minimum and maximum value of the outcome measure, the number of trajectory groups and the polynomial functions are defined as SAS macro variables later to be recalled in the SAS program.

SAS Output 6.4.4.1 Selected results of 4-group censored normal trajectory models without covariates

Group	Parameter	Estimate	Standard Error	T for H0: Parameter = 0	Prob > |T|
1	Intercept	3.53065	0.38654	9.134	0.0000
	Linear	-3.82375	0.44477	-8.597	0.0000
	Quadratic	0.45228	0.06062	7.461	0.0000
2	Intercept	3.74455	0.14610	25.630	0.0000
	Linear	-0.87271	0.09386	-9.298	0.0000
	Quadratic	0.09650	0.01296	7.446	0.0000
3	Intercept	4.51161	0.40936	11.021	0.0000
	Linear	0.68325	0.71155	0.960	0.3370
	Quadratic	-0.82005	0.29174	-2.811	0.0050
	Cubic	0.08149	0.02986	2.729	0.0064
4	Intercept	6.63424	0.27254	24.343	0.0000

```
         Linear        -1.90305       0.30320          -6.277       0.0000
         Quadratic      0.51968       0.10396           4.999       0.0000
         Cubic         -0.04342       0.00977          -4.445       0.0000

         Sigma          2.00533       0.03788          52.941       0.0000

         Group membership
1              (%)      8.37629       1.56317           5.359       0.0000
2              (%)     53.10566       3.47146          15.298       0.0000
3              (%)      8.52124       1.94793           4.375       0.0000
4              (%)     29.99681       3.23720           9.266       0.0000

BIC = -4759.06585 (N = 2569)  BIC = -4742.19156 (N = 394)  AIC = -4706.40440
```

Time-invariant and time-varying covariates are included into the model as follows.

***SAS Program 6.4.4.2**;

```
...
Data GB_Censored_model;
 Set lib.GB_Censored_Dat;

Proc SQL;
 Create table GB_Censored_model as select *, Age-mean(Age) as Age_c from
GB_Censored_model;
Quit;
%let g = 4; *define number of trajectory groups;
%let ord = 2 2 3 3; *define trajectory shape;
%let scale_min = 0; *define scale minimum;
%let scale_max = 6; *define scale maximum;
Proc Traj out = of outplot = op outstat = os outest = oe;
Var Crack_t0—Crack_t7;
INDEP t0—t7;
MODEL CNORM; *specify censored normal model;
MIN &scale_min;
MAX &scale_max;
NGROUPS &g;
ORDER &ord;
RISK Gender Ethnic Age_c Highsch College; * time-invariant covariates or
risk factors;
TCOV Trt_to— Trt_t7; *Time-varying covariates;
Run;
%TRAJPLOT(OP,OS, 'Trajectory of crack use frequency', 'Group-base censored
normal model','Crack use freq.', 'Time (Year)');
```

SAS Output 6.4.4.2 Selected results of 4-group censored normal trajectory models with time-invariant and time-varying covariates

```
                                             Standard        T for H0:
Group    Parameter       Estimate             Error      Parameter = 0    Prob > |T|
```

1	Intercept	2.57502	0.55802	4.615	0.0000
	Linear	-3.12653	0.46251	-6.760	0.0000
	Quadratic	0.36944	0.06268	5.894	0.0000
	Trt_t0	1.30230	0.58616	2.222	0.0264
2	Intercept	3.63532	0.17005	21.378	0.0000
	Linear	-0.85279	0.10131	-8.417	0.0000
	Quadratic	0.09320	0.01377	6.769	0.0000
	Trt_t0	0.16074	0.16354	0.983	0.3258
3	Intercept	4.90700	0.52704	9.311	0.0000
	Linear	-0.64973	0.99170	-0.655	0.5124
	Quadratic	-0.04077	0.51272	-0.080	0.9366
	Cubic	-0.02916	0.07230	-0.403	0.6867
	Trt_t0	-0.47388	0.52100	-0.910	0.3631
4	Intercept	6.81371	0.29685	22.953	0.0000
	Linear	-2.10476	0.31043	-6.780	0.0000
	Quadratic	0.56813	0.10342	5.493	0.0000
	Cubic	-0.04674	0.00956	-4.886	0.0000
	Trt_t0	-0.48494	0.22148	-2.190	0.0286
	Sigma	1.99788	0.03764	53.085	0.0000

Group membership

1	Constant	(0.00000)	.	.	.
2	Constant	1.67375	0.42900	3.902	0.0001
	Gender	-0.39009	0.42935	-0.909	0.3637
	Ethnic	0.70973	0.44646	1.590	0.1120
	Age_c	0.04300	0.02780	1.547	0.1220
	Highsch	0.35676	0.49282	0.724	0.4692
	College	-0.36818	0.51771	-0.711	0.4770
3	Constant	-0.77362	0.71211	-1.086	0.2774
	Gender	-0.51513	0.62713	-0.821	0.4115
	Ethnic	1.13180	0.67408	1.679	0.0933
	Age_c	0.03331	0.04319	0.771	0.4407
	Highsch	0.31760	0.79315	0.400	0.6889
	College	0.89858	0.74528	1.206	0.2280
4	Constant	0.30320	0.49710	0.610	0.5420
	Gender	-0.33767	0.45760	-0.738	0.4606
	Ethnic	2.19028	0.48172	4.547	0.0000
	Age_c	0.02026	0.02991	0.677	0.4983
	Highsch	0.06912	0.52133	0.133	0.8945
	College	-0.43996	0.55808	-0.788	0.4306

```
BIC = -4806.18903 (N = 2569)   BIC = -4771.50299 (N = 394)   AIC =
-4697.94050
```

Ever being in drug treatment in the past year had a significant negative effect on crack-cocaine use in group 4 (*Trt_t0* = −0.48494, $p = 0.0286$), but a significant positive effect on crack use in group 1 (*Trt_T0* = 1.3023, $p = 0.0264$). Being treatment has no significant effect in groups 2 and 3 in this example.

Time-invariant covariates were used to predict trajectory group membership. Because there are four trajectory groups, 3 logits in the multinomial logit model were used for membership prediction. Group 1 was treated as the reference group in each of the logits. The results show that no covariates significantly affected group membership, except that *Ethnic* had a positive effect on the odds of being in Group 4 over Group 1 (*Ethnic* = 2.1903, $p = 0.0000$).

SAS *PROC TRAJ* can also be used to analyze group-based trajectory models with continuous outcome measures that are not censored. This can be done by specifying *MIN* that is below the actual minimum outcome and a *MAX* that is above the actual maximum outcome. For example, by replacing *% let scale_min* = 0 and *% let scale_max* = 6 with *% let scale_min* = −10 and *% let scale_max* = 10 in SAS Program 6.4.4.2, we have a group-based normal model in which the outcome measure is assumed to be continuous variable with a normal distribution. However, the 4-group normal model fits the data worse (BIC = −5062) than the 4-group censored normal model (BIC = −4806). As such, we conclude that the censored normal model is preferred.

In this section, we have discussed group-based logit trajectory model, group-based ZIP trajectory model, group-based Poisson trajectory model, group-based censored normal trajectory model, and group-based normal trajectory model. These models can be further extended to "dual processes group-based trajectory model" in which two related outcome measures are modeled jointly. In addition, model-identified trajectory group membership can be used to predict a "distal" outcome measure at the end of the observation period. Interested readers are referred to Jones & Nagin (2006).

6.5 Missing values issue

Studies in many research fields suffer from missing data. In ross-sectional studies subjects may refuse to answer any questions (unit non-response) or a specific subset of questions (item non-response). In addition, investigators may not be able to find their subjects at the time of data collection. In longitudinal studies, subjects may miss one or more follow-up interviews, or may refuse to respond to some questions on any given occasion. Some subjects may drop out or withdraw from the study before its intended completion. The former types of missingness are sometimes described as "intermittent" missing while others are called dropout or attrition (Troxel et al., 1998).

Missing values complicate data analysis. First, if subjects with missing values are systematically different from "completers," analyses that ignore these differences may be biased and result in misleading inferences. Second, the existence of missing data usually implies a loss of information so that estimates will be less efficient or less precise. And finally, conventional statistical methods are designed for complete or "balanced" data. Incomplete or unbalanced data often require more advanced analytical methods (Fitzmaurice et al., 2004; Little & Schenker, 1995).

The multilevel modeling techniques presented so far mark a significant step forward in handling problems of missing data using model-based strategies. In longitudinal studies, subjects do not have to be measured at the same time, nor have the same number of measures. Subjects who are missing on some occasion would not be excluded from the study. However, validity of inferences with multilevel modeling for incomplete data rests on specific assumptions about missing data. If the specific assumption is not met, the results of multilevel models will no longer be valid. In this case, different approaches or additional care should be taken to ensure integrity of the study.

Although missing data are problematic for cross-sectional studies, missing data are far more challenging in longitudinal studies. In this section, our discussion will focus on longitudinal studies. We will review mechanisms that generate missing data, their implications for data analyses, and methods for modeling incomplete data.

6.5.1 Missing data mechanisms and their implications

Missing data "mechanisms" refers to the processes that generate missing data or the reasons why information is missing. It is critical to understand missing data mechanisms because appropriately modeling incomplete data depends on our assumptions about these mechanisms.

Despite some subtle variations, statisticians have distinguished three major missing data mechanisms: missing completely at random (MCAR), missing at random (MAR) and missing not at random (MNAR) (Rubin, 1976; Little & Rubin, 2002).

Missing completely at random (MCAR):

Data are said to be missing completely at random (MCAR) when "missingness", referring to the tendency to be missing, does not depend on either outcomes or covariates. This process implies that missing is completely random. Thus cases under MCAR can be considered as a random subset of the complete data, and nothing distinguishes the missing values from those that are observed. MCAR is the most restrictive assumption among the considered for missing data processes.

A variation of MCAR is defined as covariate-dependent missingness, which allows missingness to depend on observed covariates, but not on the observed outcome (Diggle & Kenward, 1999). Some experts, however, classify the covariate dependent missing case as a separate category (Hogan et al., 2004; Little, 1995). For our discussion here, we only consider the general version of MCAR.

Missing at random (MAR):

Missing at random (MAR) refers to a situation in which missingness depends on observed values (either outcome measures or covariates), but not to the unobserved values. In other words, MAR simply means that the missingness is conditional on observed variables, thus one can use observed variables to predict missingness. An example where longitudinal data are MAR might arise when a subject drops out of a treatment prior to a follow-up interview due to being either better or worse at the previous measurement, or exceeding a cut-off value defined by the researchers. The latter case is a called planned missing, which is under the researcher's control.

Allowing missingness to depend on observed values is far less restrictive, and is perhaps a more plausible assumption for missing data. Foster and Fang (2004, pp. 437-438) note, "*many of the forms of missingness that concern evaluators are MAR. For example, researchers often consider whether the follow-up rate differs between the treatment and control groups or whether it depends on key outcome measures at baseline. Because treatment status and baseline scores are generally observed completely, even when these relationships exist, they represent forms of*

MAR." For this reason, many experts on missing data advocate use of MAR analysis as the default approach (Hedeker & Gibbons, 2006).

Missing not at random (MNAR):

Missing not at random (MNAR) refers to a situation where the probability of missing depends on unobserved quantities, either on the potentially unobserved outcome itself or on related unobserved covariates. MNAR is also termed "informative" missingness by some authors (Diggle & Kenward, 1999; Fitzmaurice et al., 2004). MNAR informs that the distribution of the missing observations is not the same as the distribution among the completers or in the target population. For example, in quality of life (QOL) studies subjects who have lower QOL scores due to potential incidence or severity of disease may also be more likely to miss follow-ups. This results in MNAR because some important explanatory variables were missing. If the explanatory information were collected, it might be possible to reduce the missing data mechanism from MNAR to MAR, or even to MCAR (Troxel et al., 1998).

From the perspective of model estimation, the three basic missing data mechanisms can also be classified as ignorable missingness and non-ignorable missingness. Generally MAR and MCAR are ignorable while MNAR is non-ignorable (Laird, 1988). For ignorable missingness likelihood-based estimates can be constructed based on observed data, and the component of missing mechanism (the process that generates missing values) can be ignored, and the inference for full-data parameters derived from the observed data is valid (Hogan et al., 2004).

However, for non-ignorable missingness (MNAR), the missing mechanism cannot be ignored in parameter estimation. To obtain valid estimates, jointly specified models for the response and the missing data are required. The term, "non-ignorable" is used to emphasize that the missing data mechanism must be specified for correct inferences (Fitzmaurice et al., 2004).

The missing data mechanism has profound implications for analyzing incomplete longitudinal data because choice of an appropriate model and valid inference depend on assumptions about the missing mechanism. For MCAR, the observed data can be viewed as a random subsample of the entire data set. MCAR does not create any problem in data analysis.

In the case of MAR, multilevel modeling approaches will yield valid inferences. In fact, the SAS procedures used for all of the models we present in this book can appropriately deal with MAR, provided the model is correctly specified (Fitzmaurice et al., 2004).

In the case of MNAR, the methods we present in this book may not be adequate. Almost all standard methods of longitudinal analysis fail when data are MNAR (Fitzmaurice et al., 2004). To analyze data under MNAR, standard multilevel models must be enhanced. Fortunately, as we noted previously, MAR is a plausible assumption in many cases. Many experts advocate use of MAR analysis as the default approach. Even if MAR assumptions do not hold, several authors demonstrate that the departures from MAR often have only a minor impact on estimates and standard errors (Collins et al., 2001). These departures often do not invalidate the results of an MAR-based analysis (Schafer & Graham, 2002).

6.5.2 Handling missing data in longitudinal data analyses

In longitudinal studies, the MCAR condition does not present major difficulties. The MAR condition can be analyzed using likelihood based modeling approaches that we discussed in previous chapters.

Many approaches have been proposed for MNAR, or non-ignorable cases, including selection models (Heckman, 1976, 1979; Hedeker & Gibbons, 2006; Schafer & Graham, 2002; Zhou & Harel, 2007), frailty models (Hogan et al., 2004); pseudo-likelihood methods (Parzen et al.,

2007); and hybrid models (Wilkins & Fitzmaurice, 2006). These models are complex, and have yet to be integrated into commonly used statistical packages. They will not be discussed here.

Very often researchers would like to replace the missing values with estimated values. One popular method is called multiple imputations (MI). This method was originated by Rubin (1970, 1987) and elaborated by Schafer (1997). The idea of MI is straightforward; that is, replacing missing data with plausible values, and then analyzing the "complete" data sets using standard statistical procedures. The MI technique is significantly different from traditional single imputation. Single imputation method fill-in a single value for each missing value. Single imputation methods do not reflect the uncertainty about the predictions of the unknown missing values, and the resulting standard errors for the parameter estimators are biased downward. Unlike single imputation, MI is a Monte Carlo technique that simulates a random sample or distribution of missing values. This procedure results in valid statistical inferences and properly reflects uncertainty due to missing values.

Specifically, the MI procedure involves three major steps: First, the missing values are filled in multiple times, and an each complete data set is simulated. Second, each imputed complete data set is analyzed separately by using standard statistical methods (e.g., multilevel models). And finally, results are estimated from each complete data set and these results are combined to yield estimates and their confidence intervals that incorporate misssing-data uncertainty.

Ideally the model used for imputation and the model used for analysis in subsequent steps should agree with each other. If one model is more restrictive than the other, e.g., the imputation model has interaction, but the analytical model does not, then inconsistencies will arise, and inference may not be valid (Allison, 2002; Schafer, 1997). Most of the techniques available for MI assume the missing values are "ignorable" (i.e., either MCAR or MAR) (Allison, 2000; Rubin, 1987).

Many general and stand-alone statistical programs have implemented one or more methods for MI. SAS *PROC MI* and *PROC MIANALYZE* can be used together to conduct MI. However, the current version of *PROC MI* (SAS 9.2) does not build in methods for correlated longitudinal data or multilevel data. Alternative programs, such as PAN (Schafer, 2001), SOLAS can be used for MI for multilevel data. Another program can be used for MI for longitudinal data using the last value carried forward method. M (e.g., 5) sets of imputed data exported from the program can be run in SAS multilevel procedures to estimate M sets of model parameters. Then Rubin's (1987) formula can be applied to estimate the mean and standard errors of the model. For example, let b_m be the estimated model parameter in sample m of the M imputed samples, and S_m be its standard error. Then,

$$\bar{b} = \frac{\sum_{m=1}^{M} b_m}{M} \tag{6.5.2.1}$$

$$S_b = \sqrt{\frac{1}{M}\sum_m S_m^2 + \left(1 + \frac{1}{M}\right)\frac{\sum_m \left(b_m - \bar{b}\right)^2}{M-1}} \tag{6.5.2.2}$$

where the b is the parameter estimate based on the imputed data, and its standard error is S_b, which is the square root of the average of the sampling variances plus the variance of parameter estimates, multiplied by a correction factr $(1 + (1/M))$.

6.6 Statistical power and sample size for multilevel modeling

In statistical tests, we use sample data to assess whether a certain effect (e.g., effect of intervention) exists in a population by testing a null hypothesis (e.g., $H_0 : \beta = 0$) at a given α level and a given sample size. Wrongly rejecting the null hypothesis when the hypothesis is true is called a Type I error. The probability of a Type I error is called Type I error rate and designated by the Greek letter α. Wrongly accepting a null hypothesis when the hypothesis is false is called a Type II error. The probability of a Type II error is called Type II error rate and designated by the Greek letter β. The power of a statistical test is the probability that the test will reject a false null hypothesis. Power increases as α increases and also as the sample size and/or the effect size increase. In social sciences studies, a power of 0.80 or higher is usually preferred. For an introduction to power analysis, readers are referred to the well-known studies by Cohen (1988).

Sample size plays an important role in determining the power of a statistical test. Given a significance level of α and an effect size, analysts may wish to determine the sample size necessary for accurate parameter estimation, or they may wish to determine the sample size necessary to achieve a certain level of power (e.g., 0.80). Unlike simple random sample designs, there are two distinct sample sizes in multilevel sample designs: sample size at the micro level and at the macro level. The former refers to the sample size within each group (n_j) and the latter to the number of groups (J) in a two-level multilevel data set. The total sample size equals $n = \sum_{j=1}^{J} n_j$. The power of significance testing for level-1 regression coefficients depends on the total sample size, while the power of the group-level effects and the cross-level interaction effects depends more strongly on the number of groups or level-2 units than the total sample size (Hox, 1998). Adequate numbers of level-1 and level-2 units are both important, but have sufficient level-2 units or groups (J) is more important than the group size (i.e., the number of cases per group) in multilevel modeling, with respect to precision of parameter estimation and higher statistical power (Van der Leeden et al., 1997; Browne & Draper, 2000). Selecting the optimal sample sizes at both micro and macro levels for accurate estimation in multilevel modeling is a complex task. A rule of thumb suggested by Kreft (1996) is "30/30," i.e., at least 30 groups ($J \geqslant 30$) and 30 level-1 observations per group ($n_j \geqslant 30$). For accurate estimation of the group-level variance/covariance components Busing (1993) suggests that at least 100 groups ($J \geqslant 100$) are needed. Maas & Hox (2004) suggest that the number of groups needed for multilevel modeling depends on specific research interests. For instance, if analyst is interested in the fixed parameters, 10 groups may work. If the analyst is interested in contextual effects (i.e., the cross-level interactions), 30 groups are needed. If the analyst wants unbiased estimates of standard errors, then at least 50 groups are needed ($J \geqslant 50$).

6.6.1 Sample size estimation for two-level designs

Power analysis and sample size estimation for multilevel modeling is complex. However, for two-level designs, a free Windows program, PINT (Power in Two-level designs) developed by Bosker, Snijders, and Guldemond (2003)[21] performs these calculations for two-level designs. Based on studies by Snijders & Bosker (1993), the program estimates standard errors of the regression coefficients (e.g., the fixed-effect part in eq.2.2.9 in Chapter 2). The relationship be-

[21] PINT is free for download from http://stat.gamma.rug.nl/snijders/multilevel.htm.

tween the standard error of a parameter, statistical power, given an α level and effect size, is defined as (Snijders & Bosker, 2003, p.142):

$$(\text{Effect Size})/\text{S.E.} \approx \left(z_{1-\alpha} + z_{1-\beta}\right) \quad (6.6.1.1)$$

where Effect Size is defined as the difference in outcome means between two groups (e.g., intervention and control groups) divided by the square root of the variance pooled across the two groups. A small effect size is typically taken to be 0.20, a medium effect size 0.50, and a large effect size 0.80 (Cohen, 1988). S.E. denotes the standard error of a model parameter estimate of interest. $z_{1-\alpha}$ and $z_{1-\beta}$ are z-scores associated with the indicated α and β values. For a specific effect size (e.g., 0.20), a given significance level α = 0.05, and a power of 0.80 ($1-\beta$ = 0.80 or β = 0.20), the standard error can be calculated as:

$$\text{S.E.} = \frac{0.20}{1.64 + 0.84} \approx 0.0806 \quad (6.6.1.2)$$

That is, an estimated standard error of approximately 0.08 indicates a statistical power of 0.80 to detect a small effect size of 0.20 at $\alpha = 0.05$ level.

The current version of PINT estimates approximate standard errors for fixed micro and macro effects and cross-level interaction effects in a two-level multilevel linear model for a variety of combinations of N (i.e., J in our notation—number of level-2 units or groups), n (i.e., n_j in our notation—group size or number of level-1 units per group), as well as budget costs. The analyst must search for the optimal combination of the N and n, as well as $N*n$ in the output of PINT that has an estimated S.E. equal or smaller than 0.08 for the parameter estimate of interest.

Note that PINT assumes a balanced design with the same group size for all groups (i.e., $n_j = n$). This assumption is often not true in practice. However, Snijders & Bosker (2003) state that the aim of power calculations is to have insight into the magnitude of the desired sample sizes and the achieved power, and not to have perfectly precise figures for these quantities. In practice "n" can be treated as the average sample size per group.

6.6.2 Sample size estimation for longitudinal data analysis

In Chapter 4 we indicated that longitudinal or panel data with repeated measures can be considered hierarchical. Multilevel models can then be applied to analyze such data. The power of a statistical test usually depends upon three parameters: the significance criterion. (significance level), the sample size, and the effect size of the exposure variable (Cohen, 1988). However, in longitudinal studies, the number of repeated measurements plays an important role in statistical power because of a tradeoff between the sample size and the number of repeated measurements (Diggle et al., 1998). The statistical power of data analyses is strengthened by having more repeated measures in the longitudinal study.

The following formula (Diggle et al., 1998, p.31) was used to estimate the number of cases per group in a pair-wise group (e.g., intervention vs. control; males vs. females) comparison of time-averaged outcome difference when the outcome measure is numeric:

$$n = \frac{2(Z_\alpha + Z_\beta)^2 \{1 + (t-1)\rho\}}{t\Delta^2} \quad (6.6.2.1)$$

where t denotes the number of repeated measurements (or time points); $\beta = 1-P$ (P is a given power); Z_f ($f = \alpha$ or β) denotes the fth quantile of the Z distribution; the correlations between the repeated measures ρ = corr(y_{it}, y'_{it}) for all $t \neq t'$; and Δ is the effect size (i.e., the standardized mean difference between two groups), denoting the smallest meaningful time-averaged outcome difference in standard deviation units between the two groups. The following SAS program estimates sample sizes for various combinations of ρ and Δ values for a specified number of time points or repeated measurements, given $p = 0.80$ and $\alpha = 0.05$., widely used statistical power and Type I error level for social science studies.

***SAS Program 6.6.2.1**;

```
Options pagesize = 60 linesize = 90 nodate;
%let t = 5; * number of time points or repeated measurements;
Data sample_size;
   do d = 0.1 to 0.5 by 0.05;
     do rho = 0.1 to 0.6 by 0.10;
     n = round(2*(probit(1-0.05)+probit(1-0.20))**2*(1+(&t-1)*rho)/ (&t*d**2));
     output;
     end;
   end;
Proc Tabulate format = 4.0;
class rho d;
Var n;
Table rho,d*n = ''*sum = '' /rts = 9 box = 'Sample Size Estimate 1' con-
dense ;
Label d = 'Delta';
Title 'Sample size estimate - continuous outcome measure in longitudinal
study with 5 time points.';
Run;
```

In this SAS program, the number of time points or number of repeated measurements is defined as a SAS macro variable "*t*". By assigning a specific number (e.g., 5 in this example) to "*t*", the program will estimate sample size per group, corresponding to each combination of ρ and Δ values. The SAS *PROBIT(p)* function returns the pth quantile from the standard normal distribution. *p* is the probability that an observation from the standard normal distribution (i.e., *z* scores) is less than or equal to the returned quantile. The estimated number of cases per group "*n*", is tabulated using *SAS PROC TABULATE* to creates the table shown in SAS Output 6.6.2.1.

SAS Output 6.6.2.1 Sample size estimate-continuous outcome measure in longitudinal study with 5 time points

Sample Size Estimate	Delta								
	0.1	0.15	0.2	0.25	0.3	0.35	0.4	0.45	0.5
Rho									
0.1	346	154	87	55	38	28	22	17	14
0.2	445	198	111	71	49	36	28	22	18
0.3	544	242	136	87	60	44	34	27	22
0.4	643	286	161	103	71	52	40	32	26
0.5	742	330	185	119	82	61	46	37	30
0.6	841	374	210	135	93	69	53	42	34

Note that "n" is not the total sample size, but the estimated number of cases per group in the pair-wise group comparison. The total sample size should, therefore, be 2*n. For example, to detect a small effect size of $\Delta = 0.20$ at $\alpha = 0.05$ for a pair-wise group comparison in a longitudinal data in which there are 5 time points and the correlation between repeated measures is $\rho = 0.20$, the total sample size needed to have a statistical power of 0.80 would be 2*111 = 222.

The formula for binary outcome measures becomes the following (Diggle et al., 1998, p.31):

$$n = \frac{[z_{\alpha}\{2\bar{p}\,\bar{q}(1+(t-1)\rho)\}^{1/2} + z_{q}\{(1+(t-1)\rho)(p_1q_1 + p_2q_2)\}^{1/2}]^2}{td^2} \tag{6.6.2.2}$$

where p_1 and p_2 are response probabilities for groups 1 and 2, respectively; $d = (p_2 - p_1)$ is the smallest meaningful difference between p_1 and p_2; $\bar{p} = (p_1 + p_2)/2$; $\bar{q} = (1 - \bar{p})$; and ρ is the correlation between repeated measures. Assuming Group 1 is the control group and $p_1 = 0.50$, the following SAS program estimates sample size needed for each group for selected d and ρ. The estimated numbers of cases per group, "n", are shown in SAS Output 6.6.2.2.

***SAS Program 6.6.2.2**;

```
%let t = 5; *number of time points or repeated measurements;
%let p1 = 0.5;
Data sample_size;
   do d = 0.1 to 0.5 by 0.05; *d = P2-P1;
      do rho = 0.1 to 0.6 by 0.10;
      q1 = 1-&p1;
      p2 = &p1 + d;
      q2 = 1-p2;
      p_bar = (&p1 + p2)/2;
      q_bar = 1-p_bar;
      n=(probit(1-0.05)*sqrt(2*p_bar*q_bar*(1+(&t-1) *rho))+probit(1-0.20)*sqrt((1
      +(&t-1)*rho) * (&p1*q1 + p2*q2)))**2/(&t*d**2);
   output;
      end;
   end;
Proc Tabulate format = 4.0;
Class rho d;
Var n;
Table rho,d*n = ''*sum = '' /rts = 12 box = 'Sample Size Estimate 2' con-
dense;
Label d = 'Difference';
Title 'Sample size estimates for binary outcome measure in longitudinal
study';
Run;
Quit;
```

With 5 repeated measures (i.e., $t = 5$), to achieve a power $P = 0.80$ at $\alpha = 0.05$, the sample size needed to detect a small difference (e.g., $d = 0.10$) in response probability between two groups is about 2*110 = 220, assuming a correlation of 0.20 between repeated measures (see SAS Output 6.6.2.2).

For both continuous and binary outcomes sample size is positively related to correlation between repeated measures, and negatively related to effect size. In addition, the needed sample size will decrease when the number of repeated measures increases. This can be shown by running the SAS programs with different values of "t".

SAS Output 6.6.2.2 Sample size estimate-binary outcome measures in longitudinal study with 5 time points

Sample Size	Difference								
Estimate 2	0.1	0.15	0.2	0.25	0.3	0.35	0.4	0.45	0.5
Rho 0.1	85	37	20	13	8	6	4	3	2
0.2	110	48	26	16	11	8	5	4	3
0.3	134	59	32	20	13	9	7	5	4
0.4	159	69	38	24	16	11	8	6	4
0.5	183	80	44	27	18	13	9	7	5
0.6	207	91	50	31	21	14	10	8	6

In this chapter we have introduced readers to techniques for conducting multilevel analyses with data sets that have unusual or atypical distributions, including count data with excess zeros, semi-continuous outcomes, and small number of groups. We have also discussed other often encountered issues such as missing values and sample size estimation in multilevel modeling. The examples provided in this chapter, and throughout this book, are posted online at academic. hep. com. cn where sample data sets, SAS programs and other resources can be downloaded.

Endnote EN6.1

Residuals Transformation

Residual transformation is based on the rescaled or centered crude residuals. Using $\hat{U}$ to denote the rescaled level-2 crude residuals in matrix format, the empirical variance/covariance matrix of the estimated level-2 residuals is defined as:

$$S = \hat{U}'\hat{U} / J \qquad \text{(EN6.1.1)}$$

where J (i.e., the number of groups or level-2 units) rather than $(J-1)$ is used in the denominator. The level-2 residuals can be described as (Carpenter et al., 2003):

$$\hat{U}^* = \hat{U}A \qquad \text{(EN6.1.2)}$$

Where $\hat{U}^*$ is the transformed level-2 residual matrix and A is a symmetric matrix with order equal to the number of level-2 variance/covariance components. Matrix A is formed by using Cholesky factors L_S and L_Σ, which are derived from the empirical variance/covariance matrix S and the model estimated variance/covariance matrix $\hat{\Sigma}$, respectively. That is,

$$A = (L_\Sigma L_S^{-1})' \qquad \text{(EN6.1.3)}$$

Any positive definite symmetric matrix V can be written in form of its Cholesky decompositions: $V = LL'$ where L is the upper Cholesky decomposition of V (i.e., the "square root" of the upper triangular of matrix V), and L' is the lower Cholesky decomposition. Thus,

$$L_\Sigma L'_\Sigma = \hat{\Sigma} \qquad \text{(EN6.1.4)}$$

$$L_S L'_S = S \qquad \text{(EN6.1.5)}$$

The variance/covariance matrix of the transformed level-2 residuals can be written as

$$\frac{\hat{U}^{*\prime}\hat{U}^{*}}{J}=\frac{A'\hat{U}'\hat{U}A}{J}=\frac{L_{\Sigma}L_{S}^{-1}\hat{U}'\hat{U}(L_{\Sigma}L_{S}^{-1})'}{J}=\frac{L_{\Sigma}[L_{S}^{-1}\hat{U}'\hat{U}(L_{S}^{-1})']L_{\Sigma}'}{J}$$
$$=\frac{L_{\Sigma}[\hat{U}'\hat{U}/J]L_{\Sigma}'}{L_{S}L_{S}'}=L_{\Sigma}L_{\Sigma}'=\hat{\Sigma} \qquad \text{(EN6.1.6)}$$

Equations EN6.1.1—EN6.1.6 prove that the variance/covariance matrix of the transformed new level-2 residuals is equal to the model estimated variance/covariance matrix.

Transformation of level-1 residuals is a special case of the transformation shown for level-2 residuals. At level-1, the L_S and L_Σ are simply the empirical standard deviation of the crude level-1 residuals and square root of the model estimated level-1 residual variance σ^2, respectively. The transformed level-1 residuals are obtained by multiplying the crude level-1 residual by L_Σ / L_S; that is, $\hat{e}^{*}=\hat{e}(L_{\Sigma}/L_{S})$.

References

Agresti, A. 1989. Tutorial on modeling ordered categorical response data. *Psychological Bulletin*. Vol. 105: 290-301.

Agresti, A. 1996. *An Introduction to Categorical Data Analysis*. New York: John Wiley.

Agresti, A., J. G. Booth, J. P. Hobert & B. Caffo. 2000. Random-effects modeling of categorical response data. *Sociological Methodology*. Vol. 30: 27-80.

Akaike, H. 1974. A new look at the statistical model identification. *IEEE Transaction on Automatic Control*. Vol. AC-19: 716-723.

Allison, P. 2000. Multiple imputation for missing data: A cautionary tale. *Sociological Methods and Research*. Vol. 28: 301-309.

Allison, P. 2002. *Missing Data*. London: Sage Publications.

Barcikowski, R. S. 1981. Statistical power with group mean as the unit of analysis. *Journal of Educational Statistics*. Vol. 6: 267-285.

Barr, R. & R. Dreeben. 1977. Instruction in classrooms. In L. S. Shulman (Ed.), *Review of Research in Education*. Vol. 5. Itasca, IL: Peacock.

Bauer, D. & P. Curran. 2003. Distributional assumptions of growth mixture model: implication for overextraction of latent trajectory classes. *Psychological Methods*. Vol. 8: 338-363.

Bentler, P. M. 1983. Some contributions to efficient statistics for structural models. Specification and estimation of moment structures. *Psychometrika*. Vol. 48: 493-571.

Bentler, P. M. 1995. *EQS: Structural Equations Program Manual*. Encino, CA: Multivariate Software Inc.

Bingenheimerl, J. B. & S. W. Raudenbush. 2004. Statistical and substantive inferences in public health: issues in the application of multilevel models. *Annual Review of Public Health*. Vol. 25: 53-77.

Blalock, H. M. 1984. Contextual-effects models: theoretical and methodological issues. *Annual Review of Sociology*. Vol. 10: 353-372.

Blau, P. M. 1982. Structural sociology and network analysis. In P. V. Marsden & N. Lin (Eds.), *Social Structure and Network Analysis*. pp. 273-279, Beverley Hills, CA: Sage Publications.

Block, J. H. & R. B. Burns. 1976. Mastery learning. In L. S. Shulman (Ed.), *Review of Research in Education*. Vol. 4. Itasca, IL: Peacock.

Böhning, D., E. Diets, P. Schlattmann, L. Mendonca & U. Kirchner. 1999. The zero-inflated Poisson model and the decayed, missing and filed teeth index in dental epidemiology. *Journal of the Royal Statistical Society*. Series A, Vol. 162: 195-209.

Bollen, K. A. 1989. *Structural Equations With Latent Variables*. New York: John Wiley.

Booth B. M., C. Leukefeld, R. Falck, J. Wang & R. Carlson. 2006. Correlates of Rural Methamphetamine and Cocaine Users: Results from a Multi-State Community Study. *Journal of Studies on Alcohol*. Vol. 67: 493-501.

Booth, J. G. & S. Sarkar. 1998. Monte Carlo approximation of bootstrap variances. *The American Statistician*. Vol. 52: 354-357.

Bosker, R. J., T. A. B. Snijders & H. Guldemond. 2003. PINT (Power IN Two-Level Design) User's Manual, V.2.1 (http://stat.gamma.rug.nl/snijders/multilevel.htm).

Boyd, Jr., L. H. & G. R. Iversen. 1979. *Contextual Analysis: Concepts and Statistical Techniques*. Belmont, CA: Wadsworth Publishing Company.

Bozdogan, H. 1987. Model selection and Akaike's information criterion (AIC): the general theory and its analytical extensions. *Psychometrika*. Vol. 52: 345-370.

Breslow, N. R. & D. G. Clayton. 1993. Approximate inference in generalized linear mixed models. *Journal of American Statistical Association*. Vol. 88: 9-25.

Bronfenbrenner, U. 1976. The experimental ecology of education. *Educational Researcher*. Vol. 5: 5-15.

Brown, H. & R. Prescott. 1999. *Applied Mixed Models in Medicine*. New York:John Wiley.

Browne, W. J. & D. Draper. 2000. Implementation and performance issues in the Bayesian and likelihood fitting of multilevel models. *Computational Statistics*. Vol. 15: 391-420.

Browne, W. J. 1998. *Applying MCMC Methods to Multilevel Models*. Unpublished Ph.D. Thesis, University of Bath, UK.

Bryk, A. S. & S. W. Raudenbush. 1992. *Hierarchical Linear Models*. Newbury Park, CA: Sage Publications.

Burnham & Anderson. 1998. *Model Selection and Inference: A Practical Information-Theoretic Approach*. New York: Springer-Verlag.

Burstein, L. 1980. The role of levels of analysis in the specification of educational effects. In R. Dreeben & J. A. Thomas (Eds.), *The Analysis of Educational Productivity. Vol. I: Issues in Microanalysis*. Boston: Ballinger.

Busing, F. 1993. Distribution characteristics of variance estimates in two-level models: a Monte Carlo study (Tech. Rep. No. PRM 93-04). Leiden University, Department of Psychometrics and Research Methodology, Leiden, The Netherlands.

Cameron, C. A. & P. K. Trivedi. 1986. Econometric models based on count data: Comparisons and applications of some estimators and tests. *Journal of Applied Econometrics*, 1, 29-53.

Carlin, J. B. & T. A. Louis. 1996. *Bayes and Empirical Bayes Methods for Data Analysis*. London: Chapman & Hall.

Carlson, R. G., J. Wang, H. A. Siegal, R. S. Falck & J. Guo. 1994. An ethnographic approach to targeted sampling: Problems and solutions in AIDS prevention research among injection drug and crack-cocaine users. *Human Organization*. Vol. 53: 279-286.

Carpenter, J. M., H. Goldstein & J. Rasbash. 2003. A novel bootstrap procedure for assessing the relationship between class size and achievement. *Applied Statistics*. Vol. 52: 431-443.

Cheung, Y. B. 2002. Zero-inflated models for regression analysis of count data: a study of growth and development. *Statistics in Medicine*. Vol. 21: 1461-1469.

Clogg, C. C. 1995. Latent class models. In G. Arminger, C. C. Clogg & M. E. Sobel (Eds.), *Handbook Of Statistical Modeling For the Social and Behavioral Sciences*. pp. 311-359. New York: Plenum Press.

Cohen, J. 1988. *Statistical Power Analysis for the Behavioral Sciences*. Mahwah, NJ: Lawrence Erlbaum Associates.

Cohen, J. 1992. A power primer. *Psychological Bulletin*. Vol. 112: 155-159.

Collins, L., J. Schafer & C. M. Kam. 2001. A comparison of inclusive and restrictive strategies in modern missing-data procedures. *Psychological Methods*. Vol. 6: 330-351.

Collins, L. M. & S. E. Wugalter. 1992. Latent class models for stage-sequential dynamic latent variables. *Multivariate Behavioral Research*, Vol. 27: 131-157.

Collins, L. M., J. W. Graham, S. S. Rousculp & W. B. Hansen. 1997. Heavy caffeine use and the beginning of the substance use onset process: An illustration of latent transition analysis. In K. Bryant, M. Windle & S. West (Eds.), *The Science of Prevention: Methodological Advances from Alcohol and Substance Use Research*. pp. 79-99. Washington DC: American Psychological Association.

Consul, P. C. & F. Famoye. 1992. Generalized Poisson regression model. *Communications in Statistics: Theory and Methods*. Vol. 21: 89-109.

Cox, D. R. 1983. Some remarks on over-dispersion. *Biometrika*. Vol. 70: 269-274.

Cronbach, L. J. 1976. Research on classrooms and schools: Formulation of questions, design, and analysis (Occasional paper). Stanford Evaluation Consortium, Standford, CA.

Davis, J. A. 1966. The campus as a frog pond: An application of the theory of relative deprivation to career decisions of college men. *American Journal of Sociology*. Vol. 72: 17-31.

Davison, A. C. & D. V. Hinkley. 1997. *Bootstrap Methods and their Application*. Cambridge: Cambridge University Press.

De Leeuw, J. & I. G. G. Kreft. 1986. Random coefficient models for multilevel analysis. *Journal of Educational Statistics*. Vol. 11: 57-85.

Dempster, A. P. & N. Laird. 1977. Maximum likelihood estimation from incomplete data via the EM algorithm (with discussion). *Journal of the Royal statistical Society. series B*. Vol. 39: 1-38.

Dempster, A. P., D. B. Rubin & R. K. Trutakawa. 1981. Estimation in covariance components models. *Journal of the American Statistical Association*. Vol. 76 : 341-353.

Diez-Roux, A. V. 1998. Bringing context back into epidemiology: variables and fallacies in multilevel analysis. *American Journal of Public Health*. Vol. 88: 216-222.

Diez-roux, A. V. 2000. Multilevel analysis in public health Research. *Annual Review of Public Health*. Vol. 21: 171-192.

Diggle, P. & M. G. Kenward. 1999. Informative drop-out in longitudinal data analysis. *Applied Statistics*. Vol. 43: 49-93.

Diggle, P. J., K. Y. Liang & S. L. Zeger. 1998. *Analysis of Longitudinal Data*. New York: Oxford University Press Inc.

DiPrete T. A. & J. D. Forristal. 1994. Multilevel models: methods and substance. *Annual Review of Sociology*. Vol. 20: 331-357.

Drake, R., G. McHugo, R. Clark, G. Teague, H. Xie & K. Miles. 1998. Assertive community treatment for patients with co-occurring severe mental illness and substance abuse disorder: a clinical trial. *American Journal of Orthopsychiatry*. Vol. 68: 201-215.

Duan, N., W. G. Manning, C. N. Morris & J. P. Newhouse. 1983. A comparison of alternative models for the demand for medical care. *Journal of Economic and Business Statistics*. Vol. 1: 115-126.

Duncan, C., K. Jones & G. Moon. 1996. Health related behavior in context: a multilevel modelling approach. *Soc. Sci. Med.* Vol. 42: 817-830.

Efron, B. 1981. Nonparametric standard errors and confidence intervals (with discussion). *Canadian Journal of Statistics*. Vol. 9: 139-172.

Efron, B. 1982. *The Jackknife, the Bootstrap, and Other Resampling Plans*. Philadelphia: Society for Industrial and Applied Mathematics.

Efron, B. & R. Tibshirani. 1986. Bootstrap methods for standard errors, confidence intervals, and other measures of statistical accuracy. *Statistical Science*. Vol. 1: 54-77.

Efron, B. 1979. Bootstrap methods: Another look at the Jackknife. *Annals of Statistics,* Vol. 7: 1-26.

Eliason, S. R. 1993. *Maximum Likelihood Estimation*. Newbury Park, CA: Sage Publications.

Elston, D. A. 1998. Estimation of denominator degrees of freedom of f-distributions for assessing wald statistics for fixed-effect factors in unbalanced mixed models. *Biometrics*. Vol. 54: 1085-1096.

Entwisle, B., A. I. Hermalin & W. M. Mason. 1982. *Socioeconomic Determinants of Fertility Behavior In Developing Countries: Theory And Initial Results*. Washington, D. C.: National Academy Press.

Fai, A. H. T. & P. L. Cornelius. 1996. Approximate f-tests of multiple degree of freedom hypotheses in generalized least squares analyses of unbalanced split-plot experiments. *Journal of Statistical Computing and Simulation*. Vol.54: 363-378.

Falck R. S., J. Wang, H. A. Siegal & R. G. Carlson. 2003. Current physical health problems and their predictors among a community sample of crack-cocaine smokers in Ohio. *Journal of Psychoactive Drugs*. Vol. 35: 471-478.

Famoye, F. 1993. Restricted generalized Poisson regression model. *Communications in Statistics: Theory and Methods*. Vol. 22: 1335-1354.

Fitzmaurice, G. & N. Laird. 2000. Generalized linear mixture models for handling nonignorable dropouts in longitudinal studies. *Biostatistics*. Vol. 1: 141-156.

Fitzmaurice, G. 2003. Methods for handling dropouts in longitudinal clinical trials. *Statistica Neerlandica*. Vol. 57: 75-99.

Fitzmaurice, G., N. Laird & J. Ware. 2004. *Applied Longitudinal Analysis*. New York: Wiley.

Foster, E. M. & G. Y. Fang. 2004. Alternative methods for handling attrition. *Evaluation Review*. Vol. 28: 434-464.

Freund, R., R. C. Littell & W. W. Stroup. 2002. *SAS for Linear Models*, Cary, NC: SAS Institute Inc.

Gelman, A., J. B. Carlin, H. S. Stern & D. B. Rubin. 2003. *Bayesian Data Analysis*, second edition. London: CRC Press.

Glass, G. V. 1976. Primary, secondary, and meta-analysis of research. *Educational Researcher*. Vol. 5: 3-8.

Gibbons, R., D. Hedeker, C. Waternaux, H. Kraemer & J. B. Greenhouse, et al. 1993. Some conceptual and statistical issues in analysis of longitudinal psychiatric data. *Archives of General Psychiatry*. Vol. 50: 739-750.

Giraudeau, B. 1996. Negative values of the intraclass correlation coefficient are not theoretically possible (letter to the editor). *Journal of Clinical Epidemiology*. Vol. 49: 1205-1206.

Girden, E. 1992. *ANOVA Repeated Measure*. Newbury Park. CA: Sage Publication, Inc.

Goldstein, H. & J. Rasbash. 1996. Improved approximations for multilevel models with binary responses. *Journal of the Royal Statistical Society,Series A*. Vol. 159: 505-513.

Goldstein, H. 1987. *Multilevel Models in Educational and Social Research.* London: Griffin.

Goldstein, H. 1989. Restricted unbiased iterative generalized least squares estimation. *Biometrika*. Vol. 76: 622-623.

Goldstein, H. & R. McDonald. 1988. A general model for the analysis of multilevel data. *Psychometrika*. Vol. 53: 455-467.

Goldstein, H. 1991. Nonlinear multilevel models with an application to discrete response data. *Biometrika*. Vol. 78: 45-51.

Goldstein, H. 1995. *Multilevel Statistical Models.* (2nd ed.). New York: Wiley.

Goldstein, H., M. Yang, R. Omar & R. Turner, et al. 2000. Meta analysis using multilevel models with an application to the study of class size effects. *Journal of the Royal Statistical Society, Series C*. Vol. 49: 399-412.

Greene, W. 1994. *Accounting for Excess Zeros and Sample Selection in Poisson and Negative Binomial Regression Models. Working Paper EC*-94-10: Department of Economics, New York University.

Greene, W. H. 1990. *Econometric Analysis*. New York: Macmillan Publishing Company.

Greene, W. H. 2002. *LIMDEP V.* 8.0 *Reference Guide*. New York: Econometric Software, Inc.

Grogger, J. & R. Carson. 1991. Models for truncated counts. *Journal of Applied Econometrics*. Vol. 6: 225-238.

Grunwald, G. K. & R. H. Jones. 2000. Markov models for time series with mixed distribution. *Environmetrics*. Vol. 11: 327-339.

Hall, D. B. 2000. Zero-inflated Poisson and binomial regression with random effects: a case study. *Biometrics*. Vol. 56: 1030-1039.

Hannan, E. J. & B. G. Quinn. 1979. The determination of the order of an autoregression. *Journal of the Royal Statistical Society, Series B*. Vol. 41: 190-195.

Harville, D. A. 1977. Maximum likelihood approaches to variance component estimation and related problems. *Journal of the American Statistical Association*. Vol. 72: 320-338.

Heck, R. H. & S. L. Thomas. 2000. *An Introduction to Multilevel Modeling Techniques*. Mahwah, New Jersey: Lawrence Erlbaum Associates.

Heckman, J. 1976. The common structure of statistical models of truncation, sample selection, and limited dependent variables, and a sample estimator for such models. *The Annals of Economic and Social Measurement*. Vol. 5: 475-592.

Hedeker, D. & R. Gibbons. 2006. *Longitudinal Data Analysis*. New York: Wiley.

Hedeker, D. & R. Mermelstein. 1998. A multilevel threshold of change model for analysis of stage of change data. *Multivariate Behavioral Research*. Vol. 3: 427-455.

Hedeker, D., R. Mermelstein & K. Weeks. 1999. The threshold of change model: an approach to analyzing stage of change data. *Annals of Behavioral Medicine*. Vol. 21: 61-70.

Hedeker, D., R. D. Gibbons & C. Waternaux. 1999. Sample size estimation for longitudinal designs with attrition: comparing time-related contrasts between two groups. *Journal of Educational and Behavioral Statistics.* Vol. 24: 70-91.

Hermalin A. 1986. The multilevel approach: theory and concepts. The methodology for measuring the impact of family planning programs on fertility. Addendum Manual IX *Popul. Stud.* Vol. 66: 15-31.

Hermalin, A. I. & W. M. Mason. 1980. A strategy for the comparative analysis of WFS data with illustrative examples. In United Nations Fund for Population Activities. *The United Nations Program for Comparative Analysis of World Fertility Survey Data.* pp. 90-168. New York: United Nations Fund for Population Activities.

Herriott, R. E. & D. N. Muse. 1973. Methodological issues in the study of school effects. In F.N. Kerlinger (Ed.), *Review of Research in Education,* Vol. 1. Itasca, IL: Peacock.

Hinde, J. & C. Dmétrio. 2001. A score test for testing a zero-inflated Poisson regression model against zero-inflated negative binomial alternatives. *Biometrics.* Vol. 57: 219-223.

Hofmann, D. A. & M. B. Gavin. 1998. Centering decisions in hierarchical linear models: Implications for research in organizations. *Journal of Management.* Vol. 24: 623-641.

Hogan, J. W., J. Roy & K. Christina. 2004. Handling drop-out in longitudinal studies. *Statistics in Medicine.* Vol. 23: 1455-1497.

Hox, J. J. & I. Kreft. 1994. Multilevel analysis methods. *Sociol. Methods Res.* Vol. 22: 283-299.

Hox, J. J. 1994. *Applied Multilevel Analysis.* Amsterdam: TT-Publikaties.

Hox, J. J. 2002. *Multilevel Analysis, Techniques and Applications.* Mahwah, NJ: Lawrence Erlbaum Associates.

Hox, J. J. & C.J.M. Maas. 2001. The accuracy of multilevel structural equation modeling with pseudobalanced groups and small samples. *Structural Equation Modeling.* Vol. 8: 157-174.

Hox, J. J. 1998. Multilevel modeling: when and why, in: I. Balderjahn, R. Mathar & M. Schader (Eds), *Classification, Data Analysis, and Data Highways.* pp. 147-154. New York: Springer.

Huber, J. 1991. Introduction. In J. Huber (Ed.), *Macromicro Linkages in Sociology.* pp. 6-9. Newbury Park, CA: Sage Publications.

Huber, P. J. 1967. The behavior of maximum likelihood estimates under non-standard conditions, In: *Proceedings Of The Fifth Berkeley Symposium On Mathematical Statistics And Probability.* pp. 221-233. Berkeley: University of California Press.

Hur, K. 1999. *A Random-effects Zero-inflated Poisson Regression Model for Clustered Extra-zero Counts.* Unpublished Ph.D. Dessertation, University of Illinois at Chicago.

Hur, K., D. Hedeker, W. Henderson, S. Khuri & J. Daley. 2002. Modeling clustered count data with excess zeros in health care outcomes research. *Health Services and Outcomes Research Methodology.* Vol. 3: 5-20.

Hurvich, C. M. & C. L. Tsai. 1989. Regression and time series model selection in small samples. *Biometrika.* Vol. 76: 297-307.

Huyhn, H. & L. S. Feldt. 1970. Conditions under which mean squares ratios in repeated measurements designs have exact F-distributions. *Journal of the American Statistical Association.* Vol. 65: 1582-1589.

Iversen, G. R. 1991. *Contextual Analysis.* Newbury Park, CA: Sage Publications.

Jeffreys, H. 1961. *Theory of Probability* (2nd ed.). New York: Oxford University Press.

James, W. & C. Stein. 1961. Estimation with quadratic loss. In J. Neyman (Ed.), *Proceedings of the Forth Berkeley Symposium on Mathematical Statistics and Probability.* Vol. 1: 361-379. Berkeley: University of California Press.

Jennrich, R. & M. Schluchter. 1986. Unbalanced repeated-measures models with structured covariance matrices. *Biometrics.* Vol. 42: 805-820.

Jones, B. L. & D. S. Nagin. 2006. Advances in group-based trajectory modeling and a SAS procedure for estimating them. http://www.andrew.cmu.edu/user/bjones

Jones, B. L., D. S. Nagin & K. Roeder. 2001. A SAS procedure based on mixture models for estimating developmental trajectories. *Sociological Methods & Research.* Vol. 29: 374-393.

Jöreskog, K. G. 1971. Statistical analysis of sets of congeneric tests. *Psychometrika.* Vol. 36: 109-133.

Jöreskog, K. G. 1977. Structural equation models in the social sciences: Specification, estimation and testing. In P. R. Krishnaiah (Ed.), *Application of Statistics*. pp. 265-287. Amsterdam: North-Holland.

Jöreskog, K. G. & D. Sörbom. 1979. *Advances in Factor Analysis and Structural Equation Models*. Boston: Abt Books.

Jöreskog, K. G., & D. Sörbom. 1986. *LISREL VI: Analysis of Linear Structural Relationship by Maximum Likelihood and Least Squares Methods*. Mooresville, IN: Scientific Software.

Jöreskog, K. G. & D. Sörbom. 1999. *LISREL 8.30: User's Reference Guide*. Chicago: Scientific Software.

Judge, G. G., R. C. Hill, W. E. Griffiths, H. Lutkepohl & T. C. Lee. 1982. *Introduction to the Theory and Practice of Econometrics*. New York: Wiley.

Kalton, G. 1983. *Introduction to Survey Sampling*. Newbury Park, CA: Sage Publications.

Kaplan, D. & P. R. Elliott. 1997. Adidactic example of multilevel structural equation modeling applicable to the study of organizations. *Structural Equation Modeling*. Vol. 4: 1-24.

Kass, R. E. & L. Wasserman. 1995. A reference Bayesian test for nested hypotheses and its relationship to the Schwarz Criterion. *Journal of the American Statistical Association*. Vol. 90: 928-934.

Kenward, M. G. & J. H. Roger. 1997. Small sample inference for fixed effects from restricted maximum likelihood. *Biometrics*. Vol. 53: 983-997.

King, G. 1989. Event count models for international relations: generalizations and applications. *International Study Quarterly*. Vol. 33: 123-177.

Koepsell T. D., D. C. Martin, P. H. Diehr, B. M. Psaty & E. H. Wagner, et al. 1991. Data analysis and sample size issues in evaluations of community-based health promotion and disease prevention programs: a mixed model analysis of variance approach. *J. Clin. Epidemiol.* Vol. 44: 701-713.

Kononoff, P. J. & K. J. Hanford. 2006. Technical Note: Estimating Statistical Power of Mixed Models Used in Dairy Nutrition Experiments. *Journal of Dairy Science*. Vol. 89: 3968-3971.

Kreft, I. & J. de Leeuw. 1998. *Introducing Multilevel Modeling*. Thousand Oakes, CA: Sage Publications.

Kreft, I. 1996. Are Multilevel Techniques Necessary? An Overview, Including Simulation Studies. (California State University, Los Angeles, CA, 1996. Unpublished Report, available at http://ioe.ac. uk/multilevel).

Kreft, I., J. de Leeuw & L. S. Aiken. 1995. The Effects of Different Forms of Centering in Hierarchical Linear Models. *Multivariate Behavioral Research*. Vol. 30: 1-21.

Krull, J. L. & D. P. MacKinnon. 1999. Multilevel mediation modeling in group-based intervention studies. *Evaluation Review*. Vol. 23: 418-444.

Krull, J. L. & D. P. MacKinnon. 2001. Multilevel modeling of individual and group level mediated effects. *Multivariate Behavioral Research*. Vol. 36: 249-277.

Lachenbruch, P. 1976. Analysis of data with clumping at zero. *Biometrische Zeitschrift*. Vol.18: 351-356.

Lachenbruch, P. 1992. Utility of regression analysis in epidemiologic studies of the elderly. In W. R. Wallace R (Ed.), *The Epidemiologic Study of the Elderly*: Oxford University Press.

Laird, N. 1988. Missing data in longitudinal studies. *Statistics in Medicine*. Vol. 7: 305-315.

Laird, N. 1994. Discussion of "informative drop-out in longitudinal data analysis" by P. J. Diggle and M. G. Kenward. *Applied Statistics*. Vol. 43: 84.

Laird, N. M. & J. H. Ware. 1982. Random effects models for longitudinal data. *Biometrics*. Vol. 38: 963-974.

Lambert, D. 1992. Zero-inflated Poisson regression, with an application to detects in manufacturing. *Technometrics*. Vol. 34: 1-14.

Land, K. C., P. L. McCall & D. S. Nagin. 1996. A comparison of Poisson, negative binomial, and semiparametric mixed Poisson regression models with empirical applications to criminal careers data. *Sociological Methods & Research*. Vol. 24: 387-442.

Lazarsfeld, P. F. & H. Menzel. 1961. On the relation between individual and collective properties. In A. Etzioni (Ed.), *Complex Organizations: A Sociological Reader*. pp. 422-440. New York: Holt, Rinehart, and Winston.

Lee, A. & K. Wang. 2006. Zero-inflated Poisson regression modeling of correlated count data with excess zeros. *Statistical Methods in Medical Research*. Vol. 15: 47-61.

Lehman, A. F. 1988. A quality of life interview for the chronically mentally ill. *Evaluation and Program Planning*. Vol. 11: 51-62.

Li, F., T. E. Duncan, P. Harmer, A. Acock & M. Stoolmiller. 1998. Analyzing measurement models of latent variables through multilevel confirmatory factor analysis and hierarchical linear modeling approaches. *Structural Equation Modeling*. Vol. 5: 294-306.

Liang, K. & S. Zeger. 1986. Longitudinal data analysis using generalized linear models. *Biometrika*. Vol. 73: 13-22.

Liao, T. F. 1994. *Interpreting Probability Models: Logit, Probit, and Other Generalzied Linear Models*. Thousand Oaks, CA: Sage Publication.

Lindley, D. V. & A. F. M. Smith. 1972. Bayes estimates for the linear model. *Journal of the Royal Statistical Society*. Vol. B34: 1-41.

Lindstrom, M. J. & D. M. Bates. 1988. Newton-Raphson and EM algorithm for linear mixed-effects models for repeated-measures data. *Journal of the American Statistical Association*. Vol. 83: 1014-1022.

Lindstrom, M. J. & D. M. Bates. 1990. Nonlinear mixed effects models for repeated measures data. *Biometrics*. Vol. 46: 673-687.

Littell, R. C., G. A. Milliken, W. W. Stroup & R. D. Wolfinger. 1996. *SAS System for Mixed Models*. Cary, NC: SAS Institute Inc.

Littell, R. C., G. A. Milliken, W. W. Stroup, R. D. Wolfinger & O. Schabenberger. 2006. *SAS for Mixed models, Second Edition*. Carry, NC: SAS Institute Inc.

Little, R. & D. Rubin. 2002. *Statistical Analysis with Missing Data* (2nd ed.). New York: Wiely.

Little, R. 1993. Pattern-mixture models for multivariate incomplete data. *Journal of the American Statistical Association*. Vol. 88: 125-134.

Little, R. 1995. Modeling the drop-out mechanism in repeated-measures studies. *Journal of the American Statistical Association*. Vol. 90: 1112-1121.

Little, R. & N. Schenker. 1995. Missing data. In Arminger, G., C. C. Clogg & M. E. Sobel (Eds.), *Handbook of Statistical Modeling for the Social and Behavioral Sciences*. New York: Plenum Press.

Little, R., G. Milliken, W. Stroup, R. Wolfinger & O. Schabenberg. 2006. *SAS for Mixed Models*. Cary, NC: SAS Institute Inc.

Long, S. 1997. *Regression Models for Categorical and Limited Dependent Variables*. Newbury Park: CA: Sage Publications.

Longford, N. & B. Muthén. 1992. Factor analysis for clustered observations. *Psychometrika*. Vol. 65:581-597.

Longford, N. T. 1987. A fast scoring algorithm for maximum likelihood estimation in unbalanced mixed models with nested random effects. *Biometrika*, Vol. 74: 817-827.

Longford, N. T. 1993. *Random Coincident Models*. Oxford: Clarendon Press.

Lord, F. & M. Novick. 1968. *Statistical Theories of Mental Test Scores*. Massachusetts: Addison-Wesley.

Maas, C. J. M. & J. J. Hox. 2004. Robustness issues in multilevel regression analysis. *Statistica Neerlandica*. Vol. 58: 127-137.

Manning, W. G., N. Duan & W. H. Rogers. 1987. Monte Carlo evidence on the choice between sample selection and two-part models. *Journal of Econometrics*. Vol. 35: 59-82.

Mason, W., G. Wong & B. Entwisle. 1983. Contextual analysis through the multilevel linear model. In S. Leinhardt (Ed.), *Sociological Methodology* 1983-1984. pp. 72-103. San Francisco: Jossey-Bass.

McArdle, J. J. & F. Hamagami. 1996. Multilevel models from a multiple group structural equation perspective. In G. A. Marcoulides & R. E. Schumacker (Eds.), *Advanced Structural Equation Modeling: Issues and Techniques*. pp. 89-125. Mahwah, NJ: Lawrence Erlbaum Associates, Inc.

McCullagh, P. & J. A. Nelder. 1989. *Generalized Linear Models*. London: Chapman and Hall.

Merton, R. K. 1957. Continuities in the theory of reference groups and social structure. In R. Merton (Ed.), *Social Theory and Social Structure*. pp. 281-386. Glencoe, IIL: Free Press.

Miller, R. G. 1974. The Jackknife—A review. *Biometrika*. Vol. 61: 1-15.

Min, Y. & A. Agresit. 2005. Random effect models for repeated measures of zero-inflated count data. S*tatistical Modeling*. Vol. 5: 1-19.

Molenberghs, G. & G. Verbeke. 2005. *Models for Discrete Longitudinal Data*. New York: Springer.

Mooney, C. Z. & B. D. Duval. 1993. *Bootstrapping: A nonparametric approach to statistical inference*. Newbury Park: Sage Publications.

Mullahy, J. 1986. Specification and testing of some modified count data models. *Journal of Econometrics*. Vol. 33: 341-365.

Murray, D. M. & R. D. Wolfinger. 1994. Analysis issues in the evaluation of community trials: Progress toward solutions in SAS/STAT Mixed. *Journal of Community Psychology*. CSAP Special Issue: 140-154.

Murray, D. M., P. J. Hannan, R. D. Wolfinger, W. L. Baker & J. H. Dwyer. 1998. Analysis of data from group-randomized Trials with repeat observations on the same Groups. *Statistics in Medicine*. Vol. 17: 1581-1600.

Muthén, B. 1989. Latent variable modeling in heterogeneous populations. *Psychometrika*. Vol. 54: 557-585.

Muthén, B. 1991. Multilevel factor analysis of class and student achievement components. *Journal of Educational Measurement*. Vol. 28, 338-354.

Muthén, B. 1994. Multilevel covariance structure analysis. *Sociological Methods & Research*. Vol. 22: 376-398.

Muthén, B. O. & L. K. Muthen. 2000. Integrating person-centered and variable-centered analyses: growth mixture modeling with latent trajectory classes. *Alcoholism: Clinical and Experimental Research*. Vol. 24: 882-891.

Muthén, B. O. 2001. Second-generation structural equation modeling with a combination of categorical and continuous latent variables: new opportunities for latent class/latent growth modeling. *New Methods for the Analysis of Change*. A. Sayer and L. Collins. Washington D. C., American Psychological Association.

Muthén, B. & L. Muthén. 1998-2010. *Mplus User's Guide*. Los Angeles: Muthén & Muthén.

Muthén, B. & A. Satorra. 1989. Multilevel aspects of varying parameters in structural models. In R. D. Bock (Ed.), *Multilevel analysis of educational data*. pp. 87-99. San Diego: Academic Press.

Muthén, B. & A. Satorra. 1995. Complex sample data in structural equation modeling. In P. V. Marsden (Ed.), *Sociological Methodology*. pp. 267-316. Oxford, England: Blackwell.

Mwalili, S., E. Lesaffre & D. Declerck. 2006. *The Zero-inflated negative binomial regression model with correction for misclassification: an example in caries research* (No. 0462): Interuniversity Attraction Pole.

Nagin, D. 1999. Analyzing developmental trajectories: a semiparametric, Group-based approach. *Psychological Methods*. Vol. 4: 139-157.

Nagin, D. 2005. *Group-Based Modeling of Development*. Boston. Harvard University Press.

Nagin, D. & K. Land. 1993. Age, criminal careers, and population heterogeneity: specification and estimation of a nonparametric, mixed Poisson model. *Criminology*. Vol. 31: 327-362.

Needle, R., D. Fisher & N. Weatherby, et al. 1995. Reliability of self-reported HIV risk behaviors of drug users. *Psychology of Addictive Behaviors*. Vol. 9: 242-250.

Nelder, J. A. & R. W. M. Wedderburn. 1972. Generalized linear models. *Journal of the royal statistical Society*. Vol. 135: 370-384.

Neuhaus, J. M., J. D. Kalbfleisch & W. W. Hauck. 1991. A comparison of cluster-specific and population-average approaches for analyzing correlated data. *International Statistical Review*. Vol. 59: 25-35.

Olsen, M. & J. Schafer. 2001. A two-part random-effects model for semicontinuous longitudinal data. *Journal of American Statistical Association*. Vol. 96: 730-745.

Patterson, H. D. & R. Thompson. 1971. Recovery of inter-block information when block sizes are unequal. *Biometrika*. Vol. 58: 545-554.

Parzen, M., S. Lipsitz, G. Fitzmaurice, J. Ibrahim, A. Troxel & G. Molenberghs. 2007. Pseudo-likelihood methods for the analysis of longitudinal binary data subject to nonignorable non-monotone missingness. *Journal of Data Science*. Vol. 5: 1-21.

Pedhazur, E. J. 1975. Analytical methods in studies of educational effects. In F. N. Kerlinger (Ed.), *Review of Research in Education*. Vol. 3. Itasca, IL: Peacock.

Peterson, B. & F. Harrell. 1990. Partial proportional odds models for ordinal response variables. *Applied Statistics*. Vol. 39: 205-217.

Pinheiro, J. C. & D. M. Bates. 1995. Approximation to the log-likelihood function in the nonlinear mixed-models. *Journal of computational and Graphical Statistics.* Vol. 4: 12-35.

Pituch, K. A. 2001. Using multilevel modeling in large-scale planned variation educational experiments: Improving understanding of intervention effects. *Journal of Experimental Education.* Vol. 69: 347-372.

Raftery, A. E. 1996. Bayesian model selection in social research (With Discussion). In P. V. Marsden (Ed.), *Sociological Methodology.* Vol. 25: 111-163. Oxford: Basil Blackwell.

Rasbash, J., W. Browne, H. Goldstein, M. Yang, I. Plewis, M. Healy, G. Woodhouse, D. Draper, I. Langford & T. Lewis. 2000. *A User's Guide To Mlwin.* University of London, London; Multilevel Models Project.

Raudenbush, S. 1989a. Centering predictors in multilevel analysis: Choices in consequences. *Multilevel Modeling Newsletter.* Vol. 1: 10-12.

Raudenbush, S. 1989b. A response to Longford and Plewis. *Multilevel Modeling Newsletter.* Vol. 1: 8-10.

Raudenbush, S. W. 1993. A crossed random effects model for unbalanced data with applications in cross-sectional and longitudinal research. *Journal of Educational and Behavioral Statistics*, Vol. 18: 321-349.

Raudenbush, S. W. & A. S. Bryk. 2002. *Hierarchical Linear Models.* (2nd ed.). Thousand Oaks, CA: Sage Publications.

Raudenbush, S., A. Bryk, Y. F. Cheong & R. Congdon. 2000. *HLM 5, Hierarchicallinear and Nonlinear Modeling.* Chicago: Scientific Software International.

Raudenbush, S. W. & A. S. Bryk. 2002. *Hierarchical Linear Models: Applications and Data Analysis Methods* (2nd ed.). Thousand Oaks, CA: Sage Publications.

Reboussin, B.A., D. M. Reboussin, K. Y. Liang & J. C. Anthony. 1998. Latent transition modeling of progression of health-risk behavior. *Multivariate Behavioral Research.* Vol. 33: 457-478.

Ridout, M., C. Demétrio & J. Hinde. 1998. Models for count data with many zeros. *International Biometric Conference.* Cape Town, December 1998.

Ridout, M., J. Hinde & C. Dmetrio. 2001. A score test for testing a zero-inflated Poisson regression model against zero-inflated negative binomial alternatives. *Biometrics.* Vol. 57: 219-223.

Robinson, W. S. 1950. Ecological correlations and the behavior of individuals. *Sociological Review.* Vol. 15: 351-357.

Rouant, H. & D. Lépine. 1970. Comparison between treatments in a repeated-measurement design: ANOVA and multivariate methods. *British Journal of Mathematical and Statistical Psychology.* Vol. 23: 147-163.

Rubin, D. 1976. Inference and missing data. *Biometrika.* Vol. 63: 581-592.

Rubin, D. 1987. *Multiple Imputation for Nonresponse in Survey.* New York: Wiely.

Saha, A. & D. Dong. 1997. Estimating nested count data models. *Oxford Bulletin of Economics and Statistics.* Vol. 59: 423-430.

Sampson, R. J., S. W. Raudenbush & F. Earls. 1997. Neighborhoods and violent crime: A multilevel study of collective efficacy. *Science.* Vol. 227: 918-924.

SAS Institute Inc. 1999. *SAS/STAT User's Guide.* SAS On-Line Document. SAS Institute Inc.

SAS Institute Inc. 2003. *SAS/STAT User's Guide, Version 9.* SAS Institute Inc.

SAS Institute Inc. 2004. *SAS/STAT®* 9.1 *User's Guide.* SAS Institute Inc.

SAS Institute Inc. 2007. *Advanced Statistical Modeling Using NLMIXED Procedure Course Notes.* SAS Institute Inc.

SAS Institute Inc. 2008. *SAS/STATR* 9.2 *User's Guide.* Cary, NC: SAS Institute Inc.

Satterthwaite, F. E. 1946. An approximate distribution of estimates of variance components. *Biometrics Bulletin.* Vol. 2: 110-114.

Schabenberg, O. 2005. *Introducing the GLIMMIX Procedure for Generalized Linear Mixed Models.* Paper presented at the The thirty annual SAS user group international conference, Philadelphia, Pennsylvania.

Schafer, J. L. & J. W. Graham. 2002. Missing data: our view of the state of the art. *Psychological Methods.* Vol. 7: 147-177.

Schafer, J. L. & M. Olsen. 1998. Multiple imputation for multivariate missing-data problems: a data analyst's perspective. *Multivariate Behavioral Research.* Vol. 33: 545-571.

Schafer, J. L. 1997. *Analysis of incomplete multivariate data.* London: Chapman & Hall.

Schafer, J. L. 1999. Multiple imputation: a primer. *Statistical Methods in Medical Research.* Vol. 8: 3-15.

Schafer, J. L. 2001. Multiple imputation with Pan. In A. Sayer & L. Collins (Eds.), *New Methods for the Analysis of Change.* Washington, DC: American Psychological Association.

Schluchter, M. D. & J. D. Elashoff. 1990. Small-sample adjustments to tests with unbalanced repeated measures assuming several covariance structures. *Journal of Statistical Computation and Simulation.* Vol. 37: 69-87.

Schwarz, G. 1978. Estimating the dimension of a model. *Annals of Statistics.* Vol. 6: 461-464.

Searle, S. R., G. Casella & C. E. McCulloch. 1992. *Variance Components.* New York: Wiley.

Sheu, C. F. 2002. Fitting mixed-effects models for repeated ordinal outcomes with the NLMIXED procedure. *Behavior Research Methods, Instruments, and Computers.* Vol. 34: 151-157.

Shrout, P. E. & J. L. Fleiss. 1979. Intraclass correlations: Uses in assessing rater reliability. *Psychological Bulletin.* Vol. 86: 420-428.

Singer, J. D. 1998. Using SAS Proc Mixed to fit multilevel models, hierarchical models, and individual growth models. *J. of Educ. Behav. Stat.* Vol. 24: 323-355.

Snijders T. A. B. & R. J. Bosker. 1993. Standard errors and sample sizes in two-level research. *J. Educ. Stat.* Vol. 19: 237-259.

Snijders, T.A.B. & R. Bosker. 1999. *Multilevel Analysis. An Introduction to Basic and Advanced Multilevel Modeling.* Thousand Oaks, CA: Sage Publications.

Snijders, T. A. B. & R. J. Bosker. 2003. *Multilevel Analysis: An Introduction to Basic and Advanced Multilevel Modeling.* Thousand Oaks, CA: Sage Publications.

Snijders, T. A. B. & R. J. Bosker. 1994. Modeled variance in two-level models. *Sociological Methods & Research.* Vol. 22: 342-363.

Snow, R. E. 1976. Learning and individual differences. In F. N. Kerlinger (Ed.), *Review of Research in Education.* Vol. 4. Itasca, IL: Peacock.

Spady, W. G. 1973. The impact of school resources on students. In F.N. Kerlinger (Ed.), *Review of Research in Education.* Vol. 1. Itasca, IL: Peacock.

Strenio, J. L. F., H. I. Weisberg & A. S. Bryk, 1983. Empirical Bayes estimation of individual growth curve parameters and their relationship to covariates. *Biometrics.* Vol. 39: 71-86.

Swallow, W. H. & J. F. Monahan. 1984. Monte Carlo comparison of ANOVA, MIVQUE, REML, and ML estimators of variance components. *Technometrics.* Vol. 28: 47-57.

Tammemagi, M. C., J. W. Frank, M. Leblanc, H. Artsob & D. L. Streiner. 1995. Methodological issues in assessing reproducibility—A comparative study of various indices of reproducibility applied to repeat ELISA serologic tests for Lyme disease. *Journal of Clinical Epidemiology.* Vol. 48: 1123-1132.

Tobin, J. 1958. Estimation of relationships for limited dependent variables. *Econometrica.* Vol. 26: 24-36.

Tooze, A. J., G. K. Grunwald & R. H. Jones. 2002. Analysis of repeated measures data with clumping at zero. *Statistical Methods in Medical Research.* Vol. 11: 341-355.

Troxel, A., D. Fairclough, D. Curran & E. Hahn. 1998. Statistical analysis of quality of life with missing data in cancer clinical trials. *Statistics in Medicine.* Vol. 17: 653-666.

Va den Eeden, P. 1988. A two-step procedure for analyzing multi-level structured data. In W. E. Saris & I. N. Gallhofer (Eds.), *Sociometric Research*: Volume 2 Data Analysis. New York: St. Martin's Press.

Van der Leeden, R. & F. Busing. 1994. First iteration versus IGLS/RIGLS estimates in two-level models: a Monte Carlo study with ML3. Unpublished manuscript, Department of Psychometrics and Research Methodology, Leiden University.

Van der Leeden, R., F. Busing & E. Meijer. 1997. Application of bootstrap methods for two-level models (Paper presented at the Multilevel Conference, Amsterdam, April 1-2, 1997).

Verbeek, M. 2000. *A Guide to Modern Econometrics.* Hoboken: Wiley.

Von Korff M., T. Koepsell, S. Curry & P. Diehr. 1992. Multi-level research in epidemiologic research on health behaviors and outcomes. *American Journal of Epidemiology.* Vol. 135: 1077-1082.

Vuong, Q. 1989. Likelihood ratio tests for model selection and non-nested hypothesis. *Econometrica*. Vol. 57: 307-334.

Walberg, H. J. 1976. Psychology of learning environments: Behavioral structural or perceptual? In L. S. Shulman (Ed.), *Review of Research in Education*. Vol. 4. Itasca, IL: Peacock.

Wang, J., H. A. Siegal, R. S. Falck & R. G. Carlson. 1998. Needle transfer among injection drug users: a multilevel analysis. *The American Journal of Drug and Alcohol Abuse*. Vol. 24: 225-237.

Wang, J. & J. H. Fisher. 1994. Comments on "Estimating macro-relationships using micro-data: A one-stage approach." *Sociological Methods & Research*. Vol. 22: 520-531.

Wang, J. 2003. Components of difference in HIV seropositivity rate among injection drug users between low and high prevalence regions. *AIDS and Behavior*. Vol. 7: 1-8.

Wang, J., A. Rahman, H. A. Siegal & J. Fisher. 2000. Standardization and decomposition: A useful technique to behavior and public health studies. *Behavior Research Methods, Instruments, & Computers*. Vol. 32: 357-366.

Wang, J., J. R. Carpenter & M. A. Kepler. 2006. Using SAS to Conduct Nonparametric Residual Bootstrap Multilevel Modeling With a Small Number of Groups. *Computer Methods and Programs in Biomedicine, Section II: Systems and Programs*, Vol. 82: 130-143.

Wanning, W. G., N. Duan & W. H. Rogers. 1987. Monte Carlo evidence on the choice between sample selection and two-part models. *Journal of Econometrics*. Vol. 35: 59-82.

Watters, J. K. & P. Biernacki. 1989. Targeted sampling: Options for the study of hidden populations. *Social Problems*. Vol. 36: 416-430.

White, H. 1982. Maximum likelihood estimation of misspecified models. *Econometrica*. Vol. 50: 1-25.

Wilkins, K. & G. Fitzmaurice. A hybrid model for nonignorable dropout in longitudinal binary responses. *Biometrics* (in press).

Wolfinger, R. & M. O'Connell. 1993. Generalized linear models: a pseudo-likelihood approach. *Journal of Statistics and Computer Simullation*. Vol. 48: 233-243.

Wong, G. & W. Mason. 1985. The hierarchical logistic regression model for multilevel analysis. *Journal of the American Statistical Association*. Vol. 80: 513-524.

Xie, H., G. J. McHugo, A. Sengupta, D. Hedeker & R. E. Drake. 2001. An Application of the Thresholds of Change Model to the Analysis of Mental Health Data. *Mental Health Services Research*. Vol. 3: 107-114.

Xie, H., K. Hur & G. McHugo. 2000. *Using Random-Effects Zero-Inflated Poisson Model to Analyze Longitudinal Count Data with Extra Zeros*. Paper presented at the 53rd Session of the International Statistical Institute, Seoul, South Korea.

Yau, K. & A. Lee. 2001. Zero-inflated Poisson regression with random effects to evaluate and occupational injury prevention program. *Statistics in Medicine*. Vol. 20: 2907-2920.

Yau, K., K. Wang & A. Lee. 2003. Zero-inflated negative binomial mixed regression modeling of over-dispersed count data with extra zeros. *Biometrical Journal*. Vol. 45: 437-452.

Zeger, S., K. Liang & P. Albert. 1988. Models for Longitudinal data: A generalized estimating equation approach. *Biometrics*. Vol. 44: 1049-1060.

Zhou, X. H. & O. Harel. 2007. Multiple imputation: review of the theory, implementation and software. *Statistics in Medicine*. Vol. 26: 3057-3077.

Index

H

I

K

L

M

N

O

P

Q

R

S

T

Bei Fragen zur Produktsicherheit wenden Sie sich bitte an:
If you have any questions regarding product safety,
please contact:

Walter de Gruyter GmbH
Genthiner Straße 13
10785 Berlin
productsafety@degruyterbrill.com